D1087451

Representations of Integers as Sums of Squares

Emil Grosswald

Representations of Integers as Sums of Squares

Springer-Verlag
New York Berlin Heidelberg Tokyo

Emil Grosswald
Temple University
College of Liberal Arts
Philadelphia, PA 19122
U.S.A.

With 6 Illustrations

AMS Classifications: 10-01, 10B05, 10B35, 10C05, 10C15

Library of Congress Cataloging in Publication Data
Grosswald, Emil.
Representations of integers as sums of squares.
Bibliography: p.
Includes index.
1. Numbers, Natural. 2. Sequences (Mathematics)
3. Forms, Quadratic. I. Title.
QA246.5.G76 1985 512'.7 85-4664

Typeset by Asco Trade Typesetting Ltd., Hong Kong.
Printed and bound by R. R. Donnelley & Sons, Harrisonburg, Virginia.
Printed in the United States of America.

9 8 7 6 5 4 3 2 1

ISBN 0-387-96126-7 Springer-Verlag New York Berlin Heidelberg Tokyo
ISBN 3-540-96126-7 Springer-Verlag Berlin Heidelberg New York Tokyo

Preface

During the academic year 1980–1981 I was teaching at the Technion—the Israeli Institute of Technology—in Haifa. The audience was small, but consisted of particularly gifted and eager listeners; unfortunately, their background varied widely. What could one offer such an audience, so as to do justice to all of them? I decided to discuss representations of natural integers as sums of squares, starting on the most elementary level, but with the intention of pushing ahead as far as possible in some of the different directions that offered themselves (quadratic forms, theory of genera, generalizations and modern developments, etc.), according to the interests of the audience.

A few weeks after the start of the academic year I received a letter from Professor Gian-Carlo Rota, with the suggestion that I submit a manuscript for the *Encyclopedia of Mathematical Sciences* under his editorship. I answered that I did not have a ready manuscript to offer, but that I could use my notes on representations of integers by sums of squares as the basis for one. Indeed, about that time I had already started thinking about the possibility of such a book and had, in fact, quite precise ideas about the kind of book I wanted it to be.

Specifically, I had read with much pleasure a book by K. Zeller on *Summability* (Ergebnisse der Mathematik und ihrer Grenzgebiete No. 15, Springer-Verlag). What impressed me mainly was the completeness of the bibliographic references. I was moved to emulate this model and write a book on representations by sums of squares that would quote a comfortably large number of known results, occasionally with condensed proofs only, but with bibliographic references as complete as possible.

Professor Rota encouraged me to write such a text, and I proceeded. When the manuscript was completed, however, it came as a real surprise to me that, except for the attempt to have a complete bibliography, there was no resemblance whatsoever between my text and its model by Zeller.

The original draft profited greatly from suggestions made by Professors George Andrews (The Pennsylvania State University), Marvin Knopp (Temple University), and Olga Taussky-Todd (California Institute of Technology), as well as by an anonymous referee. Also, Professor Martin Kneser (University of Göttingen) read the whole manuscript at least twice, with incredible care, pointing out a large number of errors of omission as well as of commission. To all of them I express my deepest gratitude. Particular thanks are due to all colleagues, who called my attention to bibliographic items which had eluded me. I also thank Professor Rota; his encouragement was an essential element in the decision to develop my notes into the present text.

At a certain moment the original publisher appeared to have lost interest in this venture. I am happy that Springer-Verlag was receptive to the suggestion that it take over. Perhaps it is appropriate that the publishers of *Limitierungsverfahren* ... and of "Representations of integers as sums of squares" should be the same. I express my gratitude to Springer-Verlag for its support and cooperation.

Finally, I remember fondly my audience at the Technion: their keen interest was an important stimulus in the preparation of the notes that grew into this manuscript.

My visit at the Technion had been made possible by a Lady Davis Fellowship, for which I also express my gratitude.

May the reader have as much fun from this volume as the author had in writing it!

Narberth, Pennsylvania
May 22, 1984

EMIL GROSSWALD

Contents

Introduction

1. What do the relations

(i) $5^2 = 3^2 + 4^2$,
(ii) $6 = 1^2 + 1^2 + 1^2 + 1^2 + 1^2 + 1^2 = 2^2 + 1^2 + 1^2$,
(iii) $7 \neq a^2 + b^2 + c^2$

have in common? Obviously, their right hand members are all sums of squares. One way to describe those relations is as follows:

(i) The square 5^2 can be represented, in essentially one way only, as the sum of two squares.
(ii) The integer 6 can be represented in (at least) two essentially distinct ways as a sum of squares.
(iii) The integer 7 cannot be represented as a sum of three squares.

In this book, by "square" we mean the square of a rational integer (unless otherwise stated). We regard two representations as being *not essentially distinct* if they differ only by the order of the summands, or by the sign of a term, otherwise we regard them as being *essentially distinct*. To illustrate this definition, we note that $5^2 = 3^2 + 4^2 = 4^2 + 3^2 = (-3)^2 + 4^2 = 4^2 + (-3)^2 = 3^2 + (-4)^2 = (-4)^2 + 3^2 = (-3)^2 + (-4)^2 = (-4)^2 + (-3)^2$, for a total of 8 representations of 5^2 as a sum of two squares. However, any two of these representations differ only by the order of the summands, or by the sign of one of the terms, and therefore, while they differ, they are not essentially distinct.

2. Before going any further, we may ask ourselves why anybody should want to know such facts. While mathematicians rarely raise such questions, they often answer them by pointing out the usefulness of abstract mathematics in physics, engineering, and other scientific disciplines. Also, in the present instance, a case can be made for the usefulness of the study of representations of integers by sums of squares in lattice point problems, crystallography, and certain problems in mechanics. For example, (i) above is a particular instance of $d^2 = \sum_{i=1}^{k} x_i^2$, the distance formula in a k-dimensional Euclidean space. If it is required that $(d, x_1, \ldots, x_r) \in \mathbb{Z}^{r+1}$, then the number of solutions of (i) gives the number of lattice points on the sphere of integral radius d. We may remark that in this instance (as in many others), the really interesting problem is that of the

total number of solutions, rather than that of the number of essentially distinct ones.

If this example is considered slightly trivial, the numerous papers of O. Emersleben (see, e.g., [67], [68], [69]) should convince even the most skeptical of the interest of these representations by sums of squares. While it may appear from some of these papers that the theory under consideration is used merely to help solve other problems in mathematics, such as questions related to Epstein's zeta function, it becomes clear from other papers by the same author that the latter results are of direct relevance in crystallography, electrostatics, potentials of charge distributions, and even classical mechanics. Much of Emersleben's work consists in the summation over lattice points (m, n) of functions of the form $\{Q(m, n)\}^{-s} = \{am^2 + bmn + cn^2\}^{-s}$, i.e., the evaluation of Epstein zeta functions $\zeta_Q(s) = \sum_{(m,n)\neq(0,0)} Q(m, n)^{-s}$. The history of this approach (introduced, apparently, by P. Appell in 1884) is extensively reviewed by M. L. Glasser and I. J. Zucker in [83] (see especially pp. 68–96). In this review, the interested reader will find a surprising number of instances of direct applications of sums of squares. One such example is the computation of the constant Z_N that occurs in the evaluation of a certain Epstein zeta function, needed in the study of the stability of rare gas crystals, and in that of the so-called *Madelung constants* of ionic salts (see, e.g., [83] for definitions and details). Indeed, $Z_N = \sum_{n=1}^{N} r_Q(n) n^{-s/2}$, where $r_Q(n)$ stands for the number of representations of n by the quadratic form Q.

Perhaps the greatest surprise experienced in this respect by the present author occurred, when he was asked to review a paper written by a physicist with the collaboration of two mathematicians (see [15], by A. Baltes, P. K. J. Draxl, and E. R. Hilf). In it the authors study some boundary value problems arising from the Schrödinger equation in quantum physics and are led to rediscover an earlier result of the author, and A. Calloway, and J. Calloway [88] on the representation of integers as sums of three nonvanishing squares, a topic that will be discussed in detail in Chapter 6 of the present volume.

Next, as we shall see presently, one of the principal tools in the study of our representation problems will be the so called *theta functions*. The exact definition of these functions will be postponed to Chapter 8; here it is sufficient to mention that these are functions of two variables, denoted traditionally by $\theta(v|\tau)$ and that all theta functions satisfy the partial differential equation.

$$\frac{\partial^2 \theta}{\partial v^2} = c\frac{\partial \theta}{\partial \tau}.$$

We immediately recognize Poisson's heat equation (although in this instance c has the complex value $4\pi i$ rather than, as in thermodynamics, the real value $c = \gamma\rho/k, \gamma =$ specific heat, $\rho =$ density, $k =$ thermal conductivity). Thus, after a simple change of variables, the relevance of this study to the physical problem of heat becomes obvious.

One could easily add many other examples and exhibit connections of other problems to that of the representation of integers by sums of squares and, more generally, by *quadratic forms* (see Chapter 2), but we also may do well not to overstate the case of applications as a motivating element. Indeed, it rather appears that the investigators of these problems were impelled to study them by sheer curiosity, by the irresistible urge to know and to understand. One may, perhaps, be tempted to add, somewhat maliciously, that the persistence of the investigations of these and of similar problems may have been due, at least in part, to the fact that the adequate tools for their effective pursuit happened to be on hand in every generation.

First, elementary (but by no means trivial) methods were used up to the eighteenth century. Later, quadratic residues, elliptic and theta functions, complex integration and residues, algebraic number theory, and modular functions, all seem to have appeared just when needed. On the other hand, this may be an overly cynical attitude, and one may make an equally good case for the view that the pursuit of these representation problems led to the development of the appropriate tools. After all, it was the same Jacobi who contributed both to the theory of theta functions and to the problem of representations by sums of squares. As with the chicken and the egg, one can speculate which came first: Jacobi's interest in theta functions (the germ of which was known to Euler), or his interest in the sums of squares problem (which occurs in Diophantus and goes back at least to the Pythagoreans).

3. The list of mathematicians who devoted considerable effort to the study of these problems is long and contains many illustrious names. Among them we find Diophantus and Bachet, Viète and Fermat, Lagrange, Gauss, Artin, Hardy, Littlewood, Ramanujan, and many more. Among our contemporaries, the names of Siegel, Pfister, the Lehmers, and Hooley are just a modest sampling. It appears that since Fermat (at least), each generation of mathematicians has found interest in the study of these and related problems. Recently, at least since the 1962 paper [236] of Schaal, we have witnessed a renewed spurt of interest, particularly in the extension of the original problems to corresponding ones in algebraic number fields. Lagrange's contemporaries may well have felt that the last word had been said about the topic on hand by the categorical statement (often called Bachet's theorem, proved only by Lagrange, but known already by Fermat and probably even by Diophantus; see [54, Chapter VIII, p. 275, lines 12–8(b)]), that every positive integer is the sum of four squares. Today, however, we know that the field is still wide open. In fact, some of the simplest problems still await solution, and several of the best contemporary minds are actively engaged in their study.

4. This book is not meant for the specialist. A serious effort has been made to make it accessible to a wide circle of readers. Perhaps the most satisfactory scheme for the expert would be that in which one proves the most general theorems first and then obtains everything else as particular cases. This is not, however, the program chosen for the present book. On the contrary, we shall follow roughly the historical development of the subject matter. This has the advantage that the prerequisites for the understanding of the early chapters

are minimal. As more advanced tools become necessary, such as theta functions or the so-called "circle method," the theory needed will be developed here.

The success of this book will be measured, up to a point, by the number of readers who enjoy it, but perhaps more by the number who are sufficiently stimulated by it to become actively engaged in the solution of the numerous problems that are still open.

Chapter 1

Preliminaries

§1. The Problems of Representations and Their Solutions

In this book, when we speak of a quadratic form,* we mean a rational, integral quadratic form, unless the contrary is stated explicitly. Given a quadratic form Q, let $\mathbf{N}_Q$ be the set of values of Q where $\mathbf{x} \in \mathbb{Z}^k$ (i.e., $x_i \in \mathbb{Z}$ for $i = 1, 2, \ldots, k$); clearly, $\mathbf{N}_Q \subset \mathbb{Z}$. If $Q = \sum_{i=1}^{k} x_i^2$, we denote $\mathbf{N}_Q$ by $\mathbf{N}_k$. The main problems that we shall study can now be formulated as follows:

(a) Given a quadratic form Q, determine $\mathbf{N}_Q$.
(b) Given Q and $n \in \mathbf{N}_Q$, determine the *number of representations of n* by Q, i.e., the number of vectors $\mathbf{x} \in \mathbb{Z}^k$ for which $Q(\mathbf{x}) = n$.

An equivalent formulation of these problems is as follows:

(a′) Given Q and $n \in \mathbb{Z}$, determine whether the *Diophantine equation*

$$Q(x_1, \ldots, x_k) = n \tag{1.1}$$

has solutions. (We call a polynomial equation with rational integral coefficients *Diophantine* if we are interested only in solutions in integers, and if we speak of the solutions of a Diophantine equation, it will be understood that we refer only to the integral solutions of that equation.)

(b′) Given Q and a representable integer n, find the number of solutions of (1.1). (We say that n is *representable by Q*, or simply *representable* if $n \in \mathbf{N}_Q$).

Suppose that Q is positive definite. If $n < 0$, then (1.1) has no solutions, and if $n = 0$, then (1.1) has only the *trivial solution* $\mathbf{x} = 0$ (i.e., $x_1 = \cdots = x_k = 0$). However, if Q is indefinite, then $Q = 0$ may well have *nontrivial* solutions, i.e., solutions with $\mathbf{x} \neq 0$. For example,

$$2x^2 - 5y^2 + 2z^2 = 0 \tag{1.2}$$

---

* Most readers will be familiar with the basic theory of quadratic forms; complete definitions will be found in §4 of Chapter 4.

has, besides the (always present, but rarely interesting) trivial solution, also the nontrivial solution $x = y = z = 1$. This is, however, not the only nontrivial solution. In fact, we have solutions for each of the eight possible choices of signs in $\pm x = \pm y = \pm z = 1$. As mentioned in the Introduction, we consider such solutions as being *different*, but not *essentially distinct*. We recall that we already met with an example (the lattice points on a sphere with integer radius) where the total number of solutions, rather than the number of essentially distinct ones, was of interest. Surprising as it may appear, this seems to be the case in general (see, however, Chapter 7, where the emphasis is on the number of essentially distinct solutions).

Returning to equation (1.2), we note that the eight solutions $\pm x = \pm y = \pm z = 1$ do not exhaust the full set of nontrivial solutions of (1.2). Indeed, for every $m \in \mathbb{Z}$, one has the solutions $\pm x = \pm y = \pm z = m$, and one feels that these two sets of solutions are related. To formalize this, we introduce the notion of a *primitive* solution of (1.1). If $x = (x_1, \ldots, x_k)$ is a solution of (1.1), then x is called a *primitive solution* if the greatest common divisor* $(x_1, \ldots, x_k) = 1$; otherwise, x is called an *imprimitive* solution. If x is an imprimitive solution of (1.1), let $d = (x_1, \ldots, x_k)$ and write $x_i = dx_i'$ $(i = 1, 2, \ldots, k)$, whence $(x_1', \ldots, x_k') = 1$. Then $d^2 \mid n$ and $n = d^2 n_1$ with $n_1 \in \mathbb{Z}$, and $Q(x_1', \ldots, x_k') = n_1$, so that $(x_1', \ldots, x_k')$ is a primitive solution of (1.1), with n_1 instead of n. A moment's reflection shows that all solutions of (1.1) can be obtained from primitive solutions of $Q(x_1, \ldots, x_k) = n/d^2$, when d ranges over all divisors of n such that $d^2 \mid n$. Thus, if we denote by $R_Q(n)$ the number of primitive solutions of (1.1) and by $r_Q(n)$ the total number of solutions, then we have proved the following theorem:

Theorem 1. *With the above notation,* $r_Q(n) = \sum_{d^2 \mid n} R_Q(n/d^2)$.

In the particular case $Q = \sum_{i=1}^{k} x_i^2$, following the general custom, we shall denote $r_Q(n)$ by $r_k(n)$ and $R_Q(n)$ by $R_k(n)$.

§2. Methods

The authors of antiquity, the middle ages, and up to the time of Fermat used only what we would call elementary methods. This should not be misunderstood as meaning that they are simplistic, or even easy to follow. The reader who thinks otherwise is invited to read Fermat's *Oeuvres* [79]; he will be surprised by the subtlety of the reasoning. Even if we avail ourselves of the help of an expert interpreter (see, e.g., Edward's book *Fermat's Last Theorem* [63]), it often requires great concentration to understand Fermat's "elementary"

* The context is usually sufficient to indicate whether the symbol $(x_1, x_2, \ldots, x_k)$ stands for a k-tuple of integers or for their greatest common divisor; if this is not obvious, the meaning will be stated explicitly, as is done here.

considerations, and one may well marvel that a person in the seventeenth century could actually *invent* such proofs.

Shortly afterwards, with Euler, Lagrange, Legendre, and especially Gauss, the methods become more sophisticated. Specifically, since the latter part of the eighteenth century, use has been made of the quadratic reciprocity law. Next, still during Gauss's lifetime, Jacobi introduced elliptic and theta functions (see Chapters 8, 9) as tools in the study of our problems. One already finds, however, in Euler [76] (see [54], p. 277]) the statement that if in $(\sum_{n=0}^{\infty} x^{n^2})^4 = \sum_{n=0}^{\infty} r_4(n)x^n$ one has $r_4(n) \neq 0$ for all n, then Lagrange's (or Bachet's) four squares theorem immediately follows.

In principle, Jacobi's approach is extremely simple and may be considered as a method to generalize and implement Euler's idea. It will not only be invoked several times in the present book, but can also be applied to a great variety of other problems. For that reason, we shall present it in somewhat greater generality than we need right now.

Let $\mathbf{T} = \{\ldots, t_{-m}, \ldots, t_{-2}, t_{-1}, t_0, t_1, t_2, \ldots, t_m, \ldots\}$ be a finite or infinite ordered set of positive integers (in fact the proof goes through, *mutatis mutandis*, for more general numbers t_m), and consider the formal series $F(x) = \sum_{-\infty}^{\infty} x^{t_m}$. Then

$$
\begin{aligned}
F(x)^k &= \left(\sum_{-\infty}^{\infty} x^{t_m}\right)^k = \sum_{\substack{-\infty < m_j < \infty \\ j=1,2,\ldots,k}} x^{t_{m_1}+t_{m_2}+\cdots+t_{m_k}} \\
&= \sum_n x^n \sum_{t_{m_1}+t_{m_2}+\cdots+t_{m_k}=n} 1 = \sum_n b_{\mathbf{T}}(n)x^n.
\end{aligned} \tag{1.3}
$$

Clearly $b_{\mathbf{T}}(x)$ is the number of representations of n by k elements selected from the set $\mathbf{T}$. So far, the equalities (1.3) are purely formal. Let us assume, however, that, at least for some $\rho > 0$ and $|x| \leqslant \rho$, all series converge sufficiently well to permit us to justify all operations performed. Let us assume also that we have a method to expand $F(x)^k$ into a Taylor series (or even a Laurent series), say

$$F(x)^k = \sum_n a_n x^n. \tag{1.4}$$

Then, by a well-known theorem concerning the uniqueness of the expansion (1.4), we obtain immediately the desired result,

$$b_{\mathbf{T}}(n) = a_n.$$

In the particular case $\mathbf{T} = \{\ldots, (-m)^2, \ldots, (-2)^2, (-1)^2, 0, 1^2, 2^2, \ldots, m^2, \ldots\}$, we have

$$F(x) = \sum_{m=-\infty}^{\infty} x^{m^2} = 1 + 2\sum_{m=1}^{\infty} x^{m^2}$$

and (1.3) becomes

$$F(x)^k = \sum_{n=0}^{\infty} a_n^{(k)} x^n = (1 + 2\sum_{m=1}^{\infty} x^{m^2})^k = 1 + \sum_{n=1}^{\infty} r_k(n) x^n.$$

The main difficulty is, of course, to obtain (1.4).

Several methods have been devised for the computation of the $a_n^{(k)}$. We shall be interested mainly in two of them, one of which uses the theory of theta functions, and the other the theory of modular forms. Jacobi, using the theory of theta functions, represents $F(x)^k$ by Lambert series. This will be discussed in detail in Chapters 8 and 9. As a typical example, we may consider, e.g., $f(x) = \sum_{m=1}^{\infty} x^m/(1 - x^m)$. Here the coefficients of the Taylor series are easily obtained as follows: $x^m/(1 - x^m) = \sum_{k=1}^{\infty} x^{km}$; hence,

$$f(x) = \sum_{m=1}^{\infty} \sum_{k=1}^{\infty} x^{km} = \sum_{n=1}^{\infty} x^n \sum_{k \mid n} 1 = \sum_{n=1}^{\infty} d(n) x^n,$$

where $d(n)$ stands for the number of positive divisors of n. If we write $f(x) = \sum_{n=1}^{\infty} b_n x^n$ for the Taylor series of $f(x)$, then, by identification of coefficients, we obtain $b_n = d(n)$, and it is easy to understand how divisor functions will appear in similar problems. Jocobi's method works well for even values of k, but not for odd ones.

In the case $k = 3$, Gauss solved the problem (see [81, §291]) with the help of his concept of the class number of a quadratic form. Somewhat later, Dirichlet obtained formulae for Gauss's class number, in terms of the Legendre–Jacobi symbol of quadratic residuacy (for defintion see §11 of Chapter 4 and for more details see, e.g., [105]). This permits the establishment of formulae for $r_3(n)$ in terms of the Legendre–Jacobi symbol (Eisenstein [64]).

Another approach to the determination of the coefficients a_n is by use of Cauchy's integral formula

$$a_n = \frac{1}{2\pi i} \int_{|z|=\rho} \frac{F(z)^k}{z^{n+1}} dz. \tag{1.5}$$

In order to make this method effective, we have to find first, just as in Jacobi's approach, a representation for $F(x)^k$, independent of **T**. One can usually obtain such a representation without great difficulty. In fact, it turns out that if we let $z = e^{2\pi i \tau}$ and set $g(z) = F(z)^k$, then $g(z)$ has some useful properties. Specifically, if we replace τ by $(a\tau + b)/(c\tau + d)(ad - bc = 1, a, b, c, d \in \mathbb{Z})$, then $g(z)$ changes in a certain simple, easily described way. Functions with this property are called *modular*. These functions are complicated, and the effective computation of the integral in (1.5) is difficult. The expression for $g(z)$ simplifies only for $|z| \to 1^-$; unfortunately, $g(z)$ has infinitely many singularities, the set of which is dense in the unit circle. Hence, integration on $|z| = 1$ is ruled out. The "circle method" of Hardy, Ramanujan, and Littlewood (Rademacher, Vinogradov, and Davenport also contributed to it) circumvents the difficulty as follows: The contour of integration along $|z| = 1$ is replaced by a path consisting of a sequence $\mathscr{E}$ of abutting arcs inside the unit circle, but sufficiently close to it to permit a good approximation of $g(z)$ by simple functions along its new path of integration. Also, bounds can be found for the error of the approximation. If $g(z) = h(z) + \varepsilon(z)$ then $a_n = (1/2\pi i)\int_{\mathscr{E}} (h(z)/z^{n+1})\,dz + \varepsilon_n' = c_n + \varepsilon_n'$. For c_n we obtain an exact expression, and for ε_n' we can compute an upper bound. It is interesting to remark that, for $4 \leqslant k \leqslant 8$, the error terms, in fact, vanish, while this is no longer the case for $k \geqslant 9$. For $k = 3$ the method is also applicable, but only with great difficulties, due to poor convergence (see Bateman [18]), while for $k < 3$, the formally obtained results are actually false. See Chapters 11 and 12 for more details.

Modular functions can also be used directly, somewhat in the same way that Jacobi used theta functions, in order to obtain exact or approximate expressions for $r_k(n)$. The foremost exponent of this approach is Mordell, to whose papers [181–189] we refer the reader (see also [210]).

Many of these results can also be obtained from the study of hypergeometric functions. Some details concerning this approach will be given in Chapters 8 and 9. A brief sketch of this idea may be found in Andrews's book [5]. Proofs by this method of some of the theorems presented here in Chapter 9 may be found also in Section 3 of Andrews's *SIAM Review* article [6].

§3. The Contents of This Book

In Chapters 2, 3, and 4 we shall study, using elementary methods, the representations of an integer n as a sum of 2, 4, and 3 squares, respectively. In Chapter 5 we shall consider a diagonal form in three variables with arbitrary integer coefficients. Chapter 6 discusses some problems of representations by exactly k nonvanishing squares. Chapter 7 treats a certain problem of uniqueness of representations. Chapter 8 presents as much of the theory of theta functions as is needed for the problem at hand. In Chapter 9 we discuss the general problem of representations of $n \in \mathbb{Z}$ by an *even* number of squares. Chapter 10 presents various results on representations by sums of squares. Chapter 11 presents several theoretical topics needed in Chapters 12 and 13. Chapter 12 is devoted to the presentation of a certain version of the circle method. Chapter 13 sketches two alternative approaches to the problem at hand, with two examples (one for n even, one for n odd) worked out in some detail. Chapter 14 lists some recent developments, in particular on the representation problem for integers in algebraic number fields and that of positive definite functions by squares and more general quadratic forms. An Appendix lists some open problems.

This overview shows that many interesting topics have been bypassed. With few exceptions, only the simplest quadratic forms, the *diagonal forms* $Q = \sum_{i=1}^{k} x_i^2$ are here considered. More general forms were discussed already by Gauss. Indeed, one finds in [81], as well as in Bachmann's book [12], a detailed treatment of the representations by general quadratic forms, even indefinite ones. For a modern presentation see O'Meara [200]. For a more general treatment of the circle method, applicable to more general forms and allowing also the treatment of inequalities, see Davenport [49]. A superb modern presentation of quadratic forms with classical flavor is due to C. L. Siegel [243]). His results are very general, and many of the theorems of the present work can be obtained as particular cases of corresponding ones in [243]. It appears presumptuous to try to improve upon the masterly presentations of these authors, and the inclusion of the more general theorems would have at least doubled the size of the present volume. Next, no problems of representation by sums of mth powers for $m \geqslant 3$, or more general forms of degrees higher than 2, are considered (with the exceptions in §7 of Chapter 2 and §3 of Chapter 14). These problems belong properly in a discussion of *Waring's problem*, on which there exists an immense literature (see, e.g., [98] and the bibliography there; for modern contributions one may consult [160]).

This list would not be complete without the mention of the superb review article [263] on "Sums of Squares" by O. Taussky. See also [264] by the same author.

Finally, in order to keep the prerequisites for the reading of this book within the limits appropriate for the intended readership, the style of Chapter 14 differs markedly from that of the others. For a moment the author hesitated to include it at all in this book. However, the results stated are important and lead directly into contemporary research; also, they are natural generalizations of theorems stated here. Complete proofs would have required, however, some preparatory chapters on algebraic number theory and class field theory, and these could not been accommodated within the framework of this book. For that reason, much of Chapter 14 consists of definitions and examples, followed by the statements of results and commentaries, with only the barest sketches of proofs.

With the exception of Chapter 14, most of the material of this book is classical. In particular, the author's own contributions to the problems at hand have been minimal. Those included in the present volume are found in Chapters 6 and 7 (see [20], [87], and [88]). Although some proofs of well-known theorems may have been presented here in versions different from earlier ones, no claim of originality is made for any of them.

Only a handful of papers of the 1970*s* and even fewer of the 1980*s* are cited in this volume. It is entirely possible that some recent work will have far-reaching implications in the future, but it is too early for us to have the proper perspective. It often is not even possible to foresee whether a certain result, apparently without any connection with the topics here considered, may not some day have an important impact upon them. For these reasons, only the following appeared feasible. A selection of recent papers has been made, which, in the author's subjective judgment, *appear* to be relevant to the problem of representation by sums of squares or, more generally, by quadratic forms. It is listed in a separate bibliography, ordered alphabetically by the name of the author (the first listed author in the case of joint papers). Almost no references to the papers on this list occur in the text. It is quite certain that both errors of omission and of commission have been made. Some of the papers listed, even excellent and beautiful ones, may turn out to be irrelevant to the main topic of this book. More serious still, it is quite likely that, the author's efforts notwithstanding, some important papers will have been missed. The author's apologies are herewith extended to the omitted authors as well as to the readers. Papers and books, whose omission is discovered during the typesetting of this book, will be added at the end of the Bibliography as an addenda.

§4. References

An effort has been made to give bibliographic references that are as complete as possible. Nevertheless, it appeared wasteful to try to duplicate the 500 or so references found in the relevant chapters of Dickson's *History of the Theory of Numbers* [54]. Hence, especially for the literature before the eighteenth century, the reader is often directed to [54].

No particular effort has been made to establish priority in controversial cases.

§5. Problems

At the end of many chapters, the reader will find a number of problems. Some of these are fairly routine and have as their main purpose to give the reader the confidence that he has really assimilated the contents of the chapter. Some starred problems also appear. These designate either difficult problems or even unsolved ones deemed approachable. If the reader succeeds in solving one of the latter subset, he will have made a valuable contribution to mathematics. Good luck to all who try!

§6. Notation

As we did earlier, we shall use the customary symbols $\mathbb{Z}$, $\mathbb{Q}$, $\mathbb{R}$, $\mathbb{C}$, and $\mathbb{K}$ for the integers, the rationals, the reals, the complex field, and an arbitrary field. Ideals will be denoted by German letters. The symbols $(\frac{a}{p})$, $(\frac{a}{b})$ $(a, b \in \mathbb{Z}$, p a prime) are the Legendre and, more generally, the Jacobi symbols; sometimes they are denoted by (a/b). Lowercase Latin letters stand, in general, for rational integers. Sets are denoted by boldface capitals, vectors by boldface lowercase. In general p, and often q, with or without subscripts, will stand for rational primes. The greatest integer function is denoted by square brackets; this means that if $x \in \mathbb{R}$, we write $n = [x]$ for the unique integer n such that $x - 1 < n \leqslant x$.

The symbols $\sim$ and $\simeq$ are used either to indicate an equivalence relation, or an approximate equality; the meaning will be clear from the context.

Theorems, lemmas, formulae, etc., are numbered consecutively within each chapter. When Theorem 2 of Chapter 3 is cited in Chapter 3, it is called simply Theorem 2. When it is cited in any other chapter, it is called Theorem 3.2; similarly for lemmas and often for sections.

Chapter 2

Sums of Two Squares

§1. The One Square Problem

Logically, one should start with $k = 1$, i.e., the representation of a natural integer by a single (rational integral) square. In this case, however, both representation problems (i.e., problems (a), and (b) of Chapter 1) are trivial. For completeness only, we state here the result, without formal proof or further comments.

Theorem 1. *If $Q(x) = x^2$, then $Q(x) = n$ has integer solutions if and only if $n = m^2$ for some $m \in \mathbb{Z}$. If that is the case, then $r_1(n) = 2$ (namely, $x = m$ and $x = -m$).*

§2. The Two Squares Problem

Let $Q(x, y) = x^2 + y^2$; then the two representation problems of Chapter 1 are both nontrivial. Explicitly, these problems are:

(a) To characterize of the set of integers, for which the diophantine equation

$$x^2 + y^2 = n \tag{2.1}$$

has solutions in integers x and y (positive, negative, or zero).

(b) If $\mathbf{N}_Q$ is the set of integers n for which (2.1) has solutions, and if $n \in \mathbf{N}_Q$, to determine the number $r_2(n)$ of solutions of (2.1).

Diophantus (325–409 A.D.) discusses several problems connected with (2.1), often in geometric language, and the meaning of his statements is not always clear. At least some of them appear to be incorrect; one of them, however, is equivalent to the important identities

$$(a^2 + b^2)(c^2 + d^2) = (ac + bd)^2 + (ad - bc)^2 = (ac - bd)^2 + (ad + bc)^2. \tag{2.2}$$

It is, of course, trivial to verify (2.2), either by observing that all three expressions equal $a^2c^2 + a^2d^2 + b^2c^2 + b^2d^2$ or, more elegantly, by interpreting $a^2 + b^2$ as the norm $N\alpha$ of the Gaussian integer $\alpha = a + ib$, and recalling that the product of norms equals the norm of the product. In fact, the identity (2.2) is a simple particular instance of Gauss's very general theory of composition of forms (see [81; §§234–251] and, perhaps even better [63, pp. 334–339]; for the connection with the theory of ideals, see [105].

The identity (2.2) tells us that if integers m_1 and m_2 are representable as sums of two squares, then so is their product, in at least two essentially distinct ways. So, e.g., $5 = 2^2 + 1^2$, $13 = 3^2 + 2^2$, and

$$65 = 5 \cdot 13 = (2^2 + 1^2)(3^2 + 2^2) = (2 \cdot 3 + 1 \cdot 2)^2 + (2 \cdot 2 - 1 \cdot 3)^2 = 8^2 + 1^2$$

$$= (2 \cdot 3 - 1 \cdot 2)^2 + (2 \cdot 2 + 1 \cdot 3)^2 = 4^2 + 7^2.$$

§3. Some Early Work

Among the mathematicians who studied this problem were (see [54]) Mohamed Ben Alhocain (tenth century), Leonardo da Pisa (better known as Fibonacci, 1175–1230), Vieta (1540–1603), and Xylander (a sixteenth century editor and commentator on Diophantus). Bachet (1581–1638), famous for his own edition of Diophantus, made some remarks on (2.1), but it seems to have been Girard (1595–1632) who first stated the correct necessary and sufficient conditions on n for the solvability of (2.1) in integers. His (slightly paraphrased) conditions are: n has to be either a square (when $m^2 + 0^2 = m^2$), or a prime $p \equiv 1 \pmod 4$, or a product of such numbers (evident from (2.2)), or the product of one of the preceding with a power of 2 (which also follows from (2.2), because $2 = 1^2 + 1^2$). It is clear from what precedes that the crucial element of the proof is to show that an odd prime is the sum of two squares if and only if $p \equiv 1 \pmod 4$. So, e.g., $13 = 3^2 + 2^2$, but 11 is not the sum of two squares.

Shortly afterwards, Fermat (1601–1665), most likely quite independently, stated, as a condition on n, that $n \equiv 1 \pmod 4$ and that when n is divided by its largest square factor, the quotient should not contain any prime $q \equiv 3 \pmod 4$. In this statement, Fermat leaves out the even integers, but, on the other hand, makes a remark equivalent to (2.2), so that (since $2 = 1^2 + 1^2$) he evidently knew the complete theorem. In fact, he indicates (in a numerical example), how to compute $r_2(n)$.

There is no indication that Girard had (or even claimed to have) a proof for his statement, while Fermat claimed to have an "irrefutable proof" of his own, and while he never made it public, the contents of his letters (to Descartes and to Mersenne) leave hardly any doubt that he had one and that this was based on the *method of descent* of his own invention (for the meaning of this phrase and an example, see the proof of Lemma 2 below). In fact, Fermat makes similar statements for representations by several other diagonal forms, such as $x^2 + 2y^2$ or $x^2 + 5y^2$; see [63] for more details.

§4. The Main Theorems

Nevertheless, according to Gauss [81, §182]), it was Euler (1707–1783) who gave, in [75], the first known proof of a statement which is essentially equivalent to the following theorem.

Theorem 2. *The Diophantine equation* (2.1) *is solvable if and only if all prime divisors q of n with $q \equiv 3 \pmod 4$ occur in n to an even power.*

Going beyond this result, Gauss (1777–1855) in 1801 [81, §182]) by use of quadratic forms, and Jacobi (1804–1851) in 1829 [121] by use of elliptic functions, proved the following, much stronger result:

Theorem 3. *Denote the number of divisors of n by $d(n)$, and write $d_a(n)$ for the number of those divisors with $d \equiv a \pmod 4$. Let $n = 2^f n_1 n_2$, where $n_1 = \prod_{p \equiv 1 \pmod 4} p^r$, $n_2 = \prod_{q \equiv 3 \pmod 4} q^s$; then $r_2(n) = 0$ if any of the exponents s is odd. If all s are even, then $r_2(n) = 4d(n_1) = 4(d_1(n) - d_3(n))$.*

Corollary. *If all prime factors q of n_0 satisfy $q \equiv 3 \pmod 4$, then $r_2(n) = r_2(2^a n)$ $= r_2(2^a n n_0^2)$.*

As an illustration of Theorem 2, we note that 3 has no representation as a sum of two squares, but $n = 90$ has. Indeed, $90 = 2 \cdot 5 \cdot 3^2 = 9^2 + 3^2$.

Next, according to Theorem 3, $r_2(90) = 4d(5) = 4 \cdot 2 = 8$; also, $d_1(90) = 4$ ($1 \equiv 5 \equiv 9 \equiv 45 \pmod 4$) and $d_3(90) = 2$ ($15 \equiv 3 \pmod 4$), so that $4(d_1(90) - d_3(90)) = 4(4 - 2) = 8$. Finally, $90 = (\pm 9)^2 + (\pm 3)^2 = (\pm 3)^2 + (\pm 9)^2$, so that the single, essentially distinct representation $90 = 9^2 + 3^2$ leads to eight representations counted by $r_2(90)$, which are not essentially distinct.

It may well be that Jacobi had a proof of Theorem 3 long before the publication of his *Fundamenta Nova*. Indeed, in a letter to Legendre (9 September 1828; see [124]), he states formulae equivalent to

$$\left(1 + 2\sum_{k=1}^{\infty} q^{k^2}\right)^2 = \left(\sum_{k=-\infty}^{\infty} q^{k^2}\right)^2 = \sum_{k=-\infty}^{\infty} \sum_{m=-\infty}^{\infty} q^{k^2+m^2}$$

$$= \sum_{n=0}^{\infty} q^n \sum_{k^2+m^2=n} 1 = 1 + \sum_{n=1}^{\infty} r_2(n) q^n$$

$$= 1 + 4 \sum_{n=0}^{\infty} \frac{(-1)^n q^{2n+1}}{1 - q^{2n+1}}.$$

Here the first four equalities follow from (1.3) of Chapter 1. The last equality comes from the theory of elliptic functions, but elementary proofs for it exist (see [98], or Chapter 9, especially (9.10)). Theorem 3 follows immediately from the sequence of equalities

$$\sum_{n=1}^{\infty} r_2(n)q^n = 4\sum_{k=0}^{\infty} \frac{(-1)^k q^{2k+1}}{1-q^{2k+1}}$$

$$= 4\sum_{k=0}^{\infty}(-1)^k q^{2k+1}\sum_{m=0}^{\infty} q^{m(2k+1)} = 4\sum_{k=0}^{\infty}(-1)^k \sum_{m=1}^{\infty} q^{m(2k+1)}$$

$$= 4\sum_{n=1}^{\infty} q^n \sum_{2k+1\,|\,n} (-1)^k = 4\sum_{n=1}^{\infty} q^n \left\{ \sum_{\substack{2k+1\,|\,n \\ k \text{ even}}} 1 - \sum_{\substack{2k+1\,|\,n \\ k \text{ odd}}} 1 \right\}$$

$$= 4\sum_{n=1}^{\infty} q^n \left\{ \sum_{\substack{d\,|\,n \\ d\equiv 1 \ (\mathrm{mod}\ 4)}} 1 - \sum_{\substack{d\,|\,n \\ d\equiv 3 \ (\mathrm{mod}\ 4)}} 1 \right\}$$

$$= 4\sum_{n=1} q^n\{d_1(n) - d_3(n)\}.$$

Clearly, Theorem 3 implies Theorem 2, so that it is sufficient to prove Theorem 3. Nevertheless, as we have not yet proved the identity of Jacobi, it seems worthwhile to give first a direct and very simple proof of Theorem 2, to be followed by an *elementary* proof of Theorem 3.

§5. Proof of Theorem 2

We need the following lemma.

Lemma 1. *If* $p\,|\,n$ *and* $p \equiv 3 \pmod 4$, *then* (2.1) *has no primitive solution.*

Proof. Let x, y be a solution of (2.1) with $(x, y) = 1$, and let p be a prime divisor of n. If $p\,|\,n$ and $p\,|\,x$ then $p\,|\,y$, which contradicts $(x, y) = 1$. Hence, $p\,|\,n$ implies $p \nmid x$, $p \nmid y$. By Fermat's theorem, $x^{p-1} \equiv 1 \pmod p$, so that $yx^{p-1} \equiv y \pmod p$. Thus, denoting yx^{p-2} by z, we have $xz \equiv y \pmod p$, $x^2(z^2+1) \equiv y^2 + x^2 \equiv n \equiv 0 \pmod p$. Since $p \nmid x$, we obtain $z^2 \equiv -1 \pmod p$, so that $(-1/p) = +1$, which (by the complementary quadratic reciprocity law: see Section 4.11 or [105]) shows that $p \equiv 1 \pmod 4$. Hence, if (2.1) has a primitive solution, any prime divisor p of n satisfies $p \equiv 1 \pmod 4$, as claimed. □

Proof of Theorem 2. If x, y is a solution of (2.1) with $(x, y) = d$, suppose that $p \equiv 3 \pmod 4$, $p^r \,\|\, d$ (here and in what follows, the double bar means that $p^r\,|\,d$, $p^{r+1} \nmid d$). Set $x = dx_1$, $y = dy_1$, with $(x_1, y_1) = 1$, so that $n = d^2(x_1^2 + y_1^2) =$

$d^2 n_1$, say. If $p^c \parallel n$, then $p^{c-2r} \parallel n_1$. However, $x_1^2 + y_1^2 = n_1$, so that x_1, y_1 is a primitive representation of n_1. By Lemma 1, $p \nmid n_1$; hence, $c = 2r$ and is even. This proves the "only if" part of the theorem. To prove the "if" part, we use the identity (2.2). Let us also accept for a moment

Lemma 2. *The Diophantine equation* (2.1) *is solvable if n is a prime number* $p \equiv 1 \pmod 4$.

Indeed, if Lemma 2 holds, then, by (2.2), every product of primes $p \equiv 1 \pmod 4$ and of powers of 2 is representable as a sum of two squares. In particular, $n_0 = 2^f n_1 = a^2 + b^2$. If also $n_2 = \prod_{q \equiv 3 \pmod 4} q^{2t} = (\prod_{q \equiv 3 \pmod 4} q^t)^2 = m^2$, say, then $n = n_0 m^2 = (am)^2 + (bm)^2$, as claimed. It remains to prove Lemma 2. □

Proof of Lemma 2. If $p \equiv 1 \pmod 4$, then $(-1/p) = +1$, so that $x^2 + 1 \equiv 0 \pmod p$ is solvable. Hence $x^2 + y^2 = mp$ is solvable in integers x, y, and m. We now show that m may be taken equal to 1. As x and y need run only modulo p, we choose $|x| < p/2$, $|y| < p/2$, so that $x^2 + y^2 < p^2/2$ and $m < p/2$. Let m_0 be the smallest integer for which

$$x^2 + y^2 = m_0 p \tag{2.3}$$

has solutions in integers x, y. We shall show that, if $m_0 > 1$, then a solution with still smaller m_0 can be constructed, which then contradicts the definition of m_0 and proves that $m_0 = 1$, as claimed; this will finish the proof of Lemma 2 and also of Theorem 2.

Suppose that $m_0 > 1$, and assume that $m_0 \mid x$; then $m_0 \nmid y$, as otherwise $m_0^2 \mid x^2 + y^2 \Rightarrow m_0^2 \mid m_0 p \Rightarrow m_0 \mid p \Rightarrow m_0 = 1$ (because $m_0 < p/2$), contrary to $m_0 > 1$. We now choose integers c and d so that $x_1 = x - cm_0$, $y_1 = y - dm_0$, with $|x_1| < m_0/2$, $|y_1| < m_0/2$, and not both zero. Then $0 < x_1^2 + y_1^2 < 2(m_0^2/4) = (m_0/2)m_0$. As $x_1^2 + y_1^2 \equiv x^2 + y^2 \equiv 0 \pmod{m_0}$, it follows that

$$x_1^2 + y_1^2 = m_0 m_1, \qquad 0 < m_1 < m_0/2. \tag{2.4}$$

If we multiply (2.3) by (2.4), we obtain $m_0^2 m_1 p = (x^2 + y^2)(x_1^2 + y_1^2) = (xx_1 + yy_1)^2 + (xy_1 - yx_1)^2$. Here $xx_1 + yy_1 = x(x - cm_0) + y(y - dm_0) = x^2 + y^2 - m_0(cx + dy) = m_0(p - cx - dy) = m_0 X$, say. Similarly, $xy_1 - yx_1 = x(y - dm_0) - y(x - cm_0) = m_0(cy - dx) = m_0 Y$, say. Hence, $m_0^2 m_1 p = m_0^2(X^2 + Y^2)$, or $X^2 + Y^2 = m_1 p$, with $0 < m_1 < m_0/2$, as claimed, and the proofs of both Lemma 2 and of Theorem 2 are complete. □

The present scheme illustrates the idea of Fermat's "method of descent" and may well be the same that Fermat himself used in the proof of this lemma, which, on account of (2.2), is the essential part of the proof of Theorem 2.

§6. Proof of Theorem 3

We now present a proof of Theorem 3 by following, in principle, the exposition in [98]. We recall that the set of all numbers of the form $A + B\sqrt{-1} = A + Bi$ ($A, B \in \mathbb{Q}$) forms a field $\mathbb{K} = \mathbb{Q}(i)$ under addition and multiplication, and that the subset $a + bi$ ($a, b \in \mathbb{Z}$) forms a ring $\mathbf{I}$, which is, in fact, the ring of integers of $\mathbb{K}$. The integers of $\mathbb{K}$ are usually called *Gaussian integers* (see [25, p. 113, Ex. 9]). The units of $\mathbf{I}$ are ± 1, $\pm i$. The primes of $\mathbf{I}$ are (up to associates, i.e., up to multiples by units): (i) the prime divisors $1 \pm i$ of 2; (ii) the rational primes $p \equiv 3 \pmod 4$, and the integers $a + ib$ with $a, b > 0$ and $a^2 + b^2 = p$, where p is a rational prime, $p \equiv 1 \pmod 4$. Here $a^2 + b^2 = (a + ib)(a - ib)$ is called the *norm* of $\alpha = a + ib$ and is denoted by $N\alpha$, or $N(a + ib)$. We recall that factorization into Gaussian primes is unique (up to units and the order of factors) and that if $N(a + ib) = p \equiv 1 \pmod 4$, then $a \neq b$; indeed, otherwise, $p = 2a^2 \equiv 0 \pmod 2$.

Let now $n = 2^f n_1 n_2$, $n_1 = \prod_{p \equiv 1 \pmod 4} p^r$, $n_2 = \prod_{q \equiv 3 \pmod 4} q^s$. Then the (essentially unique) decomposition of n into its Gaussian prime factors reads

$$n = \{(1 + i)(1 - i)\}^f \prod_{\substack{a^2+b^2=p \\ p \equiv 1 \pmod 4}} \{(a + bi)(a - bi)\}^r \prod_{q \equiv 3 \pmod 4} q^s,$$

where it is to be understood that r and s vary with p and q. To each representation of n as a sum of two squares, say $n = u^2 + v^2 = (u + iv)(u - iv)$, one can make (taking also into account the units, all of the form i^t) the following assignments of prime factors to the two factors of n:

$$u + iv = i^t(1 + i)^{f_1}(1 - i)^{f_2} \prod \{(a + bi)^{r_1}(a - bi)^{r_2}\} \prod q^{s_1},$$

$$u - iv = i^{-t}(1 + i)^{f_2}(1 - i)^{f_1} \prod \{(a + bi)^{r_2}(a - bi)^{r_1}\} \prod q^{s_2},$$

with $t = 0, 1, 2$, or 3; $f_1 + f_2 = f$; $r_1 + r_2 = r$, $s_1 + s_2 = s$. It is clear that each allocation of prime factors to $u + iv$ determines completely the prime factors of $u - iv$ (namely those of n, not in $u + iv$), and also that to each such pair of complex conjugate factors $u \pm iv$ there corresponds a representation of n as sum of two squares. We observe that the change $i \to -i$ does not affect the real factors q, so that $s_1 = s_2$ and $s = 2s_1$ is even, as we already knew from Theorem 2. For t we have the four choices listed. For f_1 we have the $f + 1$ choices $0, 1, 2, \ldots, f$, and the choice of f_1 determines also $f_2 = f - f_1$. For r_1 we have the $r + 1$ choices $0, 1, \ldots, r$ and $r_2 = r - r_1$. Hence, we have a total of $4(f + 1)\prod(r + 1)$ possibilities to write $u + iv$, where in the last product we multiply over the primes $p^r \parallel n$, $p \equiv 1 \pmod 4$. However, not all these allocations of Gaussian primes lead to different values of $u + iv$. Indeed, when we replace a factor $1 - i$ by $1 + i$, we simply multiply the old value by $(1 + i)/(1 - i) = i$, and this corresponds simply to a different value of t. It follows that the total number of distinct factors of $u + iv$ is only

$$4 \prod_{p^r \mid n} (1 + r). \tag{2.5}$$

As $n_1 = \prod_{p \equiv 1 \pmod 4} p^r$, the factor $\prod (1 + r)$ equals $d(n_1)$, the number of divisors of n_1 (see, e.g., [86]).

Given a factor $u + iv$, we obtain the factors $u + iv$, $-v + iu$, $-u - iv$, and $v - iu$, respectively, by putting $t = 0, 1, 2, 3$. These factors correspond to the representations $u^2 + v^2, (-v)^2 + u^2, (-u)^2 + (-v)^2$, and $v^2 + (-u)^2$, respectively. These four representations are not essentially distinct, but they are all counted as different by (2.5), on account of the factor 4. We have also four other such representations of n that we obtain if we interchange u with v in the above factors. These also are counted as different by (2.5). Indeed, $n = v^2 + u^2 = (v + iu)(v - iu)$ has as first factor $v + iu$, the complex conjugate of $v - iu$ already obtained above, and similarly for the other three representations. This change corresponds to an interchange of r_1 and r_2; however, r_1 (and with it also r_2) has run through all $r + 1$ possible values; hence, (2.5) gives the number $r_2(n)$ of representations of n as a sum of two squares, in which two different representations that are not essentially distinct are counted separately.

In order to complete the proof of Theorem 3, it remains to show that $d(n_1) = d_1(n) - d_3(n)$. This may be seen in a variety of ways. We obtain all odd divisors of n as summands in the expansion of the product

$$\begin{aligned} &\prod_{\substack{p^r \parallel n \\ p \equiv 1 \pmod 4}} (1 + p + \cdots + p^r) \prod_{\substack{q^s \parallel n \\ q \equiv 3 \pmod 4}} (1 + q + \cdots + q^s) \\ &= \sum_{\substack{0 \leqslant m_i \leqslant r_i \\ 0 \leqslant k_j \leqslant s_j}} p_1^{m_1} p_2^{m_2} \cdots p_w^{m_w} q_1^{k_1} \cdots q_t^{k_t}. \end{aligned} \tag{2.6}$$

If $d \mid n$, then it is clear that one has $d \equiv 1 \pmod 4$, if and only if in its expression as a summand, we have $\sum_{j=1}^{t} k_j$ even; otherwise, $d \equiv 3 \pmod 4$. If we replace here q by -1, then the second product vanishes, if even a single exponent s is odd; if all are even, then the product in (2.6) becomes

$$\prod_{\substack{p^r \parallel n \\ p \equiv 1 \pmod 4}} (1 + p + \cdots + p^r),$$

and the terms of its expansion are precisely all the divisors of n_1. To obtain $d(n_1)$, each term has to be counted as 1. This is easily accomplished if we also replace p by 1 in (2.6); then we obtain indeed $\prod_{p^r \parallel n,\, p \equiv 1 \pmod 4} (1 + r)$. If we look now at the right hand side of (2.6) after each p is set equal to $+1$ and each q equal to -1, it is clear that each $d \mid n$, $d \equiv 1 \pmod 4$, is counted as $+1$, and each $d \mid n$, $d \equiv 3 \pmod 4$, is counted as -1. Hence, the right hand side of (2.6) becomes $d_1(n) - d_3(n)$, and the proof of Theorem 3 is complete.

§7. The "Circle Problem"

A closely related problem is that of determining the number of integral solutions x, y of the inequality

$$x^2 + y^2 \leqslant t. \tag{2.7}$$

This number is clearly the same as the sum $\sum_{n \leqslant t} r_2(n) = A(t)$, say. Once we know $A(t)$, we can immediately determine $\overline{r}_2(x) = x^{-1} \sum_{n \leqslant x} r_2(n) = A(x)/x$, the *average* value of $r_2(n)$. The problem is equivalent to that of the determination of the number $A(t)$ of lattice points (i.e., of points with integral coordinates) inside and on the circle of radius $\sqrt{t}$. This number is obviously close to the area of that circle, i.e., $A(t) = \pi t + E(t)$, where $E(t)$ is an error term. While it is relatively easy to find upper and lower bounds for the size of $E(t)$, the determination of the exact order of magnitude of $E(t)$ is still an open problem. To be more specific, we introduce the following notation (o, O due to Landau, Ω to Littlewood). Let $g(x)$ be a positive function, monotonically increasing as $x \to x_0$ (perhaps only from one side). Then, if there is a constant C such that $|f(x)| \leqslant Cg(x)$ for $x \to x_0$, we say that $f(x)$ is of order not higher than $g(x)$ and write $f(x) = O(g(x))$. If $\lim_{x \to x_0} (f(x)/g(x)) = 0$, $f(x)$ is said to be of lower order than $g(x)$ and we write $f(x) = o(g(x))$. If the last statement is false, we write $f(x) = \Omega(g(x))$ (read: f is *not* of lower order than g as $x \to x_0$). Here x_0 may be finite or infinite.

If $f(x)$ is real, inequalities such as $|f(x)| \leqslant Cg(x)$ are equivalent to the two inequalities

$$-Cg(x) \leqslant f(x) \leqslant Cg(x).$$

Sometimes, however, we are able to prove only one of these two inequalities. Then we mark the corresponding symbol O, o, or Ω with the subscript $+$ or $-$ accordingly. For example, $f = \Omega_-(g)$ means that $f(x) \geqslant -Cg(x)$ for arbitrarily small C is false, or, in other words, that $\liminf \{f(x)/g(x)\} < 0$; similarly, $f = \Omega_+(g)$ is equivalent to $\limsup \{f(x)/g(x)\} > 0$, etc.

In our last problem we now ask, for what exponents α is it true that, as $t \to \infty$, $E(t) = O(t^\alpha)$? We are interested, in particular, in $\inf \alpha = \mu$, say, so that, for every $\varepsilon > 0$, $A(t) - \pi t = O(t^{\mu+\varepsilon})$. Gauss already knew (see "De nexu inter multitudinem classium in quas formae binarias secundi gradus distribuntur earumque determinantem" [82, Vol. 2, pp. 272–275]) that $\mu \leqslant \frac{1}{2}$. Indeed, if we consider all lattice points inside the circle of radius $\sqrt{t}$ as centers of unit squares with sides parallel to the coordinate axes, then the number $A(t)$ of those lattice points equals the combined area of these squares. However, the diagonal of the unit squares is $\sqrt{2}$, so that the points of all those squares are inside the circle of radius $\sqrt{t} + \sqrt{2}/2$ and the squares cover completely the circle of radius $\sqrt{t} - \sqrt{2}/2$. It follows that

$$\pi t - \pi\sqrt{2}\sqrt{t} < \pi t\left(1 - \frac{1}{\sqrt{2t}}\right)^2 \leqslant A(t) \leqslant \pi t\left(1 + \frac{1}{\sqrt{2t}}\right)^2$$

$$= \pi t + \pi\sqrt{2}\sqrt{t} + \frac{\pi}{2}.$$

This result can be obtained also as a direct consequence of Theorem 3.

Sierpiński [248] sharpened this result to $E(t) = O(t^{1/3})$. Landau (see I in [150]) generalized this result widely. As a special case of Landau's theorem for a quadratic form Q of k variables, consider generally the number $r_Q(n)$ of the solutions of (1.1) and set $A_Q(t) = \sum_{n \leqslant t} r_Q(n)$, V_k = volume of the ellipsoid $Q(x) = 1$, and $E_Q(t) = A_Q(t) - V_k t^{k/2}$; then $E_Q(t) = O(t^c)$, where $c = k(k-1)/2(k+1)$. For $k = 2$, we obtain, in particular, Sierpiński's result $E(t) = O(t^{1/3})$.

Van der Corput [47] was the first to show that $\mu < \frac{1}{3}$; in fact, he proved that $\mu \leqslant \frac{37}{112}$. After further progress by Titchmarsh [267] ($\mu \leqslant \frac{15}{46}$) and Hua [116] ($\alpha \leqslant \frac{13}{40} + \varepsilon$), W.-lin Yin [283] and J.-run Chen [38] independently showed that $\alpha \leqslant \frac{12}{37} + \varepsilon$.

The best result to date is due to Kolesnik, who actually studied the related *Dirichlet divisor problem*, i.e., the remainder $\Delta(t) = \sum_{n \leqslant t} d(n) - \{(t \log t + (2\gamma - 1)t\}$ (here $d(n) = \sum_{d \mid n} 1$ and γ is the Euler-Mascheroni constant, $\gamma \simeq 0.577\ldots$). While the Ω-estimates for $E(t)$ and $\Delta(t)$ are not identical, it may be shown that the O-estimates are. After several earlier results, Kolesnik's last published result appears to be (see [136]) $E(t) = O(t^{(35/108)+\varepsilon})$, which, however, can still be improved by essentially the same methods [137].

In the opposite direction, Hardy [94] and Landau [149] (see also [150]) independently showed in 1915 that $\mu \geqslant \frac{1}{4}$. Hardy improved this [95] to $E(t) = \Omega_-((x \log x)^{1/4})$, $E(t) = \Omega_+(x^{1/4})$. This was further improved by Gangadharan [80] to $\Omega_+(x^{1/4}(\log_2 x)^{1/4}(\log_3 x)^{14})$ and by Corradi and Katai [48] to $E(t) = \Omega_+(x^{1/4} \exp(c(\log_2 x)^{1/4}(\log_3 x)^{-3/4}))$, where $\log_1 x = \log x$, $\log_n x = \log(\log_{n-1} x)$. On the other hand, the best Ω_--result to date seems to be that of Hafner (see [92]); (A. Selberg appears to have known, but not published, this result in the 1940s), namely

$$E(t) = \Omega_-\{(x \log x)^{1/4}(\log_2 x)^{(\log 2)/4} \exp(-B(\log_3 x)^{1/2})\}$$

for some absolute constant B.

Instead of considering the problem of representation as sum of two squares, we may consider that of representation as sum of m kth powers, where, for k odd, we ignore the signs; in other words, we consider $r_{m,k}(n)$, the number of representations of n by the sum $\sum_{j=1}^{m} |a_j|^k = n$. For $m = 2$, this leads to the representation $n = |a_1|^k + |a_2|^k$ as the sum of two kth powers. Then $B_{2,k}(t) = \sum_{n \leqslant t} r_{2,k}(n)$ gives the number of lattice points inside the simple closed curve $|x|^k + |y|^k = t$. E. Krätzel proves [140, pp. 181–192] that $B_{2,k}(t) = c_k t^{2/k} + O(t^{1/k})$, where

$$c_k = 4\int_0^1 (1 - t^k)^{1/k}\, dt = \frac{2\,\Gamma^2(1/k)}{k\,\Gamma(2/k)}.$$

In the particular case $k = 2$, this leads to Gauss's result $B_{2,2}(t) = A(t) = t + O(t^{1/2})$. B. Randol [224] improves this for $k > 2$, k even, to $B_{2,k}(t) = c_k t^{2/k} + O(t^c)$, with $c = 1/k - 1/k^2$, and shows that this is best possible (see also Section 4.15). Krätzel shows that a similar formula holds also for odd $k \geqslant 3$ and, furthermore, that $B_{2,3}(t) = c_3 t^{2/3} + O(t^{2/9}(\log t)^{2/3})$, while for $k > 3$, $B_{2,k}(t) = c_k t^{2/3} + \rho_k(t)^{1/k} + O(t^{2/3k}(\log t)^{2/3})$, with $\rho_k(t) = -8\int_0^t x^{k-1} \times (t^k - x^k)^{(1/k)-1}((t))\,dt = O(t^{1-(1/k)})$ (here $((t)) = t - [t] - \frac{1}{2}$).

§8. The Determination of $N_2(x)$

If $\mathbf{S}$ is a set of positive integers, then we denote by $\mathbf{S}(x)$ the number of integers in $\mathbf{S}$ not larger than x, i.e., $\mathbf{S}(x) = \sum_{n\in \mathbf{S}, n\leqslant x} 1$. As we denote by $\mathbf{N}_k$ the set of integers that are sums of k squares, $\mathbf{N}_k(x)$ stands for the number of such integers $n \in \mathbf{N}_k$, $n \leqslant x$. It is, of course, trivial that $\mathbf{N}_1(x) = [\sqrt{x}]$ (here $[x]$ stands for the greatest integer function, $x - 1 < [x] \leqslant x$, $[x] \in \mathbb{Z}$). In contrast to this obvious result and to the relatively easy solution of the "circle problem" (at least, if we don't insist on a good error term), the determination of $\mathbf{N}_2(x)$ is a difficult problem. It was solved in 1908 by Landau (1877–1938), and the result is [148] that

$$\mathbf{N}_2(x) = b(\log x)^{-1/2} + o(x(\log x)^{-1/2}),$$

where

$$b = \left\{\frac{1}{2}\prod_{q\equiv 3 \pmod 4} (1 - q^{-2})^{-1}\right\}^{1/2},$$

but the proof cannot be given here. The interested reader may wish to consult [146, Vol. 2] and also Hardy's *Ramanujan* [97, Lecture IV, pp. 60–63].

§9. Other Contributions to the Sum of Two Squares Problem

For completeness, let us add that work on this and related problems did not stop with the proof of Theorem 3. Indeed, Dickson [54, Chapter VI, pp. 231–257] lists results by over 100 other mathematicians who made contributions to this problem. Among them, we find the following well-known names: Goldbach, Lagrange, Legendre, Laplace, Barlow, Cauchy, Genocchi, Dirichlet, Eisenstein, Lebesgue, Hermite, Serret, Bouniakowski, Halphen, Lucas, Stieltjes, Catalan, Gegenbauer, Lipschitz, Weber, Thue, Dickson, Jacobsthal, Broccard, and many more.

§10. Problems

1. Find directly (i.e., by enumeration), and then by use of Theorem 3, the number of representations as a sum of two squares of the following integers: $n = 1, 2, 3, 4, 5, 6; 25; 100$. Compare the results.

2. Prove (2.2) formally.

3. Find $E(t) = A(t) - \pi t$ for $t = 25 - \varepsilon, 25, 25 + \varepsilon; 36 - \varepsilon, 36, 36 + \varepsilon$ ($\varepsilon > 0$, very small).

4. Prove that, for any $n \in \mathbb{Z}^+$, $d_1(n) \geqslant d_3(n)$; when exactly do we have equality here?

5. Prove Gauss's estimate of $E(t)$ by use of Theorem 3.

*6. Try to prove the conjecture $E(t) = O(t^{(1/4)+\varepsilon})$, or at least to improve the known results.

*7. Improve the error term of $\mathbf{N}_2(x)$.

*8. What can be said about the arithmetic nature of the constant $b = 0.764\ldots$ of §8 and its relations to other constants, such as π, e, $\zeta(m)$ $(= \sum_{n=1}^{\infty} 1/n^m)$, etc.?

Chapter 3

Triangular Numbers and the Representation of Integers as Sums of Four Squares

§1. Sums of Three Squares

After the discussion of representations as sums of two squares, it appears reasonable to consider next the case of representations as sums of three squares. However, as already indicated, the case of k odd is more difficult than that of k even and, what is more important, requires different methods.

To illustrate this statement, let us consider the first theorem of interest, in the case $k = 3$: that of the characterization of the set $\mathbf{N}_Q$ for $Q = \sum_{i=1}^{3} x_i^2$.

Theorem 1. *The Diophantine equation*

$$x_1^2 + x_2^2 + x_3^2 = n \tag{3.1}$$

has solutions with $x_i \in \mathbb{Z}$ if and only if n is not of the form $4^a(8m + 7)$, with $a \in \mathbf{Z}$, $a \geqslant 0$, $m \in \mathbb{Z}$.

As examples, observe that $n = 15 = 8 \cdot 1 + 7$ has no representation of the form (3.1), while $n = 17 = 2 \cdot 8 + 1 = 1^2 + 4^2 + 0^2 = 2^2 + 2^2 + 3^2$ does. Also $n = 60 = 4(8 \cdot 1 + 7)$ has no representation as a sum of three squares, while $68 = 4(2 \cdot 8 + 1) = 2^2(1^2 + 4^2 + 0^2) = 2^2 + 8^2 + 0^2 = 2^2(2^2 + 2^2 + 3^2) = 4^2 + 4^2 + 6^2$ and has two such representations. The proof that, for $n = 4^a(8m + 7)$, equation (3.1) has no solution is easy, but the converse, that if n is *not* of that form, integral solutions of (3.1) always exist, is anything but trivial and involves the rather sophisticated concepts of classes and genera of binary, and even ternary, quadratic forms (see Chapter 4).

On the other hand, the methods used to handle the case $k = 4$ are not too different from those used for $k = 2$. Therefore, letting pedagogical considerations prevail over logical ones, we shall first treat the case $k = 4$ in the present chapter and postpone the case $k = 3$ to Chapter 4. There is, however, no reason why anybody who dislikes this arrangement should not study Chapter 4 ahead of Chapter 3. The main proofs given in these two chapters are largely independent of each other and there will be no danger of running into circular reasoning.

§2. Three Squares, Four Squares, and Triangular Numbers

It appears that Diophantus was aware of the following theorem, first proven by Lagrange:

Theorem 2. *All natural integers are sums of four integral squares, i.e., for every $n \in \mathbb{Z}$, $n > 0$, the Diophantine equation*

$$x_1^2 + x_2^2 + x_3^2 + x_4^2 = n \tag{3.2}$$

has solutions with $x_i \in \mathbb{Z}$ $(i = 1, 2, 3, 4)$.

Remark. As was the case when $k = 2$, some of the squares may be zero. Thus, $25 = 5^2 + 0 + 0 + 0 = 3^2 + 4^2 + 0 + 0 = 1^2 + 2^2 + 2^2 + 4^2$ are essentially distinct representations of 25 as a sum of four squares.

There is no explicit statement of Theorem 2 in Diophantus; however, while he requires conditions for an integer to be a sum of two or of three squares, he states no conditions whatsoever for n to be a sum of four squares. His commentators Xylander and Bachet both interpret this as indicating a knowledge of Theorem 2. Bachet did state Theorem 2 explicitly and mention that he had verified it up to 325, but had no general proof of it.

Fermat claimed to have a proof of Theorem 2 and also remarked that Diophantus seemed to have known it. Somewhat earlier, Regiomontanus (Johannes Müller, 1436–1476) [54, Chapter 8] asked, among other things, for solutions of (3.2) when $n = m^2$. Descartes (1596–1650) also stated Theorem 2 and admitted that he had no proof of it. Fermat, in a letter to Carcavi [79, Vol. II, p. 433], indicated that his proof (like so many others of his) was based on the method of descent, whose application in this case, he added, required another new idea. Euler, in a letter of 1730 to Goldbach [54], mentioned that he could not prove Theorem 2, the real difficulty residing in showing that numbers of the form $n^2 + 7$ are sums of four squares. Indeed, 7 cannot be written as a sum of three squares, while if $a = \sum_{i=1}^{3} x_i^2$, then obviously $n^2 + a$ is a sum of four squares.

In another letter to Goldbach (October 1730; see [77]), Euler quoted the following claim of Fermat (made in a letter to Mersenne, 1636, if not earlier; see [79, Vol. II, p. 65]):

Theorem 3. *Every integer is the sum of three triangular numbers.*

We recall that n is said to be a *triangular number* if $n = a(a + 1)/2$, $a \in \mathbb{Z}$.

As an illustration of Theorem 3, $7 = 3 + 3 + 1$, with $3 = 2(2 + 1)/2$ and $1 = (1 + 1) \cdot 1/2$, both triangular numbers.

Euler observed that if Theorem 3 is true, then every integer $n \equiv 3 \pmod 8$

is the sum of three squares. Indeed given a nonnegative integer k, Theorem 3 enables us to write $k = \frac{1}{2}\sum_{i=1}^{3} x_i(x_i + 1)$, so that

$$8k + 3 = 4\sum_{i=1}^{3} x_i^2 + 4\sum_{i=1}^{3} x_i + 3 = \sum_{i=1}^{3} (2x_i + 1)^2. \tag{3.3}$$

Euler also observed that in order to prove Theorem 3, it is sufficient to show the following: If one defines the coefficients c_n by

$$\left(1 + \sum_{m=1}^{\infty} x^{t_m}\right)^3 = 1 + \sum_{n=1}^{\infty} c_n x^n,$$

where $t_m = m(m+1)/2$ is the mth triangular number, then $c_n \neq 0$ for all n. In fact, as seen in Chapter 1, c_n is the number of different (not necessarily essentially distinct) representations of n as a sum of three triangular numbers. This is the second instance we meet in which Euler's approach foreshadows Jacobi's method of theta functions. Indeed, if we set $y = x^{1/2}$, then

$$1 + \sum_{m=1}^{\infty} x^{t_m} = 1 + \sum_{m=1}^{\infty} y^{m^2+m} = y^{-1/4} \sum_{m=0}^{\infty} y^{(m+1/2)^2}.$$

Here the sum is just the classical theta function, denoted by Jacobi as $\frac{1}{2}\theta_2(0 \mid y)$ (see Chapter 8).

It seems that the first known proof of Fermat's Theorem 3, as of so many others, is due to Gauss [81, §293], who showed that every $n \equiv 3 \pmod 8$ is a sum of three odd squares, so that, by working backwards in (3.3), every integer k is the sum of three triangular numbers.

We shall now show that Fermat's Theorem 3 implies Theorem 2; recall that the former implies that every $n \equiv 3 \pmod 8$ is the sum of three squares. Given this, we claim that $8m + 4$ is a sum of four squares, and so also is $n = 2m + 1$. Indeed, if $n \equiv 0 \pmod 4$, with $n = \sum_{i=1}^{4} x_i^2$, then either all x_i's are even, or all are odd. In the first case, $\sum_{i=1}^{4} (x_i/2)^2 = n/4 = 2m + 1$; in the second case,

$$\left(\frac{x_1 + x_2}{2}\right)^2 + \left(\frac{x_1 - x_2}{2}\right)^2 + \left(\frac{x_3 + x_4}{2}\right)^2 + \left(\frac{x_3 - x_4}{2}\right)^2 = 4m + 2,$$

so that $4m + 2 = y_1^2 + y_2^2 + y_3^2 + y_4^2$, with two even and two odd summands. Let y_1, y_2 be even, y_3, y_4 be odd. Then

$$\left(\frac{y_1 + y_2}{2}\right)^2 + \left(\frac{y_1 - y_2}{2}\right)^2 + \left(\frac{y_3 + y_4}{2}\right)^2 + \left(\frac{y_3 - y_4}{2}\right)^2 = 2m + 1,$$

as claimed. Hence, every odd integer is the sum of four squares, and so is every integer n such that $n \equiv 4 \pmod 8$. Also, if $2m + 1 = \sum_{i=1}^{4} z_i^2$, then $(z_1 + z_2)^2 +$

$(z_1 - z_2)^2 + (z_3 + z_4)^2 + (z_3 - z_4)^2 = 4m + 2$, which proves the statement for $n \equiv 2 \pmod 4$. Since $4 = 2^2$, $4n$ is a sum of four squares if n is a sum of four squares. This finishes the proof of Theorem 2, assuming that Theorem 3 holds.

Similarly, if we accept the validity of Theorem 1, then, for $n \neq 4^a(8m + 7)$, equation (3.2) has integer solutions with $x_4 = 0$. If $n = 4^a(8m + 7)$, then $n - 4^a = 4^a(8m + 6) = \sum_{i=1}^{3} x_i^2$ and (3.2) has integer solutions with $x_4 = 2^a$.

§3. The Proof of Theorem 2

While, as observed, Theorem 2 follows almost trivially from either Theorem 1 or Theorem 3, it is easier to prove Theorem 2 directly. Indeed, Lagrange (1736–1813) gave the first proof of Theorem 2 before 1770 (see [145]), without any reference to Theorem 1 or 3. In his proof, he used a lemma that Euler had obtained some 20 years earlier [75].

Lemma 1 (Euler). *If p is an odd prime, then*

$$1 + x^2 + y^2 = mp \tag{3.4}$$

is solvable in integers x, y, m, with $0 < m < p$.

Proof. If $p \equiv 1 \pmod 4$, then $x^2 + 1 \equiv 0 \pmod p$ has a solution. x may be selected so that $|x| < p/2$, and hence $1 < 1 + x^2 < 1 + (p/2)^2$. Thus $mp = 1 + x^2 < 1 + p^2/4$, whence $m < p/2$. Equation (3.4) holds for $y = 0$ and the x found before, even for $0 < m < p/2$. If $p \equiv 3 \pmod 4$, one can no longer solve (3.4) with $y = 0$, because $(-1/p) = -1$. One can, however, still attempt to solve (3.4) with "small" values of x and y. In particular, observe that the $(p + 1)/2$ integers $0^2, 1^2, \ldots, \{(p - 1)/2\}^2$ are incongruent modulo p, and so are the $(p + 1)/2$ integers $-(1 + 0^2), -(1 + 1^2), \ldots, -(1 + ((p - 1)/2)^2)$. Now let x run through the first set and $-(1 + y^2)$ through the second set; then the set of $p + 1$ integers, consisting of the union of those two sets, cannot have all its elements incongruent to each other, because there are only p distinct residue classes modulo p. This type of reasoning is often called *Dirichlet's drawer principle* (*Schubfachprinzip*)* ($m + 1$ objects put into m drawers imply that at least one drawer contains more than a single object). Hence, for at least one value of x, $0 \leqslant x \leqslant (p - 1)/2$, and one value of y, $0 \leqslant y \leqslant (p - 1)/2$, we do have $x^2 \equiv -1 - y^2 \pmod p$, whence $x^2 + y^2 + 1 = mp$. However,

$$1 + x^2 + y^2 \leqslant 1 + 2\left(\frac{p - 1}{2}\right)^2 = \frac{(p - 1)^2 + 2}{2}$$

and so $m \leqslant (p^2 - 2p + 5)/2p < p$, as is easily seen. □

---

*Today this is often refered to also as the "pigeonhole principle."

Proof of Theorem 2. The proof of Theorem 2.2 was reduced to the easier case of $n = p$ by the existence of equation (2.2) of Chapter 2, from which it follows that if two integers are sums of two squares, then so is their product. Fortunately, a similar formula holds also in the case of four squares, and, just as in Chapter 2, one can prove the corresponding identity (equation (3.5) below) in the most elementary way, by verifying that both sides are equal to $\sum_{i=1}^{4}\sum_{j=1}^{4} x_i^2 y_j^2$. It seems that this identity appeared first in Euler (see [54, Chapter VIII, p. 277, line 2] or [76]):

$$\left(\sum_{i=1}^{4} x_i^2\right)\left(\sum_{j=1}^{4} y_j^2\right) = \left(\sum_{i=1}^{4} x_i y_i\right)^2 + (x_1 y_2 - x_2 y_1 + x_3 y_4 - x_4 y_3)^2$$
$$+ (x_1 y_3 - x_3 y_1 + x_4 y_2 - x_2 y_4)^2 \tag{3.5}$$
$$+ (x_1 y_4 - x_4 y_1 + x_2 y_3 - x_3 y_2)^2.$$

In order to prove Theorem 2, it is therefore sufficient to prove that (3.2) is solvable in integers x_i ($i = 1, 2, 3, 4$), provided that $n = p$, an odd prime, because, obviously, $2 = 1^2 + 1^2 + 0 + 0$.

From Euler's Lemma 1, it follows that

$$\sum_{i=1}^{4} x_i^2 = mp \tag{3.6}$$

is solvable even under the restrictions $x_3 = 1$, $x_4 = 0$, $0 < m < p$. We now show that, letting x_3 and x_4 be arbitrary integers, we actually may take $m = 1$ in (3.6). This proof may be compared with that of Lemma 2.2 and completes the proof of Theorem 2.

If we deny the assertion, then there exists m_0 with $1 < m_0 < p$, such that (3.6) is solvable for $m = m_0$, but not for $m < m_0$. We shall show, however, that if (3.6) is solvable for any $m_0 > 1$, then it also is solvable for some $m_1 < m_0$, which contradicts the definition of m_0. This proof is, of course, an instance of Fermat's method of descent and may well have been Fermat's own method of proof.

We make two remarks:

(i) The minimal m_0 is odd; hence $m_0 \geqslant 3$. Indeed, if $\sum_{i=1}^{4} x_i^2 = m_0 p$ is even, the number of odd summands (if any) must be even, and by pairing them together we obtain (by renumbering, if necessary) that

$$\left(\frac{x_1 - x_2}{2}\right)^2 + \left(\frac{x_1 + x_2}{2}\right)^2 + \left(\frac{x_3 - x_4}{2}\right)^2 + \left(\frac{x_3 + x_4}{2}\right)^2 = \frac{m_0}{2} p,$$

with $m_1 = m_0/2 < m_0$.

(ii) not all x_i's can be divisible by m_0, because, if they all were then m_0^2

would divides $\sum_{i=1}^4 x_i^2$, or $m_0^2 | m_0 p, m_0 | p$, whence (recall $m_0 < p$) $m_0 = 1$, contrary to our assumption.

Let b_i be the integer closest to x_i/m_0; then $x_i = m_0 b_i + y_i$, with $|y_i| < m_0/2$, and, by (ii), not all y_i are zero. Hence,

$$0 < \sum_{i=1}^{4} y_i^2 < 4\left(\frac{m_0}{2}\right)^2 = m_0^2. \tag{3.7}$$

Also, from $y_i \equiv x_i \pmod{m_0}$ follows

$$\sum_{i=1}^{4} y_i^2 \equiv \sum_{i=1}^{4} x_i^2 = m_0 p \equiv 0 \pmod{m_0}. \tag{3.8}$$

From (3.7) and (3.8) it follows that, for some $0 < m_1 < m_0$, $\sum_{i=1}^4 y_i^2 = m_0 m_1$. If we multiply this by $\sum_{i=1}^4 x_i^2 = m_0 p$ and use (3.5), we obtain that $\sum_{i=1}^4 X_i^2 = m_0^2 m_1 p$. By (3.5), $X_i \equiv 0 \pmod{m_0}$ $(i = 1, 2, 3, 4)$. Indeed,

$$X_1 = \sum_{i=1}^{4} x_i y_i = \sum_{i=1}^{4} x_i(x_i - b_i m_0) \equiv \sum_{i=1}^{4} x_i^2 \equiv m_0 p \equiv 0 \pmod{m_0},$$

$$\begin{aligned} X_2 &= x_1 y_2 - x_2 y_1 + x_3 y_4 - x_4 y_3 \\ &= x_1(x_2 - b_2 m_0) - x_2(x_1 - b_1 m_0) + x_3(x_4 - b_4 m_0) - x_4(x_3 - b_3 m_0) \\ &= m_0(-x_1 b_2 + x_2 b_1 - x_3 b_4 + x_4 b_3) \equiv 0 \pmod{m_0}, \end{aligned}$$

and similarly for X_3 and X_4. Let us now set $X_i = m_0 x_i'$; then

$$\sum_{i=1}^{4} X_i^2 = m_0^2 \sum_{i=1}^{4} (x_i')^2 = m_0^2 m_1 p,$$

or

$$\sum_{i=1}^{4} (x_i')^2 = m_1 p, \quad \text{with } 1 < m_1 < m_0,$$

contrary to the minimality of m_0. This contradiction finishes the proof that (3.2) is solvable if n is any prime, and consequently, by (3.5), that (3.2) is solvable if n is any product of primes, i.e., any integer, as claimed. □

The present proof is much shorter and simpler than Lagrange's original one.

Soon after Lagrange, Euler published two proofs of his own, simpler than Lagrange's (one of them is essentially the previous one). Other proofs are due to Legendre (1752–1833). Gauss proved Theorem 2 by use of Theorem 1, which had been proved by then. He also gave an interpretation (published only posthumously) of (3.5) in terms of norms of Gaussian integers [82, Vol. 8, p. 3].

§4. Main Result

Soon afterwards (1828), Jacobi [122], by the use of elliptic and theta functions, proved the definitive theorem on this representation problem:

Theorem 4. *Let $\sigma'(n)$ stand for the sum of the divisors of n that are not multiples of 4, $\sigma'(n) = \sum_{d \mid n,\, d \not\equiv 0 \pmod 4} d$. Then*

$$r_4(n) = 8\sigma'(n). \tag{3.9}$$

It follows from (3.9) that the divisors $d \mid n$, $d \equiv 0 \pmod 4$, are irrelevant for $r_4(n)$. This fact is also an immediate consequence of the following easy

Theorem 5. *For every $n \in \mathbb{Z}$, $n > 0$,*

$$r_4(2n) = r_4(8n).$$

Corollary. *If n is odd, then for every nonnegative $a \in \mathbb{Z}$,*

$$r_4(2 \cdot 4^a n) = r_4(2n).$$

The corollary follows immediately from Theorem 5. Hence, it is sufficient to justify the latter. We do it by exhibiting a 1–1 correspondence between the representations of $2n$ and of $8n$ as sums of four squares.

If $2n = \sum_{i=1}^4 x_i^2$, then $8n = \sum_{i=1}^4 (2x_i)^2$. On the other hand, if $8n = \sum_{i=1}^4 y_i^2$, then $\sum_{i=1}^4 y_i^2 \equiv 0 \pmod 8$, so that all y_i's have to be even, say $y_i = 2x_i$. Then

$$8n = \sum_{i=1}^{4} (2x_i)^2 = 4 \sum_{i=1}^{4} x_i^2 \quad \text{and} \quad 2n = \sum_{i=1}^{4} x_i^2,$$

as claimed.

Caution. It is *not* true, in general, that $r_4(n) = r_4(4n)$ (e.g., $r_4(1) = 8$, $r_4(4) = 24$); indeed, if n is odd, then $4n \equiv 4 \pmod 8$ and $4n = \sum_{i=1}^4 x_i^2$ is possible with all four x_i's odd, $x_i^2 \equiv 1 \pmod 8$, and above proof does not go through.

As a first example of Theorem 4, let $n = 11$. Clearly, for any integer that is itself not a multiple of 4, $\sigma'(n) = \sigma(n)$. Hence, $\sigma'(n) = 11 + 1 = 12$, and Theorem 4 states that $r_4(11) = 8 \cdot 12 = 96$. In fact, 11 has essentially only one representation as sum of four squares, namely $11 = 1^2 + 1^2 + 3^2 + 0^2$. The four summands have 12 distinct permutations, and each nonvanishing square permits two choices of signs, $(\pm 1)^2$ and $(\pm 3)^2$, for a total of $2^3 = 8$ different choices of signs and a total of $12 \cdot 8 = 96$ different (although not essentially distinct) representations of 11.

Next, $4 = (\pm 1)^2 + (\pm 1)^2 + (\pm 1)^2 + (\pm 1)^2 = (\pm 2)^2 + 0^2 + 0^2 + 0^2$; here the first representation is counted as $2^4 = 16$ different ones, according to the choices of signs; in the second representation, the summand $(\pm 2)^2$ may occupy any one of four positions, and, as we may choose either $+2$ or -2, the representation counts for $4 \cdot 2 = 8$ distinct ones. This leads to a total of $16 + 8 = 24$ representations. According to Theorem 4, we consider only the divisors 1 and 2 (4 itself is eliminated), so that $\sigma'(4) = 1 + 2 = 3$. By (3.9), we find that $r_4(4) = 8 \cdot \sigma'(4) = 8 \cdot 3 = 24$. Similarly, $1 = (\pm 1)^2 + 0^2 + 0^2 + 0^2$ counts for 8 representations (4 permutations and 2 choices of signs for each).

Shortly after his first proof of Theorem 4, Jacobi (in 1834; see [123]) also gave an elementary proof of it.

§5. Other Contributions

Among the numerous later contributions on this problem listed by Dickson [54, Chapter VIII, pp. 285–303], are those by Barlow, Cauchy, Bouniakowski, Libri, Eisenstein, Tchebychef, Hermite, Genocchi, Dirichlet, Sylvester, Liouville, Pollock, H. J. S. Smith, Lebesgue, Glaisher, Réalis, Catalan, T. Pepin, Lipschitz, M. A. Stern, E. Humbert, Gegenbauer, Dickson, von Sterneck, Bolzano, Bachmann (expository work), Mordell, and many others. It is only natural that many of these contributions should be repetitions, rediscoveries, or variations of already published proofs. E. Landau showed [149] (compare with Section 2.7) that $\sum_{n \leqslant x} r_4(n) = \frac{1}{2}\pi^2 x^2 + R(x)$. Here the first term is, of course, the volume of the 4-dimensional sphere of radius $x^{1/2}$, and $R(x)$ is an error term and satisfies $R(x) = O(x^{1+\varepsilon})$. We infer, in particular, that $\bar{r}_4(x) = x^{-1} \sum_{n \leqslant x} r_4(n)$, the average number of representations of an integer n $(1 \leqslant n \leqslant x)$ by four squares, is asymptotically equal to $\frac{1}{2}\pi^2 x$; see also Section 9.6.

§6. Proof of Theorem 4

We shall see an elegant proof of Theorem 4 in Chapter 9. Nevertheless, it seems worthwhile to indicate here also an elementary proof of it, in the spirit of the proof of Theorem 2.3. Certain computational steps in this proof will be given in a somewhat condensed form. While the motivation for the proof is to be found in the theory of elliptic functions, the reader who does not know anything about them will not be at any disadvantage, if he is willing to forgo the motivation.

Lemma 2 (Jacobi [121]). *Let $u_r = u_r(x) = x^r/(1 - x^r)$; then*

$$\left(1 + 2\sum_{n=1}^{\infty} x^{n^2}\right)^2 = \left(\sum_{n=-\infty}^{\infty} x^{n^2}\right)^2 = 1 + \sum_{n=1}^{\infty} r_2(n)x^n = 1 + 4\sum_{r=1}^{\infty}(-1)^r u_{2r+1}.$$

Proof of Lemma 2. We already know the first two equalities from Chapter 1. Without appeal to elliptic functions, we prove the last equality by observing that

$$\sum_{r=1}^{\infty}(-1)^r u_{2r+1} = \sum_{r=1}^{\infty}(-1)^r \frac{x^{2r+1}}{1 - x^{2r+1}}$$

$$= \sum_{r=1}^{\infty}(-1)^r \sum_{m=1}^{\infty} x^{m(2r+1)} = \sum_{k=1}^{\infty} x^k \sum_{2r+1 \mid k} (-1)^r$$

$$= \sum_{k=1}^{\infty} x^k \left\{ \sum_{\substack{d \mid k \\ d \equiv 1}} 1 - \sum_{\substack{d \mid k \\ d \equiv 3}} 1 \right\},$$

where the congruences on d are modulo 4; by Theorem 2.3, the expression between the curly braces equals $\frac{1}{4}r_2(k)$, and the proof is complete. □

Remark. If the reader happens to know elliptic and theta functions, he will recognize that Lemma 2 is an immediate consequence of the fact that the extreme terms are both equal to $2K/\pi$ ($K =$ the quarter period of the elliptic integral with $v = \pi/2$, $u = K$, so that ns $u = 1$; see [1], 16.38.5 and 16.23.10, respectively).

To prove Theorem 4, we again recall from Chapter 1 that

$$\left(1 + 2\sum_{m=1}^{\infty} x^{m^2}\right)^4 = \left(\sum_{m=-\infty}^{\infty} x^{m^2}\right)^4 = 1 + \sum_{n=1}^{\infty} r_4(n)x^n.$$

By Lemma 2,

$$1 + \sum_{n=1}^{\infty} r_4(n)x^n = \left(1 + 4\sum_{r=1}^{\infty}(-1)^r u_{2r+1}\right)^2. \tag{3.10}$$

Let us also accept for a moment the following lemma.

Lemma 3. *With the previous notation,*

$$\left(1 + 4\sum_{r=1}^{\infty}(-1)^r u_{2r+1}\right)^2 = 1 + 8\sum_{n \not\equiv 0 \pmod 4} n u_n.$$

By combining (3.10) with Lemma 3, we obtain

$$1+\sum_{n=1}^{\infty} r_4(n)x^n = \left(1+4\sum_{r=1}^{\infty}(-1)^r u_{2r+1}\right)^2$$

$$= 1 + 8 \sum_{r\not\equiv 0 \ (\mathrm{mod}\ 4)} r u_r = 1 + 8 \sum_{r\not\equiv 0 \ (\mathrm{mod}\ 4)} \frac{r x^r}{1-x^r}$$

$$= 1 + 8 \sum_{r\not\equiv 0 \ (\mathrm{mod}\ 4)} r \sum_{m=1}^{\infty} x^{rm}$$

$$= 1 + 8 \sum_{n=1}^{\infty} x^n \sum_{\substack{r\mid n \\ r\not\equiv 0 \ (\mathrm{mod}\ 4)}} r = 1 + 8\sum_{n=1}^{\infty}\sigma'(n)x^n,$$

by the definition of $\sigma'(n)$. The identification of the coefficients of x^n now finishes the proof of Theorem 4, except for the proof of Lemma 3.

§7. Proof of Lemma 3

In order not to interrupt the proof with distracting technical details, we note here the following identities, due to Ramanujan (1887–1920) (see [222] and [223]):

$$\sum_{m=1}^{\infty} u_m(1+u_m) = \sum_{m=1}^{\infty}\frac{x^m}{1-x^m}\left(1+\frac{x^m}{1-x^m}\right) = \sum_{m=1}^{\infty}\frac{x^m}{(1-x^m)^2}$$

$$= \sum_{m=1}^{\infty} x^m \sum_{n=1}^{\infty} n x^{(n-1)m} = \sum_{m=1}^{\infty}\sum_{n=1}^{\infty} n x^{nm}$$

$$= \sum_{n=1}^{\infty} n \sum_{m=1}^{\infty} x^{nm} = \sum_{n=1}^{\infty}\frac{n x^n}{1-x^n} = \sum_{n=1}^{\infty} n u_n.$$

We shall need later only the final result obtained, namely

$$\sum_{m=1}^{\infty} u_m(1+u_m) = \sum_{n=1}^{\infty} n u_n. \tag{3.11}$$

Next,

$$\sum_{m=1}^{\infty}(-1)^{m-1}u_{2m}(1+u_{2m}) = \sum_{m=1}^{\infty}(-1)^{m-1}\frac{x^{2m}}{(1-x^{2m})^2}$$

$$= \sum_{m=1}^{\infty}(-1)^{m-1}\sum_{r=1}^{\infty} r x^{2mr} = \sum_{r=1}^{\infty} r \sum_{m=1}^{\infty}(-1)^{m-1}x^{mr}$$

$$= \sum_{r=1}^{\infty} \frac{rx^{2r}}{1+x^{2r}} = \sum_{r=1}^{\infty} r\left\{\frac{x^{2r}}{1-x^{2r}} - \frac{2x^{4r}}{1-x^{4r}}\right\}$$

$$= \sum_{r=1}^{\infty} \frac{rx^{2r}}{1-x^{2r}} - \sum_{r=1}^{\infty} \frac{2rx^{4r}}{1-x^{4r}}$$

$$= \sum_{n=1}^{\infty} \frac{(2n-1)x^{2(2n-1)}}{1-x^{2(2n-1)}},$$

because all terms of the sum $\sum_{r=1}^{\infty} (rx^{2r})/(1-x^{2r})$ with even r are canceled by terms of the sum $\sum_{r=1}^{\infty} 2r \cdot x^{4r}/(1-x^{4r})$. It follows that

$$\sum_{m=1}^{\infty} (-1)^{m-1} u_{2m}(1+u_{2m}) = \sum_{n=1}^{\infty} (2n-1)u_{4n-2}. \tag{3.12}$$

The interchange in the order of summation is easily justified.

Let now $L = L(x,\theta) = \frac{1}{4}\cot(\theta/2) + \sum_{n=1}^{\infty} u_n \sin n\theta$, $T_1 = T_1(x,\theta) = \frac{1}{16}\cot^2(\theta/2) + \sum_{n=1}^{\infty} u_n(1+u_n)\cos n\theta$, $T_2 = T_2(x,\theta) = \frac{1}{2}u_1(1+\cos\theta) + \sum_{n=2}^{\infty} nu_n(1-\cos n\theta)$. Then we claim that

$$L^2 = T_1 + T_2. \tag{3.13}$$

To see this, observe that $L^2 = \frac{1}{16}\cot^2(\theta/2) + S_1 + S_2$, where

$$S_1 = \frac{1}{2}\sum_{n=1}^{\infty} u_n \cot\frac{\theta}{2} \sin n\theta$$

$$= \sum_{n=1}^{\infty} u_n\left(\frac{1}{2} + \sum_{m=1}^{n-1} \cos m\theta + \frac{1}{2}\cos n\theta\right)$$

and

$$S_2 = \sum_{m=1}^{\infty}\sum_{n=1}^{\infty} u_m u_n \sin m\theta \sin n\theta$$

$$= \frac{1}{2}\sum_{m=1}^{\infty}\sum_{n=1}^{\infty} u_m u_n\{\cos(m-n)\theta - \cos(m+n)\theta\}.$$

Hence, by rearranging the series in L^2 (permitted by the absolute convergence of $\sum_n nu_n$ and $\sum_m \sum_n u_m u_n$), it follows that $L^2 = \frac{1}{16}\cot^2(\theta/2) + C_0 + \sum_{k=1}^{\infty} C_k \cos k\theta$. Here C_0 contains the terms free of cosines, i.e., $C_0 = \frac{1}{2}\sum_{n=1}^{\infty} (u_n + u_n^2) = \frac{1}{2}\sum_{n=1}^{\infty} nu_n$, by (3.11). The computations for C_k are elementary, but somewhat lengthy, and lead to $C_k = u_k(1 + u_k - k/2)$ (see for details [98, Chapter 12] or [97]). It follows that

$$L^2 = \frac{1}{16}\cot^2\frac{\theta}{2} + \frac{1}{2}\sum_{n=1}^{\infty} nu_n + \sum_{k=1}^{\infty} u_k\left(1 + u_k - \frac{k}{2}\right)\cos k\theta$$

$$= \frac{1}{16}\cot^2\frac{\theta}{2} + \sum_{k=1}^{\infty} u_k(1 + u_k)\cos k\theta + \frac{1}{2}\sum_{k=1}^{\infty} ku_k(1 - \cos k\theta)$$

$$= T_1 + T_2,$$

as claimed. In particular, for $\theta = \pi/2$, we obtain

$$T_1 = \frac{1}{16} - \sum_{m=1}^{\infty} (-1)^{m-1} u_{2m}(1 + u_{2m}) = \frac{1}{16} - \sum_{m=1}^{\infty} (2m - 1)u_{4m-2},$$

by (3.12), and $T_2 = \frac{1}{2}\sum_{m=1}^{\infty} (2m-1)u_{2m-1} + 2\sum_{m=1}^{\infty} (2m-1)u_{4m-2}$ and $T_1 + T_2 = \frac{1}{16} + \frac{1}{2}\sum_{n=1,\, n\not\equiv 0 \,(\mathrm{mod}\, 4)}^{\infty} nu_n$; also for $\theta = \pi/2$, $L = \frac{1}{4}\cot(\pi/4) + \sum_{n=1}^{\infty} (-1)^n u_{2n+1}$. If we substitute this in (3.13), we obtain

$$\frac{1}{16}\left(1 + 4\sum_{n=1}^{\infty} (-1)^n u_{2n+1}\right)^2 = \frac{1}{16} + \frac{1}{2}\sum_{n=1}^{\infty} nu_n,$$

which proves Lemma 3 and completes the proof of Theorem 4.

§8. Sketch of Jacobi's Proof of Theorem 4

Jacobi's own, elegant proof is based, as in the case of $r_2(n)$, on an identity from the theory of elliptic functions:

$$1 + \sum_{n=1}^{\infty} r_4(n)x^n = \left\{\sum_{m=-\infty}^{\infty} x^{m^2}\right\}^4 = \left(1 + 2\sum_{m=1}^{\infty} x^{m^2}\right)^4$$
$$= 1 + 8\sum_{m=1}^{\infty} \frac{mx^m}{1 + (-1)^m x^m}, \tag{3.14}$$

where (see Chapter 1) only the last equality still needs justification. Assuming it, we observe that for m odd,

$$\sum_{m \text{ odd}} \frac{mx^m}{1 + (-1)^m x^m} = \sum_{m \text{ odd}} mx^m(1 + x^m + x^{2m} + \cdots + x^{km} + \cdots)$$

$$= \sum_{n=1}^{\infty} x^n \sum_{\substack{m \mid n \\ m \text{ odd}}} m = \sum_{n=1}^{\infty} x^n(d_1(n) + d_3(n)),$$

while for m even,

$$\sum_{m \text{ even}} \frac{mx^m}{1 + x^m} = \sum_{m \text{ even}} m(x^m - x^{2m} + x^{3m} - \cdots)$$

$$= \sum_{n=1}^{\infty} x^n \left(\sum_{\substack{m \text{ even} \\ n/m \text{ odd} \\ m \mid n}} m - \sum_{\substack{m \text{ even} \\ n/m \text{ even} \\ m \mid n}} m \right).$$

We shall now evaluate the coefficient of x^n in this series. If $n = 2^k v$ with v odd and $m \mid n$, then $m = 2^a d$ with $d \mid v$, $1 \leqslant a \leqslant k$. In the first sum, with n/m odd, only the divisors $2^k d$ occur. In the second sum only the divisors $2^a d$ with $1 \leqslant a \leqslant k - 1$ occur. This means that, for that particular n, the value of the expression in parentheses is

$$\sum_{\substack{d \mid n \\ d \text{ odd}}} d\{2^k - (2^{k-1} + 2^{k-2} + \cdots + 2^2 + 2)\} = \sum_{\substack{2d \mid n \\ 2d \equiv 2 \pmod 4}} (2d)$$

$$= \sum_{\substack{m \mid n \\ m \equiv 2 \pmod 4}} m = d_2(n).$$

Hence, by (3.14),

$$1 + \sum_{n=1}^{\infty} r_4(n)x^n = 1 + 8 \sum_{n=1}^{\infty} x^n(d_1(n) + d_3(n) + d_2(n))$$

$$= 1 + 8 \sum_{n=1}^{\infty} x^n \sum_{\substack{d \mid n \\ d \not\equiv 0 \pmod 4}} d,$$

and Theorem 4 follows by identification of the coefficients. The proof of Jacobi's elliptic function identity, assumed here, will be given in §4 of Chapter 9 (see also Problem 4 in this chapter).

§9. Problems

1. Find all essentially distinct representations of the integers 1, 2, 3, 4, 5, 6, 7, 8, 9, 10, 100, 144:

   (a) as sums of three squares (if any exist);
   (b) as sums of three triangular numbers;
   (c) as sums of four squares.

   To each such representation as a sum of squares correspond a certain number of representations that differ from it and among each other only by a permutation of terms or the sign of a squared term. Indicate how many different representations correspond to each essentially distinct one.

2. Use Theorem 4 to compute $r_4(n)$ for the values of n in Problem 1, and compare the results.

3. Determine the number of primitive representations by sums of four squares for each of the values of n in Problem 1, and verify the values of $r_4(n)$ by use of Theorem 1.1.

4. Prove Jacobi's identity (3.14).
(Hint: On account of (3.10), it suffices to prove

$$\sum_{\substack{r=1 \\ r \not\equiv 0 \pmod 4}}^{\infty} \frac{rx^r}{1-x^r} = \sum_{m=1}^{\infty} \frac{mx^m}{1+(-1)^m x^m}.$$

By expanding both sides, we obtain $\sum_{n=1}^{\infty} a_n x^n = \sum_{n=1}^{\infty} b_n x^n$, where

$$a_n = \sum_{\substack{r \mid n \\ r \not\equiv 0 \pmod 4}} r \quad \text{and} \quad b_n = \sum_{\substack{m \mid n \\ m \text{ odd}}} m + \sum_{\substack{m \mid n \\ m \text{ even} \\ n/m \text{ odd}}} m - \sum_{\substack{m \mid n \\ m \text{ even} \\ n/m \text{ even}}} m.$$

The equality $a_n = b_n$ is immediately verified for $n \equiv 1, 2$, or $3 \pmod 4$. If $m = 2^a n$, $a \geqslant 2$, one may proceed as in §8).

5. Can the error term in $\sum_{n \leqslant x} r_4(n) = (\pi^2/2)x^2 + R(x)$ be improved to $R(x) = o(x)$? Why or why not?

Chapter 4

Representations as Sums of Three Squares

§1. The First Theorem

As already remarked in Chapter 3, it is easy to verify the "only if" part of Legendre's Theorem 3.1, which we recall here in its complete form.

Theorem 1. *The Diophantine equation*

$$x_1^2 + x_2^2 + x_3^2 = n \tag{4.1}$$

has solutions in integers x_i $(i = 1, 2, 3)$ *if and only if* n *is not of the form* $4^a(8k + 7)$ *with* $a, k \in \mathbb{Z}$. For all n, $r_3(4^a n) = r_3(n)$.

On the other hand, the proof that (4.1) always has solutions for $n \neq 4^a \times (8k + 7)$ is rather difficult. This is due, to a large extent, to the fact that in the present case, we do not have at our disposal an identity analogous to (2.2) of Chapter 2, or (3.5) of Chapter 3. In fact, it is *not true* that if n and m do each have representations as sums of three squares, then also their product nm has such a representation. For example, $3 = 1^2 + 1^2 + 1^2$ and $5 = 2^2 + 1^2 + 0^2$, but $3 \cdot 5 = 15 \equiv 7 \pmod 8$, and 15 cannot be represented as a sum of three squares. In fact, for any $n \equiv 3 \pmod 8$ and any $m \equiv 5 \pmod 8$, the product satisfies $nm \equiv 7 \pmod 8$ and nm is not the sum of three squares.

§2. Proof of Theorem 1, Part I

For any integer x, one always has $x^2 \equiv 0, 1$, or $4 \pmod 8$. In general, we may set $x_i = 2^{a_i} y_i$, with $2 \nmid y_1 y_2 y_3$, so that $y_i^2 \equiv 1 \pmod 8$. Without loss of generality, let $0 \leqslant a = a_1 \leqslant a_2 \leqslant a_3$. Then $n = \sum_{i=1}^{3} x_i^2 = 4^a(y_i^2 + 4^b y_2^2 + 4^c y_3^2) = 4^a n_1$, say, with $0 \leqslant b \leqslant c$. If $b \geqslant 2$, then $n_1 \equiv 1 \pmod 8$. Otherwise, according to the possible values of the pair (b, c), namely $(0, 0)$, $(0, 1)$, $(0, c)$, $(1, 1)$, $(1, c)$ with $c \geqslant 2$, we find $n_1 \equiv 3, 6, 2, 1, 5 \pmod 8$, respectively. If $a \geqslant 1$ and $n_1 \equiv 1 \pmod 8$, we also can write $n = 4^a(8k + 1) = 4^{a-1}(8k_1 + 4)$ and, if $a \geqslant 2$, $n = 4^{a-2}(8k_2 + 0)$. It follows that if n admits a representation as a sum of three squares, then it is necessarily of the form $n = 4^a(8k + m)$, where $0 \leqslant m \leqslant 6$. Also, $n \equiv 0 \pmod 4)$

occurs only if all summands are even; in particular, if $4 \mid n$, then (4.1) has no primitive solutions. Now if $n = x_1^2 + x_2^2 + x_3^2$, then $4n = (2x_1)^2 + (2x_2)^2 + (2x_3)$, and conversely, if $4n = y_1^2 + y_2^2 + y_3^2$, then all three y_i's are even, say $y_i = 2x_i$, so that $n = x_1^2 + x_2^2 + x_3^2$ and there is a 1 to 1 correspondence between the representations by sums of three squares of n and of $4n$. Hence, $r_3(n) = r_3(4n)$ and, by induction on a, $r_3(n) = r_3(4^a n)$.

§3. Early Results

Diophantus (see [54, p. 259]) gave a condition for the solvability of (4.1) if $n \equiv 1 \pmod 3$, which is essentially equivalent to $n \neq 24k + 7$. Bachet found this condition insufficient and required instead that, besides $n \neq 24k + 7$, also $n \neq 96k + 28$ $(= 4(24k + 7))$ should hold. These conditions are not quite as far off the mark as it may seem, because, in the quasigeometric formulation of the problem by Diophantus, the coefficient of k had to be divisible by 3. Fermat observed that even Bachet's conditions were insufficient and gave many counterexamples, of which the simplest is $112 \equiv 1 \pmod 3$, $112 = 4^2 \cdot 7$, not excluded by either Diophantus or Bachet. Fermat himself formulated the correct conditions in a precise, if rather cumbersome form. In a letter [79, Vol. II, p. 66; Vol. III, p. 287] to Mersenne (1636), he states that no integer of the form $8k + 7$ is the sum of three squares, his proof being based on $x^2 \equiv 0, 1$, or $4 \pmod 8$, as above. Fermat also considered the problem of representation of an integer as a sum of three *nonvanishing* squares (in two letters to Mersenne, 1636; see [79, Vol. II, pp. 29, 57]), a problem we shall consider in Chapter 6.

Particular cases of representations as sums of three squares were considered (sometimes partially solved, other times only verified numerically up to a few hundred) by Descartes, Matsunago (ca. 1730), Euler, Goldbach, Lagrange, Legendre, and others (see [54, Chapter VII]).

A proof that every integer not of the form $4^a(8k + 7)$ is indeed a sum of three squares was given by Legendre [154] in 1798. Shortly afterwards, in 1801, Gauss [81, §291], going way beyond Legendre, actually obtained a formula for the number of primitive representations of an integer as a sum of three squares. However, before we discuss this topic, we want to complete the proof of Theorem 1.

To facilitate the presentation in this, as well as in some of the following chapters, we recall some well-known facts concerning quadratic forms. For more information the reader may want to consult [12], [28], [81], [125], and [200].

§4. Quadratic Forms

A homogeneous polynomial of degree r in k variables is commonly referred to as a *form* of degree r. For example, $x^2 + y^2$, $2x^2 + 5xy - 6y^2 - z^2$, and more generally $Q(x_1,\ldots,x_n) = \sum_{i,j=1}^n a_{ij}x_ix_j$ are forms of degree two, or *quadratic forms*; $x^3 - 3xyz + yzw - 5y^3$, $x^3 + 2xy^2 + y^3$, and more generally $Q(x_1,\ldots,x_n) = \sum_{i,j,m=1}^n a_{ijm}x_ix_jx_m$ are forms of degree three, or cubic forms, etc.

We shall be concerned mainly with quadratic forms with rational integral coefficients, which we shall call *rational integral forms*. This means that $Q(x_1,\ldots,x_n) = \sum_{i,j=1}^n a_{ij}x_ix_j$ holds with $a_{ii} \in \mathbb{Z}$ and, for $i \neq j$, $a_{ij} + a_{ji} \in \mathbb{Z}$ ($\mathbb{Z}$ = set of rational integers; similarly, we shall use the customary notation $\mathbb{Q}$, $\mathbb{R}$, and $\mathbb{C}$ for the rationals, the real numbers, and the complex numbers, respectively). It is no restriction to assume that $a_{ij} = a_{ji}$, and we shall do so in what follows. The determinant of the coefficient matrix $A = (a_{ij})$ is called the *discriminant* of Q and is denoted by $d(Q)$. If we consider $x_1, \ldots, x_n$ as the components of a column vector $\mathbf{x}$, then the quadratic form $Q(x_1,\ldots,x_n)$ is easily identified with the inner product $(\mathbf{x}, A\mathbf{x}) = \mathbf{x}' \cdot A\mathbf{x}$, where $\mathbf{x}'$ is the transpose of $\mathbf{x}$, i.e., $\mathbf{x}$ written as a row vector, and $(\mathbf{x}, \mathbf{y}) = \mathbf{x}' \cdot \mathbf{y} = \sum_{i=1}^n x_iy_i$, the inner product of $\mathbf{x}$ and $\mathbf{y}$. Only in Chapter 14 will the more general case occur, where $a_{ii} \in \mathbb{I}$, $a_{ij} + a_{ji} \in \mathbb{I}$, with $\mathbb{I}$ some integral domain, more general than $\mathbb{Z}$.

Let $C = (c_{ij})$ be a matrix with integral coefficients. If

$$x_i = \sum_{j=1}^n c_{ij}y_j \tag{4.2}$$

and if $|C|$, the determinant of C, equals 1, then

$$y_j = \sum_{i=1}^n d_{ji}x_i \tag{4.2$'$}$$

with the d_{ji}'s also integers. If we replace the x_i's in the quadratic form $Q = Q(x_1,\ldots,x_k)$ by their values (4.2), we obtain another quadratic form $Q_1 = Q_1(y_1,\ldots,y_k)$, also with integral coefficients. Forms that are related like Q and Q_1, i.e., that are transformed into each other by substitutions like (4.2) or (4.2′), are said to be *equivalent* to each other, in symbols, $Q \sim Q_1$. One immediatly verifies that this is indeed an equivalence relation. It also is easy to check that equivalent forms have the same discriminant. To give an example, let $Q = x_1^2 + x_1x_2 + x_2^2$ and define y_1, y_2 by $x_1 = y_1$, $x_2 = y_1 + y_2$; then the determinant $\left|\begin{smallmatrix}1 & 0\\ 1 & 1\end{smallmatrix}\right| = 1$, so that, indeed, $y_1 = x_1$, $y_2 = x_2 - x_1 = -x_1 + x_2$, again with integer coefficients and with determinant $\left|\begin{smallmatrix}1 & 0\\ -1 & 1\end{smallmatrix}\right| = 1$. If we substitute these expressions in Q, we obtain $Q_1 = y_1^2 + y_1(y_1 + y_2) + (y_1 + y_2)^2 = 3y_1^2 + 3y_1y_2 + y_2^2$. We verify that indeed

$$d(Q) = \begin{vmatrix} 1 & \frac{1}{2} \\ \frac{1}{2} & 1 \end{vmatrix} = 1 - \tfrac{1}{4} = \tfrac{3}{4} \quad \text{and} \quad d(Q_1) = 3 - \tfrac{9}{4} = \tfrac{3}{4}$$

are equal. The proof that $Q \sim Q_1 \Rightarrow d(Q) = d(Q_1)$ is not difficult. Under the equivalence relation $\sim$, the binary quadratic forms (i.e., forms of two variables) of a given discriminant d split into a number of classes. It is a nontrivial theorem that this number is finite. The concept of *class number* is already clearly presented in Gauss's *Disquisitiones Arithmeticae* [81]. In fact, this concept is not restricted to binary forms, and we shall need it for both binary and ternary forms.

Gauss makes a distinction between the case $|C| = 1$, in which he speaks of *proper* equivalence, and the case $|C| = -1$, when he calls the equivalence *improper*. From $CD = I$ it is clear that in either case $|C| = |D|$. This shows, in particular, that if Q is properly equivalent to Q_1, then also Q_1 is properly equivalent to Q. From this easily follows that proper equivalence is indeed an equivalence relation. In what follows we shall be interested only in the case of proper equivalence and for simplicity will suppress the adjective "proper." The (finite) number of classes defined by this equivalence relation is denoted by h (or by $h(d)$, if necessary).

We shall denote the column vector of components $x_1, \ldots, x_n$ by $\mathbf{x}$ whenever convenient and, as already done, use the simplest of the symbols Q, or $Q(\mathbf{x})$, or $Q(x_1, \ldots, x_n)$ that can be used without ambiguity.

Generalizing slightly, let $D = (d_{ji})$ be the matrix of the coefficients in (4.2′). Then (4.2′) may be written as $\mathbf{y} = D\mathbf{x}$, and (4.2) as $\mathbf{x} = C\mathbf{y}$. These relations remain meaningful even for nonintegral entries in C and D. In general, however, we shall have to assume that $|C| \neq 0$; indeed, $D = C^{-1}$, so D would not be defined if $|C| = 0$. We observe that from (4.2), (4.2′) it follows that $\mathbf{x} = C\mathbf{y} = CD\mathbf{x}$, so that $(CD - I)\mathbf{x} = 0$, with I the unit matrix. The last equation can hold for every vector $\mathbf{x}$ only if $CD = I$, so that C and D are inverse matrices, $D = C^{-1}$, as claimed. The situation is symmetric, so that in this case also $|D| \neq 0$.

Given Q with matrix A, we have $Q(\mathbf{x}) = \mathbf{x}'A\mathbf{x} = \mathbf{y}'D'AD\mathbf{y} = \mathbf{y}'B\mathbf{y} = Q_1(y_1, \ldots, y_n)$, and we infer that the matrices of equivalent quadratic forms are related like A and B, i.e., $B = D'AD$, with a matrix D such that $|D| \neq 0$. Such matrices are also said to be *equivalent matrices*. One may verify that if one does not insist on the integrality of the coefficients, then every quadratic form of an arbitrary number of variables is equivalent over Q (in fact, over any field over which $Q(x)$ is defined) to a *diagonal* form, i.e., to a form with $a_{ij} = 0$ for $i \neq j$.

If Q is a *diagonal* form with $a_{ii} > 0$ $(i = 1, 2, \ldots, k)$, then since $a_{ii} \in \mathbb{Z}$, Q is in fact a sum of squares (e.g., $2x^2 + y^2 = x^2 + x^2 + y^2$). Any form equivalent over Q to a sum of squares is said to be *positive definite*. It is clear that the sets of values of equivalent forms are the same. In particular, positive definite forms (like sums of squares) cannot take negative values and vanish only when all variables are equal to zero. Similarly, a form Q is said to be *negative definite* if $Q \sim -\sum_{i=1}^{k} a_{ii}x_i^2$, $a_{ii} > 0$. If Q is neither positive, nor negative definite, it is said to be *indefinite*.

There is also a borderline case, when the given form is equivalent to a diagonal form in fewer than k variables. Such a form is sometimes called *degenerate*. So, e.g., under $x = u + v$, $y = v$, i.e., $u = x - y$, $v = y$, one has

$x^2 - 2xy + y^2 = u^2$ and Q contains a single square. The form can vanish for $x = y \neq 0$, when $u = 0$ and the form either vanishes or is positive. Such forms are sometimes also called semidefinite, but we shall not need them further.

§5. Some Needed Lemmas

The proof of Theorem 1 uses some properties of binary and ternary forms, which we shall present as a sequence of lemmas. In the present and some of the following sections, we shall often find it convenient to use Gauss's notation, somewhat different from our previous one. Specifically, we shall require that all coefficients a_{ij} (not only $a_{ij} + a_{ji} = 2a_{ij}$) of the quadratic form $Q = \sum_{i,j=1}^{k} a_{ij}x_ix_j$ be integers. It then follows that the coefficients of the mixed terms, $a_{ij} + a_{ji} = 2a_{ij}$, are all even. A quadratic form $Q = \sum_{i,j=1}^{2} a_{ij}x_ix_j$ with $a_{ij} \in \mathbb{Z}$ is said to be *primitive* if $(a_{11}, a_{12}, a_{22}) = 1$; it is said to be *properly primitive* if the stronger condition $(a_{11}, 2a_{12}, a_{22}) = 1$ holds. Clearly, in a properly primitive form, at least one of the coefficients a_{11} or a_{22} must be odd. Also, instead of the usual discriminant of the quadratic form $a_{11}x_1^2 + 2a_{12}x_1x_2 + a_{22}x_2^2$, $D = 4a_{12}^2 - 4a_{11}a_{22} = 4(a_{12}^2 - a_{11}a_{22})$, we shall often use the determinant

$$d_2 = \begin{vmatrix} a_{11} & a_{12} \\ a_{21} & a_{22} \end{vmatrix} = a_{11}a_{22} - a_{12}^2, \quad \text{with } D = -4d_2.$$

More generally, and regardless of the parity of b, if $Q = ax^2 + bxy + cy^2$, we set $D = b^2 - 4ac$. Clearly, only $D \equiv 0$ or $1 \pmod 4$ is possible, while d_2 is not subject to such restrictions.

In §4 we have called the determinant of the coefficients by its customary name, namely *discriminant*. This is a well-defined entity for forms in any number of variables. However, in the particular case of *binary forms* (i.e., of forms in two variables) the traditional meaning of *discriminant* has always been D. To avoid confusion, in the present chapter, we shall henceforth reserve the name *discriminant* for the traditional D, and shall call d_2 the *determinant* of the form. The apparent confusion is alleviated, at least in part, by the fortunate fact that, at least in the cases of interest here, the number h of equivalence classes of positively definite, primitive binary forms corresponding to a given discriminant is the same with either definition of the discriminant* except when $d_2 \equiv 3 \pmod 8$. In that case, when $-D = d_2 \equiv 3 \pmod 8$, we have $h(d_2) = 3h(-D)$. One may verify, e.g., that for $d_2 = 11$, $h = 3$, while for $D = -11$, $h = 1$.† In general, the context will make it clear which "discriminant" is

---

*Obviously, the forms with a given d_2 are not the same as those with $D = -d_2$, because for any given form, $D = -4d_2$.

†Gauss was fully aware of this fact and specifically points it out at the end of §291 in [81]: "... $k = h$ for $M \equiv 1, 2, 5$, or $6 \pmod 8$, but $k = h/3$ for $M \equiv 3 \pmod 8$, with the single case $M = 3$ excepted (where $h = k = 1$)."

meant. In a few cases, when both concepts appear in the same section, we shall distinguish them as *discriminant* and *determinant* according to the above convention. For m-ary forms, with $m > 2$, we shall use only $d_m = \text{determinant}(a_{ij})$, as defined in Chapter 1. With this terminology the following lemmas hold.

Lemma 1. *A binary quadratic form $a_{11}x_1^2 + 2a_{12}x_1x_2 + a_{22}x_2^2$ is positive definite if and only if the two principal minors are positive.*

As

$$d_2 = \begin{vmatrix} a_{11} & a_{12} \\ a_{21} & a_{22} \end{vmatrix},$$

Lemma 1 requires that $a_{11} > 0$ and $d_2 = a_{11}a_{22} - a_{12}^2 > 0$.

Proof of Lemma 1. To have Q definite, it is necessary that Q should not change sign. This will be true if $a_{11}\lambda^2 + 2a_{12}\lambda + a_{22}$ has no real zeros, i.e., if $D/4 = a_{12}^2 - a_{11}a_{22} < 0$, which means $d_2 > 0$. Being definite, Q is positive, or negative definite, according to the sign of any of its values, in particular of $Q(1,0) = a_{11}$. This proves the necessity of the condition. The sufficiency follows, using also $a_{11} > 0$, from $a_{11}Q = (a_{11}x_1 + a_{12}x_2)^2 + d_2x_2^2$. □

Lemma 2. *In each class of binary, positive definite forms, there is a form with*

$$2|a_{12}| \leqslant a_{11} \leqslant a_{22}; \tag{4.3}$$

such a form is called reduced. *In a reduced form,*

$$a_{11} \leqslant \frac{2}{\sqrt{3}}\sqrt{d_2}. \tag{4.3$'$}$$

Corollary 1. *Every positive definite binary quadratic form of discriminant $d_2 = 1$ is equivalent to a sum of two squares.*

Proof of the Corollary. By (4.3′), $a_{11} = 1$; by (4.3), $a_{12} = 0$; and by $1 = d_2 = a_{11}a_{22} - a_{12}^2 = a_{22}$, also $a_{22} = 1$. □

Proof of Lemma 2. Let a be the smallest positive integer that can be represented by $a'_{11}x_1^2 + 2a'_{12}x_1x_2 + a'_{22}x_2^2$; then, for certain integers x'_1, x'_2,

$$a = a'_{11}(x'_1)^2 + 2a'_{12}x'_1x'_2 + a'_{22}(x'_2)^2. \tag{4.4}$$

Here x'_1, x'_2 are coprime, as otherwise, if $(x'_1, x'_2) = m > 1$, then $a/m^2 < a$ would also be representable. Since $(x'_1, x'_2) = 1$, it follows that $x'_1u - x'_2s = 1$ is solvable for $u, s \in \mathbb{Z}$. If u_0, s_0 is any solution, the general solution is $u = u_0 + tx'_2$, $s = s_0 + tx'_1$ with $t \in \mathbb{Z}$. Let

$$\mathbf{x} = \begin{pmatrix} x_1 \\ x_2 \end{pmatrix}, \qquad \mathbf{y} = \begin{pmatrix} y_1 \\ y_2 \end{pmatrix}$$

and let

$$M = \begin{pmatrix} x'_1 & s \\ x'_2 & u \end{pmatrix};$$

then for the determinant of M we have $|M| = 1$, and if $\mathbf{x} = M\mathbf{y}$, then $Q(\mathbf{x}) = Q(M\mathbf{y}) = Q_1(\mathbf{y})$, with $Q \sim Q_1$ in the same class, say $Q_1 = a_{11}y_1^2 + 2a_{12}y_1y_2 + a_{22}y_2^2$. By direct substitution of $M\mathbf{y}$ for $\mathbf{x}$ in Q, we obtain $a_{11} = a$, as given by (4.4), and $a_{12} = g + ta$, where g depends on the elements of $A = (a_{ij})$ and of M, but not on t. Hence, the arbitrary integer t may be selected so that $|a_{12}| = |g + ta| \leqslant a/2$. Also, $a_{22} = Q_1(0, 1)$ is representable; hence $a_{22} \geqslant a = a_{11}$, and (4.3) is proved. Finally, $a_{11}^2 \leqslant a_{11}a_{22} = a_{12}^2 + d_2 \leqslant \frac{1}{4}a_{11}^2 + d_2$, and (4.3′) follows. □

Lemma 3. *A ternary quadratic form $Q = \sum_{i,j=1}^{3} a_{ij}x_ix_j$ is positive definite if and only if all three principal minors of the matrix $A = (a_{ij})$ are positive.*

As in the case of binary forms, this means that

$$d_3 = \begin{vmatrix} a_{11} & a_{12} & a_{13} \\ a_{21} & a_{22} & a_{23} \\ a_{31} & a_{32} & a_{33} \end{vmatrix} > 0, \qquad b = \begin{vmatrix} a_{11} & a_{12} \\ a_{21} & a_{22} \end{vmatrix} > 0, \quad \text{and} \quad a_{11} > 0.$$

Proof. By "completing the square," $a_{11}Q = L^2 + Q_1$, where $L = \sum_{i=1}^{3} a_{1i}x_i$ and $Q_1 = Q_1(x_2, x_3)$. By direct computation, if d_2 stands for the discriminant of Q_1, then $d_2 = a_{11}d_3$ and the coefficient of x_1^2 in Q_1 is b. Here Q is positive definite if and only if Q_1 is positive definite. The sufficiency is clear. For the necessity, assume that $Q_1(x'_2, x'_3) \leqslant 0$. Then also $Q_1(x''_2, x''_3) \leqslant 0$ with $x''_2 = a_{11}x'_2$, $x''_3 = a_{11}x'_3$, and we may choose $x''_1 = a_{11}^{-1}(a_{12}x''_2 + a_{13}x''_3) \in \mathbb{Z}$, so that $L = 0$, and $Q(x''_1, x''_2, x''_3) \leqslant 0$, contrary to the requirement that Q be positive definite. By Lemma 1, Q_1 is positive definite if and only if $d_2 = a_{11}d_3 > 0$ and $b > 0$; but $a_{11} = Q(1, 0, 0) > 0$, and this implies $d_3 > 0$, thus proving the lemma. □

Remark. The general conditions for the positive definiteness of an m-ary quadratic form $Q = \sum_{i,j=1}^{m} a_{ij}x_ix_j$ in terms of the principal minors of A are now almost obvious. The proof may be molded on the previous proof of Lemma 3, by induction on m and completion of the square, and by observing that

$a_{11} = Q(1,0,\ldots,0) > 0$ and $a_{11}Q = L^2 + Q_1$, with $L = \sum_{i=1}^{m} a_{1i}x_i$ and $Q_1 = \sum_{i,j=2}^{m} b_{ij}x_ix_j$. We shall not need this more general theorem.

Lemma 4. *Each class of positive definite ternary quadratic forms* $Q = \sum_{i,j=1}^{3} a_{ij}x_ix_j$ *contains a* reduced *form with*

$$0 < a_{11} \leqslant \tfrac{4}{3}\sqrt[3]{d}, \qquad 2|a_{12}| \leqslant a_{11}, \qquad 2|a_{13}| \leqslant a_{11}. \tag{4.5}$$

Remark. This is a particular case of the *Hermite reduction*; a more general case is mentioned in §12 of Chapter 14 (for a general treatment see, e.g., [36] of the Bibliography).

Proof of Lemma 4. Given $Q = \sum_{i,j=1}^{3} a'_{ij}x'_ix'_j$, let a be the smallest representable integer, say $a = Q(m_{11}, m_{21}, m_{31})$. It simplifies the proof if, as a preparatory step, we find $Q_1 \sim Q$, $Q_1 = \sum_{i,j=1}^{3} a_{ij}x_ix_j$, such that $a_{11} = a$; indeed, then $Q_1(1,0,0) = a$. If we set $M = (m_{ij})$, with m_{i1} in the first column, then M can be completed (in an infinite number of ways—one disposes of many parameters), so that $|M| = 1$. We easily verify that* $Q(\mathbf{x}') = Q(M\mathbf{x}) = Q_1(\mathbf{x})$, with $a_{11} = Q(m_{11}, m_{21}, m_{31}) = Q_1(1,0,0)$. Next, we construct a matrix

$$N = \begin{pmatrix} 1 & v & w \\ 0 & & \\ 0 & B & \end{pmatrix},$$

with v, w integers to be selected later and B a 2×2 matrix with $|B| = 1$. Clearly, $|N| = 1$, so that, if we set $\mathbf{x} = N\mathbf{y}$, then $Q_1(\mathbf{x}) = Q_1(N\mathbf{y}) = Q_2(\mathbf{y})$ and $Q \sim Q_1 \sim Q_2$ are in the same class, with $Q_2 = \sum_{i,j=1}^{3} b_{ij}y_iy_j$, say, and $b_{11} = a_{11}$. By completing the square and using $a_{11} = b_{11}$, we obtain

$$a_{11}Q_1(\mathbf{x}) = L_1^2 + \tilde{Q}_1(x_2, x_3), \qquad L_1 = \sum_{i=1}^{3} a_{1i}x_i,$$

$$a_{11}Q_2(\mathbf{y}) = L_2^2 + \tilde{Q}_2(y_2, y_3), \qquad L_2 = \sum_{i=1}^{3} b_{1i}y_i.$$

Let

$$\mathbf{y}_0 = \begin{pmatrix} y_2 \\ y_3 \end{pmatrix};$$

then, as we know, $\tilde{Q}_2(\mathbf{y}_0) = \tilde{Q}_1(B\mathbf{y}_0)$, with $d_2(\tilde{Q}_1) = a_{11}d_3 = b_{11}d_3$, where the coefficient of y_2^2 is equal to $b_{11}b_{22} - b_{12}^2 = b$, say. As seen in the proof of

* Here $\mathbf{x}'$ stands, exceptionally, not for the transpose of $\mathbf{x}$, but for the vector of components x'_i.

Lemma 2, B may be selected so that $b < (2/\sqrt{3})(d_2(\tilde{Q}_1))^{1/2}$. This means here $b_{11}b_{22} - b_{12}^2 \leqslant (2/\sqrt{3})\sqrt{b_{11}d_3}$. Also, b_{12} and b_{13} are linear forms in a_{11}, with coefficients v and w, respectively. Hence, these may be selected so that $|b_{1j}| \leqslant \frac{1}{2}a_{11} = \frac{1}{2}b_{11}$ for $j = 2, 3$. Finally, $b_{22} = Q_2(0, 1, 0)$ is representable; hence $b_{22} \geqslant a_{11}$, and we obtain the sequence of inequalities

$$b_{11}^2 \leqslant b_{11}b_{22} = (b_{11}b_{22} - b_{12}^2) + b_{12}^2 \leqslant \frac{2}{\sqrt{3}}\sqrt{b_{11}d_3} + \tfrac{1}{4}b_{11}^2.$$

Taking the first and last member, $\frac{3}{4}b_{11}^2 \leqslant (2/\sqrt{3})b_{11}^{1/2}d_3^{1/2}$, $(3\sqrt{3}/8)b_{11}^{3/2} \leqslant d_3^{1/2}$, $\frac{27}{64}b_{11}^3 \leqslant d_3$, and by changing the b's into a's, we complete the proof of (4.5). □

Corollary 2. *Every positive definite ternary quadratic form of discriminant $d_3 = 1$ is equivalent to a sum of three squares.*

Proof. From (4.5) it follows that $0 \leqslant a_{11} \leqslant \frac{4}{3}$; hence, $a_{11} = 1$ and $|a_{1j}| \leqslant \frac{1}{2}a_{11} \leqslant \frac{1}{2}\,(j = 2, 3)$ imply $a_{12} = a_{13} = 0$; hence, $Q = x_1^2 + Q_1$, with $Q_1 = \sum_{i,j=2}^{3} a_{ij}x_ix_j$. By Lemma 2, there exists a 2×2 matrix B with $|B| = 1$ and such that if

$$\mathbf{x}_0 = \begin{pmatrix} x_2 \\ x_3 \end{pmatrix}, \qquad \mathbf{y}_0 = \begin{pmatrix} y_2 \\ y_3 \end{pmatrix}, \quad \text{and} \quad \mathbf{x}_0 = B\mathbf{y}_0,$$

then $Q_1(\mathbf{x}_0) = Q_1(B\mathbf{y}_o) = y_2^2 + y_3^2$. If we now set

$$N = \begin{pmatrix} 1 & 0 & 0 \\ 0 & & \\ 0 & & B \end{pmatrix}, \qquad \mathbf{x} = \begin{pmatrix} x_1 \\ x_2 \\ x_3 \end{pmatrix}, \qquad \mathbf{y} = \begin{pmatrix} y_1 \\ y_2 \\ y_3 \end{pmatrix},$$

with $\mathbf{x} = N\mathbf{y}$, then $Q(\mathbf{x}) = Q(N\mathbf{y}) = y_1^2 + Q_1(B\mathbf{y}_0) = y_1^2 + y_2^2 + y_3^2$. □

§6. Proof of Theorem 1, Part II

If $n = 4^a n_1$, $4 \nmid n_1$ and n_1 is the sum of three squares, say $n_1 = \sum_{i=1}^{3} x_i^2$, then $n = \sum_{i=1}^{3}(2^a x_i)^2$ is also a sum of three squares. Hence, it is sufficient to consider only the case $n \not\equiv 0 \pmod 4$. We know that if $n \equiv 7 \pmod 8$, then (4.1) has no solution; therefore, it is sufficient to consider the cases $n \not\equiv 0, 4, 7 \pmod 8$. The idea of the proof consists in showing first that, under those conditions, there exists a positive definite ternary quadratic form $Q = \sum_{i,j=1}^{3} a_{ij}x_ix_j$, of discriminant 1, which represents n. Next, we invoke Corollary 2 to complete the proof.

The equation

$$Q(x) = \sum_{i,j=1}^{3} a_{ij}x_i x_j = n \tag{4.6}$$

has nine parameters: the six coefficients a_{ij} and the three values x_i. The conditions to be met are, besides (4.6),

$$d_3 = |a_{ij}| = 1\ (>0), \qquad b = \begin{vmatrix} a_{11} & a_{12} \\ a_{21} & a_{22} \end{vmatrix} > 0, \quad \text{and} \quad a_{11} > 0.$$

The problem is indeterminate; this is clear *a priori*, because if (4.6) holds for some form Q with $d_3 = 1$, then it will hold for any other form $Q_1 \sim Q$. In fact, while it is true that equivalent forms represent the same integers, the converse is not true. Indeed, the same integer may be represented by nonequivalent forms of the same discriminant. Such forms are said to belong to the same *genus*. For example, the two forms $Q_1 = x^2 + 161y^2$ and $Q_2 = 9x^2 + 2xy + 18y^2$ both have $d_2 = 161$. Also, $Q_1(1,1) = 162 = Q_2(0,3)$ and $Q_1(3,0) = 9 = Q_2(1,0)$. However, the two forms are not equivalent. Indeed, if we set $x = ax_1 + by_1$, $y = cx_1 + dy_1$, then $Q_1(x,y) = Ax_1^2 + 2Bx_1y_1 + Cy_1^2$, with $A = a^2 + 161c^2$, $B = ab + 161cd$, $C = b^2 + 161d^2$. If we identify this with $Q_2(x_1, y_1)$, we have to solve the system of equations

$$a^2 + 161c^2 = 9, \qquad ab + 161cd = 1, \qquad b^2 + 161d^2 = 18$$

in integers a, b, c, d with $ad - bc = 1$. The first equation requires $c = 0$, $a = \pm 3$; the second then yields $b = a^{-1} = \pm\frac{1}{3}$, and the last one $d = \pm\frac{1}{3}$. With appropriate choices of signs, we find indeed $ad - bc = 1$, but $b, d \notin \mathbb{Z}$.

It is clear that, given a form, its whole equivalence class belongs to a given genus; consequently, a genus consists of a certain number of equivalence classes of forms. We shall not need the complete theory of genera—not even their precise definition. For their definition by Gauss, see [81, §231]; for a modern presentation, see, e.g., [105, §48, Satz 145, §53]. We quote, nevertheless, the important, nontrivial theorem (for a proof, see, e.g., [105, §48, Satz 145, §53]), that each of the g genera of a given discriminant contains the same number $k = h/g$ of equivalence classes. If the discriminant $D = b^2 - 4ac$ of the form $ax^2 + bxy + cy^2$ is divisible by t distinct primes, then $g = 2^{t-1}$, except for $-D = 4n$, with $n \equiv 3 \pmod 4$, when $g = 2^{t-2}$, and for $-D = 4n$, $n \equiv 0 \pmod 8$, when $g = 2^t$—cases that will not concern us. This is essentially equivalent to Gauss's rule of counting only the odd prime divisors of n. By following carefully the number of different possible choices of the parameters at our disposal, one could actually obtain, by such an analysis, the exact number $r_3(n)$ of solutions of (4.1), essentially as Gauss did.

At present, however, we shall content ourselves with completing the proof of Theorem 1. To make this as simple as possible, we freeze, with the advantage of hindsight, the values of three of the nine available parameters and take $a_{13} = 1$, $a_{23} = 0$, $a_{33} = n$. This reduces Q to

$$Q = a_{11}x_1^2 + 2a_{12}x_1x_2 + 2x_1x_3 + a_{22}x_2^2 + nx_3^2.$$

Clearly, $Q(0,0,1) = n$; hence, we also take $x_1 = x_2 = 0, x_3 = 1$. It only remains to determine the three coefficients a_{11}, a_{12}, a_{22} subject to the conditions

$$a_{11} > 0, \qquad b = a_{11}a_{22} - a_{12}^2 > 0, \qquad d_3 = 1.$$

As

$$d_3 = \begin{vmatrix} a_{11} & a_{12} & 1 \\ a_{21} & a_{22} & 0 \\ 1 & 0 & n \end{vmatrix} = nb - a_{22},$$

the last condition is equivalent to $a_{22} = bn - 1$, while (for $n \geqslant 2$) the first condition is automatically satisfied on account of the other two and may be ignored. Indeed, $a_{22} = nb - 1 \geqslant 2b - 1 > 0$ and $a_{11}a_{22} = a_{12}^2 + b \geqslant b > 0$, which, by $a_{22} > 0$, implies $a_{11} > 0$.

The main difficulty is the proper choice of b in order to insure that $a_{11} = a_{22}^{-1}(a_{12}^2 + b) \in \mathbb{Z}$. This condition means that $-b \equiv a_{12}^2 \pmod{a_{22}}$. As a_{12} is an arbitrary integer, the condition really reduces to the selection of $-b$ as a quadratic residue modulo a_{22}, i.e., modulo $nb - 1$. The easiest way to accomplish this is to choose b so that $nb - 1 = p$ or $nb - 1 = 2p$, with p a prime, and to verify that then $(\frac{-b}{p}) = +1$ or $(\frac{-b}{p}) = (\frac{-b}{2}) = +1$, respectively. It is convenient to treat the cases of n even and n odd separately. In either case, we shall make us of Dirichlet's theorem [28] that if $(k, m) = 1$, then arithmetic progression $kr + m$ $(r = 0, 1, 2, \ldots)$ contains infinitely many primes.

(i) Let $n \equiv 2$ or $6 \pmod 8$. From $(4n, n - 1) = 1$ it follows from Dirichlet's theorem that there is an integer m such that $4nm + (n - 1) = p$ with p a prime. We then select $b = 4m + 1$, so that $p = bn - 1$, and observe that $b \equiv p \equiv 1 \pmod 4$. Also,

$$\left(\frac{-b}{p}\right) = \left(\frac{b}{p}\right) = \left(\frac{p}{b}\right) = \left(\frac{bn-1}{b}\right) = \left(\frac{-1}{b}\right) = 1.$$

With this b, the remaining coefficients of Q are easily obtained. Indeed, $a_{22} = bn - 1 = p > 0$, and $-b \equiv a_{12}^2 \pmod p$ is solvable, yielding a_{12}. Consequently, $a_{11} = (b + a_{12}^2)/a_{22}$ is indeed an integer, and Q is completely determined.

(ii) If $n \equiv 1$, 3, or $5 \pmod 8$, we choose c such that $cn - 1 \equiv 2 \pmod 4$. Then $(4n, (cn - 1)/2) = 1$ and, by Dirichlet's theorem, we may choose m so that $4nm + (cn - 1)/2 = p$, a prime. We now set $b = 8m + c$, so that $2p = (8m + c)n - 1 = bn - 1$. We may verify that, with n in any of the three residue classes modulo 8, $-b$ is a quadratic residue modulo $2p$. For example, if $n \equiv 1 \pmod 8$, let $c = 3$. Then $b \equiv c \equiv 3 \pmod 8$, $p \equiv (3n-1)/2 \equiv 1 \pmod 4$, and $(\frac{-b}{p}) = (\frac{b}{p}) = (\frac{p}{b})$; also, $(\frac{-2}{b}) = (\frac{-1}{b})(\frac{2}{b}) = (-1)^2 = +1$. It follows that

$$\left(\frac{-b}{p}\right) = \left(\frac{-b}{p}\right)\left(\frac{-2}{b}\right) = \left(\frac{p}{b}\right)\left(\frac{-2}{b}\right) = \left(\frac{-2p}{b}\right)$$

$$= \left(\frac{1-bn}{b}\right) = \left(\frac{1}{b}\right) = 1,$$

as claimed. The cases $n \equiv 3$ or $5 \pmod 8$ are handled similarly. From here on, we proceed as in (i). Indeed, clearly $a_{22} = bn - 1 = 2p$, $-b \equiv u^2 \pmod p$ is solvable, and so is $-b \equiv u^2 \pmod 2$.

Hence, $-b \equiv a_{12}^2 \pmod{2p}$ has a solution a_{12}^2 such that $b + a_{12}^2$ is divisible by $2p = a_{22}$, $a_{11} = (a_{12}^2 + b)/a_{22} \in \mathbb{Z}$, and the proof is complete.

§7. Examples

The proof of Theorem 1 that we just saw, while fully constructive, is rather long, and some of its details may be confusing when first seen. For that reason, it may be worthwhile to give two completely worked-out examples, in which we follow step by step the proof just given.

Let $n = 30 \equiv 6 \pmod 8$. We choose m such that $4 \cdot 30m + (30 - 1) = 120m + 29$ is prime. We may choose $m = 0$. Then $p = 29 = a_{22}$, $b = (p+1)/n = 1$. For a_{12} we choose the smallest solution of $-1 \equiv u^2 \pmod{29}$, i.e., $u = 12$. Then $a_{12} = 12$ and $a_{11} = (b + a_{12}^2)/a_{22} = (1 + 12^2)/29 = 5$. The quadratic form is therefore $Q = 5x_1^2 + 24x_1x_2 + 2x_1x_3 + 29x_2^2 + 30x_3^2$. We verify that, indeed,

$$d_3 = \begin{vmatrix} 5 & 12 & 1 \\ 12 & 29 & 0 \\ 1 & 0 & 30 \end{vmatrix} = 30b - 29 = 30 - 29 = 1.$$

It remains to find a matrix M such that $\mathbf{x} = M\mathbf{y}$ and $Q(\mathbf{x}) = Q(M\mathbf{y}) = \sum_{i=1}^{3} y_i^2$. By proceeding as in the proof of Lemma 3, we complete the square and obtain $5Q = (5x_1 + 12x_2 + x_3)^2 + Q_1$, where $Q_1 = x_2^2 - 24x_2x_3 + 149x_3^2$. It is clear that $Q_1(1,0) = 1$ is the smallest integer representable by Q_1. We then try to minimize $|L(x_1, 1, 0)| = |5x_1 + 12|$. Hence, we choose $x_1 = -2$ and have that the minimum of $5Q$ occurs for $x_1 = -2$, $x_2 = 1$, $x_3 = 0$. This minimum is $2^2 + 1 = 5$, so that we expect $Q(-2,1,0) = 1$. This is indeed the case, as $5(-2)^2 + 24(-2)\cdot 1 + 0 + 29 \cdot 1^2 + 30 \cdot 0 = 20 - 48 + 29 = 1$. We now complete the column $\left(\begin{smallmatrix} -2 \\ 1 \\ 0 \end{smallmatrix}\right)$ to a matrix M, with $|M| = 1$. A simple such matrix is

$$M = \begin{pmatrix} -2 & 1 & 0 \\ 1 & -1 & 0 \\ 0 & 0 & 1 \end{pmatrix}, \qquad \text{say.}$$

Then, if $\mathbf{x} = M\mathbf{y}$, we have $Q(\mathbf{x}) = Q(M\mathbf{y}) = \tilde{Q}(\mathbf{y}) = y_1^2 - 6y_1y_2 + 10y_2^2 - 4y_1y_3 + 30y_3^2 + 2y_2y_3 = (y_1 - 3y_2 - 2y_3)^2 + Q_1(y_2, y_3)$. Here $Q_1 = y_2^2 - 10y_2y_3 + 26y_3^2$, $\min Q_1(y_2, y_3) = Q_1(1,0) = 1$. Hence, we form $B = \left(\begin{smallmatrix} 1 & s \\ 0 & u \end{smallmatrix}\right)$, and $|B| = 1$ requires $u = 1$ and $s \in \mathbb{Z}$, otherwise arbitrary. We define $\left(\begin{smallmatrix} z_2 \\ z_3 \end{smallmatrix}\right)$ by

$$\begin{pmatrix} y_2 \\ y_3 \end{pmatrix} = \begin{pmatrix} 1 & s \\ 0 & 1 \end{pmatrix} \begin{pmatrix} z_2 \\ z_3 \end{pmatrix},$$

substitute in Q_1, and set the coefficient of $z_2 z_3$ equal to zero. This requires $s = 5$ and $B = \left(\begin{smallmatrix} 1 & 5 \\ 0 & 1 \end{smallmatrix}\right)$. Now

$$N = \begin{pmatrix} 1 & v & w \\ 0 & 1 & 5 \\ 0 & 0 & 1 \end{pmatrix}.$$

With

$$\mathbf{y} = \begin{pmatrix} y_1 \\ y_2 \\ y_3 \end{pmatrix}, \qquad \mathbf{z} = \begin{pmatrix} z_1 \\ z_2 \\ z_3 \end{pmatrix},$$

set $\mathbf{y} = N\mathbf{z}$ and obtain $y_1 = z_1 + vz_2 + wz_3$, $y_2 = z_2 + 5z_3$, $y_3 = z_3$. We substitute these in $\tilde{Q}(\mathbf{y})$ and obtain $\tilde{Q}(\mathbf{y}) = L^2 + z_2^2 + z_3^2$, with $L = z_1 + (v - 3)z_2 + (w - 17)z_3$. For $v = 3$, $w = 17$, we have obtained $Q(\mathbf{x}) = \tilde{Q}(\mathbf{y}) = \tilde{Q}(N\mathbf{z}) = z_1^2 + z_2^2 + z_3^2$. In order to obtain $Q(\mathbf{x}) = 30$, we needed $x_1 = x_2 = 0$, $x_3 = 1$. Under $\mathbf{x} = M\mathbf{y}$, either from $\mathbf{y} = M^{-1}\mathbf{x}$, or directly, we obtain $y_1 = y_2 = 0$, $y_3 = 1$ and, finally, $z_3 = y_3 = 1$, $z_2 = y_2 - 5z_3 = -5$, $z_1 = y_1 - 3z_2 - 17z_3 = -3(-5) - 17 \cdot 1 = -2$. We have obtained the solution of (4.1):

$$(+1)^2 + (-5)^2 + (-2)^2 = 30.$$

We recall that, by permuting the three summands and by considering all eight possible choices of signs, this solution counts for 48 different (although not essentially distinct) representations of 30.

To take another example, let $n = 11 \equiv 3 \pmod 8$. With $c = 1$, $(cn - 1)/2 = 5$ and we may choose again $m = 0$, $p = 5$, $2p = 10 = bn - 1$, $bn = 11$, $b = 1$. The congruence $-1 \equiv u^2 \pmod 5$ has the solutions $2 + 5r$, and to have $-1 \equiv u^2 \pmod{2 \cdot 5}$, we may choose $r = -1$, i.e., $u = \pm 3$. With $a_{12} = 3$, $a_{22} = 2p = 10$, and $b = 1$, we obtain $a_{11} = (a_{12}^2 + b)/a_{22} = (9 + 1)/10 = 1$, so that

$$Q = x_1^2 + 6x_1x_2 + 2x_1x_3 + 10x_2^2 + 11x_3^2.$$

We verify that all required conditions hold: $Q(0,0,1) = 11$, $a_{11} = 1 > 0$, $b = \left|\begin{smallmatrix} 1 & 3 \\ 3 & 10 \end{smallmatrix}\right| = 1 > 0$, and

$$d_3 = \begin{vmatrix} 1 & 3 & 1 \\ 3 & 10 & 0 \\ 1 & 0 & 11 \end{vmatrix} = b \cdot 11 + \begin{vmatrix} 3 & 10 \\ 1 & 0 \end{vmatrix} = 11 - 10 = 1.$$

Having obtained a positively definite form Q with $d_3 = 1$ that represents $n = 11$, we continue, in order to obtain a reduced form which is precisely a sum of three squares. We follow the same method of completing the square. However, here $a_{11} = 1$ and is, of course, already the smallest representable integer, $1 = Q(1,0,0)$, so that we don't need the preliminary transformation to make $a_{11} = a$. We have $Q = (x_1 + 3x_2 + x_3)^2 + Q_1(x_2, x_3)$, $Q_1 = x_2^2 - 6x_2x_3 + 10x_3^2$. Clearly, $1 = Q_1(1,0)$ is the smallest integer representable by Q_1; hence, $B = \left(\begin{smallmatrix} 1 & s \\ 0 & u \end{smallmatrix}\right)$, and again, $u = 1$, while s is so far arbitrary. Substituting $x_2 = y_2 + sy_3$, $x_3 = y_3$ in Q_1, we have $Q_1(x_2, x_3) = y_2^2 + 2y_2y_3(s-3) + y_3^2(s^2 - 6s + 10)$. To eliminate the mixed term y_2y_3, set $s = 3$; then, as expected, $s^2 - 6s + 10 = 1$, so that $Q_1(By_0) = y_2^2 + y_3^2$. Hence, let

$$N = \begin{pmatrix} 1 & v & w \\ 0 & 1 & 3 \\ 0 & 0 & 1 \end{pmatrix}$$

and, with the earlier notation, set $\mathbf{x} = N\mathbf{y}$. Then $L = (y_1 + vy_2 + wy_3) + 3(y_2 + 3y_3) + y_3 = y_1 + (v+3)y_2 + (w+10)y_3$. We now choose $v = -3$ and $w = -10$; then $L = y_1$ and $Q(\mathbf{x}) = Q(N\mathbf{y}) = y_1^2 + y_2^2 + y_3^2$. As $Q(0,0,1) = 11$, we have $x_1 = 0 = y_1 + vy_2 + wy_3 = y_1 - 3y_2 - 10y_3$, $x_2 = 0 = y_2 + 3y_3$, and $x_3 = 1 = y_3$, so that $y_2 = -3y_3 = -3$ and $y_1 = 3y_2 + 10y_3 = -9 + 10 = 1$. We verify that, indeed, $1^2 + (-3)^2 + 1^2 = 11$. The terms admit three different permutations and each may be taken as the square of a positive or of a negative integer, for eight different possible choices of signs. The solution of (4.1) just found corresponds, therefore, to $8 \cdot 3 = 24$ representations counted by $r_3(11)$.

§8. Gauss's Theorem

The number $R_3(n)$ of primitive solutions of (4.1) has been determined by Gauss. The theorem may be formulated as follows.

Theorem 2. *Let h stand for the number of classes of primitive binary quadratic forms, corresponding to the discriminant $D = -n$ if $n \equiv 3 \pmod 8$, $D = -4n$ if*

$n \equiv 1, 2, 5,$ or $6 \pmod 8$), and let $\delta_n = 1$ except for $\delta_1 = \frac{1}{2}$ and $\delta_3 = \frac{1}{3}$. Then $R_3(n)$ is given by

$$R_3(n) = \begin{cases} 12h\delta_n & \text{if } n \equiv 1, 2, 5, \text{ or } 6 \pmod 8, \\ 24h\delta_n & \text{if } n \equiv 3 \pmod 8). \end{cases} \tag{4.7}$$

Gauss's own formulation is somewhat different. We recall that $h = gk$, where h is the number of classes, $g = 2^{t-1}$ is the number of genera, k is the number of classes in each genus, and t is the number of distinct prime factors of D. If $d_2 = n \equiv 1, 2, 5$, or $6 \pmod 8$, then $D = -4n$ and, if n contains t odd prime factors, then D contains $t + 1$ primes, and $g = 2^{(t+1)-1} = 2^t$. Hence, the first line of (4.7) becomes $12 \cdot 2^t k\delta_n = 3 \cdot 2^{t+2} k\delta_n$. If $n \equiv 3 \pmod 8$, then $D = -n = -d_2$, and if d_2 contains t primes (all odd), then $g = 2^{t-1}$. However, as already mentioned, the class number $h' = h(d_2) = 3h(-D)$, so that $24h\delta_n = 8h'\delta_n = 8gk'\delta_n = 2^3 \cdot 2^t k'\delta_n$, and if we write, with Gauss, k for k', the second line of (4.7) becomes $R_3(n) = 2^{3+t}k\delta_n$. (In fact, Gauss does not use the symbol δ_n, but discusses the cases $n = 1$ and $n = 3$ separately.) We thus obtain

Theorem 2′ (Gauss). *With the above meanings of t and k, for $n \neq 1, 3$,*

$$R_3(n) = \begin{cases} 3 \cdot 2^{t+2}k & \text{if } n \equiv 1, 2, 5, \text{ or } 6 \pmod 8, \\ 2^{t+2}k & \text{if } n \equiv 3 \pmod 8). \end{cases} \tag{4.7′}$$

To take a few examples, first let $n = 1$. We already know that all positive definite primitive binary quadratic forms of discriminant 1 are equivalent; hence $h = 1$, and *a fortiori*, $k = g = 1$. By (4.7), $R_3(1) = 12 \cdot 1 \cdot \frac{1}{2} = 6$. Indeed, $1 = (\pm 1)^2 + 0^2 + 0^2$ stands for six representations (three permutations of the terms and two choices of sign). Similarly, for $n = 2$ we have $h = 1$. Then (4.7) yields $R_3(2) = 12 \cdot 1 \cdot 1 = 12$, and (4.7′) (with $t = 0$, as 2 is the only prime divisor of n and is not odd) also yields $R_3(2) = 3 \cdot 2^2 \cdot 1 = 12$. Indeed, $2 = (\pm 1)^2 + (\pm 1)^2 + 0^2$, with three permutations and four choices of signs. Also for $n = 3$, $h = 1$, so that, by (4.7), $R_3(3) = 24 \cdot 1 \cdot \frac{1}{3} = 8$, and, by (4.7′), with $k = t = 1$, $R_3(3) = 2^3 \cdot 1 = 8$ (Gauss discusses this case in detail, to explain why k is not 3). Again, we verify that $3 = (\pm 1)^2 + (\pm 1)^2 + (\pm 1)^2$ counts for eight representations, corresponding to the different choices of sign.

As a less trivial example, let $n = 5$. Here we have the two inequivalent reduced forms $x^2 + 5y^2$ and $2x^2 + 2xy + 3y^2$, so that $h = 2, g = 2, k = h/g = 1$, and we have, by (4.7), $R_3(5) = 12 \cdot 2 = 24$, or by (4.7′), $3 \cdot 2^{1+2} \cdot 1 = 24$. We verify directly that $5 = (\pm 2)^2 + (\pm 1)^2 + 0^2$, with six permutations and four choices of sign for each, so that indeed $R_3(5) = 6 \cdot 4 = 24$.

Finally, for $n = 38$ we have $h = 6$, $g = 2$, $k = 3$ (see, e.g., [28] or [43]), so that, since $38 \equiv 6 \pmod 8$, we have $R_3(38) = 12 \cdot 6 = 72$ by (4.7), and $R_3(38) = 3 \cdot 2^{1+2} \cdot 3 = 72$ by (4.7′). This is verified directly by $38 = (\pm 1)^2 + (\pm 1)^2 + (\pm 6)^2 = (\pm 2)^2 + (\pm 3)^2 + (\pm 5)^2$; indeed, the first sum corresponds to $3 \cdot 8 = 24$ representations and the second to $6 \cdot 8 = 48$ representations, and $24 + 48 = 72$.

§9. From Gauss to the Twentieth Century

In 1847 Eisenstein (1823–1852), by using Dirichlet's class number formulae for h, showed (as had been hinted at already by Dirichlet (1805–1859)) that a consequence of Theorem 2 is the following theorem.

Theorem 3. *With previous notation, with $[x]$ equal to the greatest integer not in excess of x, and with $(\frac{r}{n})$ the Jacobi symbol, one has, for squarefree n,*

$$R_3(n) = \begin{cases} 24 \sum_{r=1}^{[n/4]} \left(\frac{r}{n}\right) & \text{if } n \equiv 1 \pmod 4 \\ 8 \sum_{r=1}^{[n/2]} \left(\frac{r}{n}\right) & \text{if } n \equiv 3 \pmod 8. \end{cases} \tag{4.7''}$$

So far, we have considered only the primitive representations $R_3(n)$. In order to handle completely the problem of $r_3(n)$, the total number of solutions of (4.1), we need, besides Theorem 2, also Theorem 1.1, which we recall here as

$$r_3(n) = \sum_{d^2 | n} R_3\left(\frac{n}{d^2}\right). \tag{4.8}$$

As an example, consider $n = 108$. Then $r_3(108) = R_3(108) + R_3(27) + R_3(12) + R_3(3)$. As $108 \equiv 12 \equiv 0 \pmod 4$, we have $R_3(108) = R_3(12) = 0$. We already know that $R_3(3) = 8$. For $R_3(27)$, we use (4.7) with $h = 1$ (or directly, by $27 = (\pm 5)^2 + (\pm 1)^2 + (\pm 1)^2$) and find $R_3(27) = 24$. Hence, $r_3(108) = 24 + 8 = 32$. This is verified immediately: $108 = (\pm 2)^2 + (\pm 2)^2 + (\pm 10)^2 = (\pm 6)^2 + (\pm 6)^2 + (\pm 6)^2$.

It is clear that Legendre's Theorem 1 is an immediate corollary of Gauss's Theorem 2. In 1850, Dirichlet gave a new proof of Theorem 1 (essentially the one given here; see [56]). Liouville (1809–1882) indicated a large number of combinatorial identities connected with representations of integers as sums of three squares; we shall discuss them in §14. In 1850, Kronecker (1823–1891), by the use of elliptic functions, found [143] an expression for $r_3(n)$ which may be shown (nontrivially—see Problem 6) to be equivalent to Theorem 2.

Among the other contributors to the solution of this and related problems, Dickson [54, Chapter VII] lists Genocchi, Catalan, Lebesgue (if (4.1) has solutions for some n, it also has for n^2), Halphen (equation (4.1) is solvable for $n = p \equiv 3 \pmod 8$), Neuberg, Glaisher, Hermite, Bachmann (solutions of (4.1) with $x_1 \neq x_2 \neq x_3 \neq x_1$), von Sterneck, Schubert (the case $n = m^2$ in (4.1)),

Humbert (equation (4.1) in $Q(\sqrt{5})$), Landau (see §15 below), Sierpiński, Mathieu, Aubry (if $17 < n = p \equiv 5 \pmod{12}$, then (4.1) has solutions with $x_1 \neq x_2 \neq x_3 \neq x_1$), Mordell, and many more. Some of the contributions not specified above in parentheses after the authors' names refer to particular cases of (4.1); others treat related problems. Often these problems are quite interesting, but too special and complicated for quotation here. Many of the above contributions also consist in new proofs of the formula for $r_3(n)$ or of Theorem 1. Finally, some belong more properly to the *Terry–Escott problem*: to find nontrivial solutions of equations of the form $\sum_{i=1}^{k} x_i^m = \sum_{j=1}^{r} y_j^m$ (here usually $k = 3$, $r = 2$, 3, or 4).

In our rapid historical survey we mention again P. T. Bateman [18], who obtained the formula for $r_3(n)$ by a modification of the method that Mordell and Hardy (1877–1947) had successfully used for the cases of $k \geqslant 5$ squares. Hardy had actually stated that the method could be extended easily to $k = 4$; the case $k = 3$ presents, however, special difficulties. As already mentioned (see Chapter 1), the results obtained formally by this approach are actually false for $k = 1$ and $k = 2$.

§10. The Main Theorem

We now state formally the main theorem of this chapter, in the form obtained in [18].

Theorem 4. *The number $r_3(n)$ of solutions of $x_1^2 + x_2^2 + x_3^2 = n$ is given by*

$$r_3(n) = \frac{16}{\pi}\sqrt{n}L(1,\chi)q(n)P(n), \tag{4.9}$$

where $n = 4^a n_1$, $4 \nmid n_1$,

$$q(n) = \begin{cases} 0 & \text{if } n_1 \equiv 7 \pmod 8, \\ 2^{-a} & \text{if } n_1 \equiv 3 \pmod 8, \\ 3 \cdot 2^{-a-1} & \text{if } n_1 \equiv 1, 2, 5, \text{ or } 6 \pmod 8, \end{cases}$$

$$P(n) = \prod_{\substack{p^{2b} \mid n \\ p \text{ odd}}} \left(1 + \sum_{j=1}^{b-1} p^{-j} + p^{-b}\left\{1 - \left(\frac{-n/p^{2b}}{p}\right)\frac{1}{p}\right\}^{-1}\right)$$

($P(n) = 1$ for squarefree n), and $L(s,\chi) = \sum_{m=1}^{\infty} \chi(m)m^{-s}$, with $\chi(m) = (-4n/m)$.*

---

* Attention is called to the factor 4, which occurs in $\chi(m)$ even if $-n \equiv 1 \pmod 4$; its effect is to eliminate the terms with even m.

On the basis of (4.8), Theorem 4 is equivalent to the much simpler

$$R_3(n) = \pi^{-1} G_n \sqrt{n} L(1, \chi), \tag{4.10}$$

with

$$G_n = \begin{cases} 0 & \textit{for } n \equiv 0, 4, 7 \pmod 8 \\ 16 & \textit{for } n \equiv 3 \pmod 8 \\ 24 & \textit{for } n \equiv 1, 2, 5, 6 \pmod 8). \end{cases}$$

As Mordell (1888–1972) (see [180]) ruefully admits, there seems to be no simple proof of the theorems of the present chapter. This is, perhaps, not too surprising if we consider the fact that even the statements depend on the rather deep concepts of class number, number of genera of quadratic forms, etc.

In some way or another, all "elementary" proofs seem to be adaptations of Gauss's original (and very readable, although long) proof of Theorem 2′ in [81]; see, e.g., Bachmann [12].

We have already verified the equivalence of Theorem 2 and Theorem 2′. Before we consider any proofs, we show that on account of known representations of the class number, Theorem 3 is a corollary of Theorem 2, that (4.9) follows from (4.8) and (4.10), and that (4.10) is equivalent to (4.7). Then it will be sufficient to justify any one of the Theorems 2, 2′, or 4 and all will have been proved. First, however, we recall a few number theoretic results.

§11. Some Results from Number Theory

Let k be a rational integer. A function $\chi(m)$, defined on the integers, is called a *Dirichlet character* if it satisfies the following (not independent) conditions for all $m \in \mathbb{Z}$: if $(m, k) = 1$ then $|\chi(m)| = 1$; if $(m, k) > 1$ then $\chi(m) = 0$; $\chi(m + k) = \chi(m)$; $\chi(m)\chi(n) = \chi(mn)$; $\chi(1) = 1$. The function

$$\chi_0(m) = \begin{cases} 1 & \text{for } (m, k) = 1 \\ 0 & \text{for } (m, k) > 1 \end{cases}$$

is a character, called the *principal* character. If $\chi \neq \chi_0$, then $|\sum_{m=a}^{b} \chi(m)| \leqslant \phi(k)/2$ ($\phi(k)$ = Euler's function), so that $\sum_{m=1}^{\infty} \chi(m) m^{-s}$ converges for σ (= Re s) > 0; if $\chi = \chi_0$, it converges for all $\sigma > 1$. The function defined by this series in $\sigma > 1$ (or in $\sigma > 0$, respectively) is denoted by $L(s, \chi)$.

We recall that, for $Q = \sum_{i,j=1}^{2} a_{ij} x_i x_j = a x_1^2 + b x_1 x_2 + c x_2^2$, we are led to consider either the *discriminant* $D = b^2 - 4ac = 4(a_{12}^2 - a_{11}a_{22})$ or the *determinant* $d_2 = a_{11}a_{22} - a_{12}^2$, $D = -4d_2$. If $a_{ij} \in \mathbb{Z}$, then $D \equiv 0 \pmod 4$, if only a, $b, c \in \mathbb{Z}$ is required and b is odd, then $D \equiv 1 \pmod 4$. In a positive definite form we have $d_2 > 0$, $D < 0$. If $n \in \mathbb{Z}$ and $-n$ has to be used as a discriminant, then this is possible only for $n \equiv 0$ or 3 (mod 4). If $n \equiv 1$ or 2 (mod 4), then $-n \equiv 3$ or 2 (mod 4) and we have to take instead $D = -4n$ as discriminant, while $d_2 = n$ is possible for any $n \in \mathbb{Z}$. It follows that, for $n \not\equiv 0, 4$, or 7 (mod 8) (i.e., in all cases of interest to us), the number of classes of positive definite primitive binary quadratic forms is unchanged, regardless of which point of view we take, if also $n \not\equiv 3 \pmod 4$. If $n \equiv 3 \pmod 4$ (i.e., in the cases of interest to us, if $n \equiv 3 \pmod 8$), the classes of determinant n are 3 times the number of classes of discriminant $-n$.

Finally, while we have assumed knowledge of the Legendre–Jacobi symbol, we recall here the quadratic reciprocity law as it applies to the Kronecker symbol (see [105] for proofs). Just as for the Legendre–Jacobi symbol, $(\frac{n}{m}) = 0$ if $(n, m) > 1$; if $(n, m) = 1$ and $2 \nmid nm$, then $(\frac{n}{m}) = (\frac{m}{n})(-1)^{(m-1)(n-1)/4}$. Next, for m odd, we have $(\frac{-1}{m}) = (-1)^{(m-1)/2}$ and $(\frac{2}{m}) = (-1)^{(m^2-1)/8}$. Also,

$$\left(\frac{a}{2}\right) = \left(\frac{a}{-2}\right) = \begin{cases} 0 & \text{if } a \equiv 0 \pmod 4 \\ +1 & \text{if } a \equiv 1 \pmod 8 \\ -1 & \text{if } a \equiv 5 \pmod 8; \end{cases}$$

otherwise, the symbol is not defined. Whenever $(\frac{a}{2})$ is defined, we have $(\frac{a}{2}) = (\frac{2}{a})$.

Given $n \in \mathbb{Z}$, let $D = -n$ if $n \equiv 0, 3 \pmod 4$ and $D = -4n$ otherwise. Set $\chi(m) = (\frac{D}{m})$. Then, if h is the number of classes of primitive binary quadratic forms $ax^2 + 2bxy + cy^2$ with $n = ac - b^2$, Dirichlet showed that

$$h = \frac{u_n\sqrt{n}}{\pi} L(1, \chi), \tag{4.11}$$

with $u_n = 1$ if $n \equiv 3 \pmod 4$, $u_n = 2$ if $n \equiv 1, 2, 5$, or 6 (mod 8).

A discriminant D can be factored in exactly one way so that $D = -fg^2$, with f squarefree if $f \equiv 1 \pmod 4$, and otherwise with $4 \mid f$, $4^2 \nmid f$, $f/4$ squarefree. If in this factorization $g = 1$, then D is called a *fundamental discriminant*. If $n = fg^2$, then, for $n \equiv 3 \pmod 4$ we have $L(1, \chi) = L(1, (\frac{-fg^2}{m}))$, while if $n \equiv 1$ or 2 (mod 4), then $L(1, \chi) = L(1, (\frac{-4fg^2}{m}))$. For these two cases, we record the following identities:

$$\begin{aligned} L(1, \chi) &= \prod_{p \mid g}\left(1 - \left(\frac{-f}{p}\right)\frac{1}{p}\right) L\left(1, \left(\frac{-f}{m}\right)\right) && \text{if } n \equiv 3 \pmod 4, \\ L(1, \chi) &= \prod_{p \mid g}\left(1 - \left(\frac{-4f}{p}\right)\frac{1}{p}\right) L\left(1, \left(\frac{-4f}{p}\right)\right) && \text{if } n \equiv 1 \text{ or } 2 \pmod 4. \end{aligned} \tag{4.12}$$

In particular, let n be squarefree, set $\chi(m) = (\frac{-4n}{m})$ and if $n \equiv 3 \pmod 4$, let

$\chi_1(m) = (\frac{-n}{m})$; otherwise, let $\chi_1(m) = \chi(m)$. Then it follows from (4.12) that for $n \equiv 3 \pmod 4$,

$$L(1,\chi) = \left(1 - \left(\frac{-n}{2}\right)\frac{1}{2}\right) L(1,\chi_1) = \tfrac{3}{2} L(1,\chi_1). \tag{4.13}$$

If $n \not\equiv 3 \pmod 4$, then $L(1,\chi) = L(1,\chi_1)$.

For future use, we record also Dirichlet's formula, valid for any fundamental discriminant f:

$$h = \frac{v_f}{2 - (\frac{f}{2})} \sum_{r=1}^{t} \left(\frac{f}{r}\right), \qquad t = \left[\frac{f}{2}\right], \tag{4.14}$$

with $v_f = 1$ except for $v_{-4} = 2$, and $v_{-3} = 3$.

Proofs for all statements of this section may be found, e.g., in [146, Vol. 1], especially in Part IV.

§12. The Equivalence of Theorem 4 with Earlier Formulations

Let $n \equiv 3 \pmod 8$, $n \neq 3$ and squarefree, $f = -n$. Then, by (4.14),

$$h = \frac{1}{2 - (\frac{-3}{2})} \sum_{r=1}^{[n/2]} \left(\frac{-n}{r}\right).$$

Here one has $(\frac{-3}{2}) = -1$, $(\frac{-n}{r}) = (\frac{-1}{r})(\frac{n}{r}) = (-1)^{(r-1)/2}(\frac{r}{n})(-1)^{(r-1)(n-1)/4} = (\frac{r}{n})$ for odd r. Also, for even r one has, if $r = 2^a r_1$, $(\frac{-n}{r}) = (\frac{-n}{2})^a(\frac{-n}{r_1}) = (\frac{2}{-n})^a(\frac{r_1}{n}) = (\frac{2}{n})^a(\frac{r_1}{n}) = (\frac{r}{n})$. Consequently, by (4.14), $h = \frac{1}{3}\sum_{r=1}^{[n/2]}(\frac{r}{n})$, and by (4.7), $R_3(n) = 24h = 8\sum_{r=1}^{[n/2]}(\frac{r}{n})$, i.e., (4.7″) for $n \equiv 3 \pmod 8$. In the case $n \equiv 1 \pmod 4$, the computations are similar, but somewhat more complicated. Indeed, we have $f = -4n$, by (4.14) with $n \equiv 1 \pmod 4$, and $h = \frac{1}{2}\sum_{r=1}^{2n}(\frac{-4n}{r})$. Here the sum equals

$$\sum_{\substack{r=1\\ r\text{ odd}}}^{2n} \left(\frac{-n}{r}\right) = \sum_{\substack{r=1\\ r\text{ odd}}}^{2n} (-1)^{(r-1)/2}\left(\frac{n}{r}\right) = \sum_{\substack{r=1\\ r\text{ odd}}}^{2n} (-1)^{(r-1)/2}\left(\frac{r}{n}\right) = \sum_{\substack{r=1\\ r\text{ odd}}}^{n-2} + \sum_{\substack{r=n+2\\ r\text{ odd}}}^{2n-1}.$$

In the second sum, set $r = 2n - s$ and add, obtaining

$$2\sum_{\substack{r=1\\ r\text{ odd}}}^{n-2} (-1)^{(r-1)/2}\left(\frac{r}{n}\right) = 2\sum_{r=1}^{(n-1)/2} (-1)^{[r/2]}\left(\frac{r}{n}\right) = 4\sum_{r=1}^{(n-1)/4}\left(\frac{r}{n}\right)$$

(the last equality is nontrivial) and $h = 2\sum_{r=1}^{[n/4]}(\frac{r}{n})$. Substituting this in (4.7), we obtain $R_3(n) = 24\sum_{r=1}^{[n/4]}(\frac{r}{n})$, i.e., (4.7″) for $n \equiv 1 \pmod 4$. (See [64].*)

We now show that on account of (4.11), (4.10) is equivalent to (4.7) and that (4.9) follows from (4.8) and (4.10). We first consider the case of squarefree n, so that $r_3(n) = R_3(n)$. We have to show that (4.10) is equivalent to (4.7). For $n \equiv 1$, 2, 5, or 6 (mod 8), $R_3(n) = (24/\pi)\sqrt{n}L(1,\chi) = (24/\pi)\sqrt{n}L(1,\chi_1) = (12.2/\pi)\sqrt{n}L(1,\chi_1) = 12h$, by (4.11). If $n \equiv 3 \pmod 8$, then, by (4.13), $L(1,\chi) = \frac{3}{2}L(1,\chi_1)$ and $R_3(n) = 16\pi^{-1}\sqrt{n}\cdot\frac{3}{2}L(1,\chi_1) = (24/\pi)\sqrt{n}L(1,\chi_1) = 24h$, according to (4.11); hence, (4.10) is equivalent to (4.7). Next, for squarefree n, in (4.9), $q(n) = 1$ if $n \equiv 3 \pmod 8$ and $q(n) = \frac{3}{2}$ if $n \equiv 1$, 2, 5, or 6 (mod 8), with $P(n) = 1$ in both cases. Consequently, in this case, (4.9) is equivalent to (4.10).

If $n = 4^a n_1$, $4 \nmid n_1$, then (see §2), $r_3(4^a n_1) = r_3(n_1)$. Hence, if χ_1 is as defined before (4.13) and n_1 is squarefree, $n_1 \equiv 3 \pmod 4$, then $r_3(n) = r_3(n_1) = (16/\pi)\sqrt{n_1}\,L(1,\chi) = (16/\pi)\sqrt{n}\cdot 2^{-a}L(1,\chi) = (16/\pi)\sqrt{n}q(n)L(1,\chi_1)$, with $q(n) = 2^{-a}$. Similarly, if n_1 is squarefree and $n_1 \equiv 1$, 2, 5, or 6 (mod 8), then $r_3(n) = r_3(n_1) = (16/\pi)\sqrt{n_1}\,L(1,\chi) = (16/\pi)\sqrt{n}\cdot 2^{-a}\cdot\frac{3}{2}L(1,\chi_1)$, by (4.13). Consequently, $r_3(n) = (16/\pi)\sqrt{n}L(1,\chi_1)q(n)$, with $q(n) = 3\cdot 2^{-a-1}$. If $n_1 \equiv 7 \pmod 8$, then $r_3(n) = 0$, and that finishes the proof of (4.9) for $n = 4^a n_1$ with n_1 squarefree.

To complete the proof of the equivalence of (4.9) with (4.7), coupled with (4.8), let us assume that $n_1 = p^2 n_2$, with p a prime and squarefree $n_2 \equiv 1$, 2, 5, or 6 (mod 8). Then, by (4.8) (or by Theorem 1.1), $r_3(n_1) = R_3(n_1) + R_3(n_2)$. By (4.10), $R_3(n_1) = 24\pi^{-1}\sqrt{n_1}\,L(1,\chi_1)$, with $\chi_1(m) = (\frac{-4n_2p^2}{m})$, and $R_3(n_2) = 24\pi^{-1}\sqrt{n_2}\,L(1,\chi_2)$, with $\chi_2(m) = (\frac{-4n_2}{m})$. By (4.12), $L(1,\chi_1) = (1 - (\frac{-4n_2}{p})\frac{1}{p}) \times L(1,\chi_2)$, so that

$$L(1,\chi_2) = \frac{p}{p - (\frac{-4n_2}{p})}L(1,\chi_1),$$

and

$$R_3(n_2) = 24\pi^{-1}\sqrt{n_2}\,L(1,\chi_2) = 24\pi^{-1}\sqrt{\frac{n_1}{p^2}}\frac{p}{p - (\frac{-4n_2}{p})}L(1,\chi_1)$$

$$= 24\pi^{-1}\frac{1}{p - (\frac{-n_1/p^2}{p})}L(1,\chi_1).$$

*Note: The formulae are incorrectly quoted in [54]. So, e.g., for $n \equiv 7 \pmod 8$, the formula in [54] yields $R_3(n) > 0$, while, in fact, $R_3(n) = 0$. Also, the condition n squarefree is needed.

Adding,

$$r_3(n) = R_3(n_1) + R_3(n_2)$$

$$= 24\pi^{-1}\sqrt{n_1}\,P(n_1)L(1,\chi_1) = 16\pi^{-1}\sqrt{n_1}\,q(n_1)P(n_1)L(1,\chi_1)$$

with

$$q(n_1) = \tfrac{3}{2} \quad \text{and} \quad P(n_1) = 1 + \frac{1}{p - \left(\frac{-n_1/p^2}{p}\right)},$$

and this is precisely the case $n = n_1 = p^2 n_2$, n_2 squarefree, $n_2 \equiv 1, 2, 5$, or 6 (mod 8) of Theorem 4. The proof for the case $n_1 \equiv 3 \pmod 8$ is entirely similar. The proof for $n_1 = p^{2b} n_2$ may be completed by induction on b and the general proof by induction on the number of square prime divisors of n_1. Finally, if $n = 4^a n_1$, then $r_3(n) = r_3(n_1)$, and if we replace $\sqrt{n_1}$ in the formula for $r_3(n_1)$ by $\sqrt{n}\cdot 2^{-a}$, we obtain, as already seen, the stated value for $q(n)$. The proof of the equivalence of (4.9) with (4.7) combined with (4.8) is complete.

In what precedes, we have consistently taken $\delta_n = 1$, i.e., we have ignored the exceptional values $n = 1$ and $n = 3$. Theorem 4, however, holds also in those cases. Indeed, for $n = 1$,

$$\chi(m) = \left(\frac{-4n}{m}\right) = \left(\frac{-4}{m}\right) = \begin{cases} 1 & \text{if } m \equiv 1 \pmod 4, \\ 0 & \text{if } m \equiv 0 \pmod 2, \\ -1 & \text{if } m \equiv 3 \pmod 4, \end{cases}$$

$L(1,\chi) = 1 - \frac{1}{3} + \frac{1}{5} - \cdots = \pi/4$, and $r_3(1) = (24/\pi)(\pi/4) = 6$. Similarly, for $n = 3$ we have $\chi(m) = (\frac{-4\cdot 3}{m}) = (\frac{-12}{m})$, $L(1,\chi) = \frac{3}{2}L(1,(\frac{-3}{m}))$, $L(1,(\frac{-3}{m})) = \pi/3\sqrt{3}$, so that $L(1,\chi) = \frac{3}{2}(\pi/3\sqrt{3}) = \pi/2\sqrt{3}$ and $r_3(3) = (16/\pi)\sqrt{3}(\pi/2\sqrt{3}) = 8$.

§13. A Sketch of the Proof of (4.7′)

The following is the barest sketch of Gauss's proof. Let $Q_1(t,u) = at^2 + 2btu + cu^2$, and let $Q_2(\mathbf{x}) = Q_2(x_1, x_2, x_3) = \sum_{i,j=1}^{3} a_{ij}x_i x_j$. We say that Q_1 is *represented* by Q_2 if there exist integers m_i, n_i such that, if we set $x_i = m_i t + n_i u$ and substitute these in Q_2, then $Q_2(\mathbf{x}) = Q_1(t,u)$. In what follows, we shall be interested mainly in the form $Q_2 = \sum_{i=1}^{3} x_i^2$, but much of what we obtain applies to any positive definite ternary form.

Clearly, if $Q = t^2 + nu^2$, then $Q(0,1) = n$, so that Q represents n. In fact, as seen, forms with $-D = ac - b^2 = n$, but not equivalent to Q, may also represent n. These forms belong to a certain number k of classes of discriminant n and, in general, form one of the g genera corresponding to that discriminant. However, in Gauss's notation (where the coefficient of tu has to be even), for $n \equiv 3 \pmod 8$, only a third of classes in the *principal genus* (i.e., the genus that contains $t^2 + nu^2$) represent n. We now select one form (for definiteness, e.g., the reduced form) $Q^{(j)}(t, u)$ in each one of those k classes and determine all sets of integers $m_i^{(j)}, n_i^{(j)}$ such that, when we substitute

$$x_i = \pm(m_i^{(j)}t + n_i^{(j)}u), \qquad i = 1, 2, 3, \quad j = 1, 2, \ldots, k, \tag{4.15}$$

in Q_2, then Q_2 becomes $Q^{(j)}(t, u)$. Gauss showed that the number of such sets of solutions is 2^t, where t is the number of odd prime divisors of n. Hence, we obtain $2^t k$ representations of forms $Q^{(j)}$ that (for appropriate integers t, u) take the value n. These lead, by (4.15), to integer values for the x_i's. By considering the six possible permutations of the x_i's and the 2^3 choices of signs indicated in (4.15), we obtain all together $48 \cdot 2^t k = 3 \cdot 2^{t+4} k$ representations. However, simultaneous changes of signs in all three relations, with simultaneous change of t into $-t$ and u into $-u$, will lead to the same values for the x_i's; hence, we obtain only $3 \cdot 2^{t+3} k$ representations of n as a sum of three squares, that differ at least by the order of the summands or by the sign of one of the x_i. The proofs that the number of solution sets $m_i^{(j)}, n_i^{(j)}$ equals 2^t, and that there are no further duplications in the representations obtained, are anything but simple (they take the better part of 200 pages in [81]). This essentially finishes the proof of (4.7′) for $n \equiv 1, 2, 5$, or $6 \pmod 8$. By recalling that for $n \equiv 3 \pmod 8$ Gauss's k is actually $3k$, the proof of (4.7′) is complete. The particular cases $n = 1$ and $n = 3$ need a special discussion (see [81, §291]) and are not covered by (4.7′) as written.

§14. Liouville's Method

Liouville gave a proof of (4.10) by the use of elliptic functions. However (see [165] and [270]), his identities can be proved in an entirely elementary fashion. While it is difficult to obtain the full statement of (4.10), and even more difficult to motivate the approach, without the consideration of elliptic functions, the elementary character of the considerations makes them intriguing, and in any case one is able to prove Theorem 1 by use of Liouville's identities.

The most condensed presentation of this method is presumably that in [270] and takes almost 50 pages. While the weakness of the results obtained does not justify a complete presentation here, it appears worthwhile to state at least the result, in order to suggest the flavor of this approach.

Theorem 5. *Given an integer n, denote by $T(n)$ the total number of solutions of at least one of the two equations*

$$4n + 1 = d\delta + (d + \delta - 2)\delta'',$$

$$4n + 1 = d\delta + (d + \delta + 2)\delta''$$

in positive, odd integers d, δ, δ'', *such that* $d + \delta \equiv 0 \pmod 4$. *Then* $T(n)$ *is always positive and we have* $r_3(8m + 3) = 2T(8m + 3)$, $r_3(4m + 1) = 3T(4m + 1)$, $r_3(4m + 2) = 3T(4m + 2)$, $r_3(4m) = r_3(m)$, $r_3(4^a(8m + 7)) = 0$.

Remark. The positivity of $T(n)$ is clear: if n is odd, then $d = 2n + 1, \delta = \delta'' = 1$ satisfy $d + \delta \equiv 0 \pmod 4$ and also the first equation; similarly, if n is even, then $d = 2n - 1, \delta = \delta'' = 1$ satisfy $d + \delta \equiv 0 \pmod 4$ and also the second equation.

§15. The Average Order of $r_3(n)$ and the Number of Representable Integers

We finish this chapter with two related questions.

Problem A. From Theorem 4 it is clear that the behavior of the function $r_3(n)$ is very irregular. On the one hand, it seems to increase roughly like $\sqrt{n}$; on the other hand, it vanishes for arbitrarily high values of n, e.g., whenever $n \equiv 7 \pmod 8$. In such cases it is often worthwhile to consider the average value of the function $r_3(n)$, say $\overline{r}_3(n)$, just as we have done for $r_2(n)$ and $r_4(n)$. Hence, let $S(n) = \sum_{m=1}^{n} r_3(m)$; then $\overline{r}_3(n) = S(n)/n$. This is easily computed by the same method as for $r_2(n)$ (see §2.7). Let us consider the set of lattice points in 3-dimensional space. If $m = \sum_{i=1}^{3} x_i^2$, $x_i \in \mathbb{Z}$, then (x_1, x_2, x_3) are the coordinates of a lattice point on the sphere of radius $\sqrt{m}$, and the set of all lattice points on this sphere corresponds to the different (in general not essentially distinct) solutions of $x_1^2 + x_2^2 + x_3^2 = m$. Their number is counted precisely by $r_3(m)$. We now think of cubes of volume 1, centered at each of these lattice points. These cubes completely fill a portion of space, without overlapping. Each lattice point corresponds to the solution of some equation (4.1). It follows that $S(n)$ equals the sum V of the volumes of all these cubes with centers inside or on the sphere of radius $\sqrt{n}$. As the distance between the center of a unit cube and a vertex equals $\sqrt{3}/2$, it follows that V is enclosed by the sphere of radius $R = \sqrt{n} + \sqrt{3}/2$ and completely contains the sphere of radius $r = \sqrt{n} - \sqrt{3}/2$. We conclude that $\frac{4}{3}\pi r^3 \leqslant S(n) \leqslant \frac{4}{3}\pi R^3$. If we replace r and R with their values, elementary computations lead to

$$|S(n) - \tfrac{4}{3}\pi n^{3/2}| \leqslant 2\sqrt{3}\pi n\left(1 + \frac{\sqrt{3}}{2}n^{-1/2} + \tfrac{1}{4}n^{-1}\right).$$

Consequently, we obtain the not unexpected result that $\overline{r}_3(n) = \frac{4}{3}\pi\sqrt{n} + R(n)$, with a bounded error term $R(n)$. This result may be compared with $\overline{r}_2(n) \sim \pi$

and with $\bar{r}_4(n) \sim \frac{1}{2}\pi^2 n$. (See §§2.7, 3.5, and also 9.7, for other results concerning the behavior of $r_k(n)$.)

As for the error term $E_3(n) = |S(n) - \frac{4}{3}\pi n^{3/2}|$, Vinogradov [272] has shown that it satisfies $E_3(n) = O(n^{u+\varepsilon})$, with $\mu = \frac{19}{28}$, and this improves considerably the previous result $E_3(n) = O(n)$. A simple proof of the somewhat weaker result $E_3(n) = O(n^{3/4}\log n)$ is due to M. Bleicher and M. I. Knopp [26]. In the same paper the authors also show that $E_3(n) = \Omega(n^{1/2}\log_2 n)$, presumably the best Ω-result available.

It may be worthwhile to add that recently B. Randol [224] has obtained a very general related result. It is profitable to consider the error terms of Vinogradov and Bleicher and Knopp, as well as the corresponding error terms in the case of two variables (see §2.7), in the light of Randol's results. We shall combine his two theorems into a single one, in a slightly modified formulation.

Theorem (Randol). *Let $T(\mathbf{z})$ be a positive function defined in Euclidean n-space, $T(\mathbf{z}) \in \mathbf{C}^\infty$, homogeneous of weight w. Let $C = \{\mathbf{z} \mid T(\mathbf{z}) \leqslant 1\}$, $\partial C = \{\mathbf{z} \mid T(\mathbf{z}) = 1\}$, V = volume of C, $N(x)$ = number of lattice points $\mathbf{z}_j$ such that $T(\mathbf{z}_j) \leqslant x$ $(x > 0)$; and set $E(x) = N(x) - Vx^{n/w}$. Then (under a very mild, technical condition on the boundary ∂C), $E(x) = O(x^\gamma)$, where*

$$\gamma \leqslant \gamma_1 = \frac{n(n-1-(n-1)/w)}{wn-n+1} = \frac{n(n-1)(w-1)}{w(wn-n+1)}.$$

In particular, if $T(\mathbf{z}) = \sum_{j=1}^n z_j^{2k}$ (so that $w = 2k$) and if we set $A = (w-1) \times (n-1)/w^2$, $B = n(n-1)/w(n+1)$, the previous result can be improved to $\gamma = \gamma_2 \leqslant \max(A, B)$. Moreover, if $A > B$, this result is best possible, i.e., $E(x) = \Omega(x^{\gamma_2})$. For $n = 2$, $k > 1$, the last result can be improved to $\gamma = \gamma_3 = (w-1)/w^2$, and this is best possible.

Comments. If we consider the circle problem, with $n = w = 2$, then $\gamma_1 = 2\cdot 1\cdot 1/2(4-2+1) = \frac{1}{3}$, while for $n = 3$, $w = 2$ we have $\gamma_1 = 3\cdot 2\cdot 1/2(6-3+1) = \frac{3}{4}$. These are the exponents of Sierpiński (see §2.7) and (essentially) Bleicher and Knopp, previously considered. (For $n = 4$, however, we easily verify that $\gamma_1 = \frac{6}{5}$, $A = \frac{3}{4}$, $B = \frac{6}{5}$, so that $\gamma_1 = \gamma_2 = \frac{6}{5}$, a result poorer than that of Landau in §3.5.) In all these cases, $T(\mathbf{z})$ is a sum of squares, so that $k = 1$, and we compute, for $n = 2$, that $A = \frac{1}{4}$, $B = \frac{1}{3}$, $\gamma_2 = \frac{1}{3}$; for $n = 3$, that $A = \frac{1}{2}$, $B = \frac{3}{4}$. It follows that in both these cases, $\gamma_1 = \gamma_2$. For $n = 2$, $k > 1$, however, we obtain the stronger result of §2.7 (observe the change of notation!).

Problem B. Perhaps of more interest is the problem of determining the number of integers $n \leqslant x$ for which $x_1^2 + x_2^2 + x_3^2 = n$ is solvable (see [148]; also [146]). If we denote their number by $\mathbf{N}_3(x)$, then $\mathbf{N}_3(x) = \sum 1$, with the summation extended over the integers $a \geqslant 0$, $k \geqslant 0$ for which $n \neq 4^a(8k+7)$, $n \leqslant x$. If we set $\mathbf{M} = \{n \mid n = 4^a n_1,\ n_1 \equiv 7 \pmod 8\}$, then $\mathbf{M}(x) = [x] - \mathbf{N}_3(x) = \sum_{4^a(8k+7)\leqslant x} 1 = \sum_{a\geqslant 0}\{[(x\cdot 4^{-a} - 7)/8] + 1\} =$

$\sum_{a \geqslant 0} [(x \cdot 4^{-a} + 1)/8]$, where the square bracket vanishes if $x \cdot 4^{-a} + 1 < 8$, i.e., if $a > [(\log(x/7))f] = y$, say. Here we abbreviate the numerical constant $(\log 4)^{-1}$ by f. Also, all quantities θ_i that will occur, satisfy $0 \leqslant \theta_i \leqslant 1$. It follows that, with the above notation,

$$\mathbf{M}(x) = \sum_{a=0}^{y} \left(\left(\frac{x}{8} \cdot 4^{-a} + \frac{1}{8} \right) - \theta_1 \right)$$

$$= \sum_{a=0}^{y} \left(\frac{x}{8} \cdot 4^{-a} - \frac{7}{8} + \theta_2 \right)$$

$$= \frac{x}{8} \frac{1 - 4^{-y-1}}{1 - 4^{-1}} - \left(\frac{7}{8} - \theta_2 \right)(y + 1).$$

By observing that $4^{-y} = \exp\{(-\log 4)(f \log(x/7) - 1 + \theta_3)\} = (7/x) \cdot 4^{1-\theta_3}$, we obtain

$$\mathbf{M}(x) = \frac{x}{24} \cdot 4(1 - 4^{-y-1}) - \left(\frac{7}{8} - \theta_2 \right) \left(f \log \frac{x}{7} + \theta_3 \right)$$

$$= \frac{x}{6} - \frac{x}{24} \cdot \frac{7}{x} \cdot 4 \cdot 4^{-\theta_3} - \tfrac{7}{8} f \log \frac{x}{7} - \frac{7}{8}\theta_3 + \theta_2 \left(f \log \frac{x}{7} + \theta_3 \right)$$

$$= \frac{x}{6} - \frac{7}{8} f \log \frac{x}{7} + R_0(x),$$

$$R_0(x) = \theta_2 \left(f \log \frac{x}{7} + \theta_3 \right) - \frac{7}{8}\theta_3 - \frac{7}{6} \cdot 4^{-\theta_3}.$$

As $0 \leqslant \theta_i \leqslant 1$, we have $R_0(x) \leqslant f \log(x/7) + \theta_3/8 - \frac{7}{6} \cdot 4^{-\theta_3} \leqslant f \log(x/7) + \frac{1}{8} - \frac{7}{24} = f \log(x/7) - \frac{1}{6}$ and $R_0(x) \geqslant -(\frac{7}{8}\theta_3 + \frac{7}{6} \cdot 4^{-\theta_3}) \geqslant -\frac{7}{6}$. It follows that $\mathbf{M}(x) = x/6 - \frac{7}{8} f \log(x/7) - \frac{7}{6} + R(x)$, where $0 \leqslant \frac{7}{6} + R_0(x) = R(x) \leqslant f \log(x/7) + 1$. An easy computation now yields

Theorem 6. *Let* $\mathbf{M}$ *be the set of natural integers that are not the sum of three squares, and set* $\mathbf{M}(x) = \sum_{n \in \mathbf{M}, n \leqslant x} 1$; *then*

$$\mathbf{M}(x) = \frac{x}{6} - \frac{7}{8 \log 4} \log x + C + R(x),$$

with

$$C = \frac{7 \log 7}{8 \log 4} - \frac{7}{6} \quad \textit{and} \quad 0 \leqslant R(x) \leqslant \frac{\log(x/7)}{\log 4} + 1.$$

Corollary 3. $\lim_{x\to\infty} \mathbf{M}(x)/x = \frac{1}{6}$.

The number $\mathbf{N}_3(x)$ of integers $n \leqslant x$ and that are sums of three squares is

$$\mathbf{N}_3(x) = [x] - \mathbf{M}(x) = x - \frac{x}{6} + \frac{7\log x}{8\log 4} - C - R(x) - x + [x]$$

$$= \frac{5x}{6} + \frac{7}{8\log 4}\log x - \frac{\log x}{\log 4} - 1 + \frac{\log 7}{\log 4}$$

$$- \frac{7\log 7}{8\log 4} + \frac{7}{6} - \left(R(x) + \theta - \frac{\log(x/7)}{\log 4} - 1\right)$$

$$= \frac{5x}{6} - \frac{\log x}{8\log 4} + C_1 + R_1(x),$$

where $C_1 = (\log 7)/(8\log 4) + \frac{1}{6} \simeq 0.3421\ldots$, $0 \leqslant \theta = x - [x] < 1$, and

$$R_1(x) = -R(x) - \theta + \frac{\log(x/7)}{\log 4} + 1 \leqslant \frac{\log(4x/7)}{\log 4}, \qquad R_1(x) \geqslant -\theta > -1.$$

It follows that if we set $R_2(x) = R_1(x) + 1$, $C_2 = C_1 - 1$, then

$$0 \leqslant R_2(x) \leqslant \frac{\log(x/7)}{\log 4} + 2 = \frac{\log(16x/7)}{\log 4},$$

$C_2 = (\log 7)/(8\log 4) - \frac{5}{6} \simeq -0.65787\ldots$, and we may state

Corollary 4. *With the above notation,*

$$\mathbf{N}_3(x) = \frac{5x}{6} - \frac{\log x}{8\log 4} + C_2 + R_2(x).$$

From Corollary 4 immediately follows

Corollary 5. $\lim_{x\to\infty} \mathbf{N}_3(x)/x = \frac{5}{6}$.

§16. Problems

1.

(a) Find all essentially distinct representations as sums of three squares of the following integers: 1 to 10, 25, 100, 153.

(b) For each of the representations found under (a), determine the number of representations counted by $r_3(n)$, and compute $r_3(n)$ for the values of n indicated under (a).

2. Check, on the sample of Problem 1, that $r_3(4n) = r_3(n)$.

3. Indicate at least one representation as a sum of three triangular numbers for the integers listed in Problem 1.

4. Find the number of classes of binary positive definite, properly primitive quadratic forms of discriminants equal to 5, 6, 7, 11, 13, 14, 19. Observe that this number is the same, regardless of whether we consider the form written as $ax^2 + bxy + cy^2$ and use $D = b^2 - 4ac$ (< 0), or consider it as $ax^2 + 2bxy + cy^2$ and use $d_2 = ac - b^2$ with $-D = d_2$ (or $= 4d_2$, accordingly), except for the cases $d_2 \equiv 3 \pmod 8$, $d_2 \neq 3$.

(Hint: Construct sets of inequivalent, reduced forms for each discriminant).

5.

(a) Obtain representations as a sum of three squares for $n = 13$ and $n = 19$, by following the steps used in the examples of §7.
(b) Use the results of Problem 4 in order to determine $r_3(19)$ and $r_3(13)$ by use of Theorem 2.
(c) Find $r_3(13)$ and $r_3(19)$ by direct decompositions.
(d) Use Theorem 3 to compute $r_3(13)$ and $r_3(19)$.
(e) Use Theorem 4 to compute $r_3(13)$ and $r_3(19)$.
(f) Use Theorem 4 to compute $r_3(250)$.
(g) Compare all results.

(One may want to know that $L(1,(-4\cdot 13/m)) = \pi/\sqrt{13}$ and $L(1,(-4\cdot 19/m)) = 3\pi/2\sqrt{19}$.)

6. Let $G(n)$ be the number of classes of binary quadratic forms $ax^2 + 2bxy + cy^2$ of determinant $ac - b^2 = n$, and denote by $F(n)$ the number of such classes in which at least one of the extreme coefficients is odd. Kronecker proved [143] that $R_3(n) = 24F(n) - 12G(n)$. Prove this theorem by showing that it is equivalent to Theorem 2.

Hint: Use $a_{11}a_{22} - a_{12}^2 = n$, and show that $G(n) = F(n)$ if $n \equiv 1$ or $2 \pmod 4$; $G(n) = 2F(n)$ if $n \equiv 7 \pmod 8$; and $3G(n) = 4F(n)$ if $n \equiv 3 \pmod 8$, except if $n = 3(2m+1)^2$, when $3G(n) = 4F(n) + 2$. In the case $n \equiv 7 \pmod 8$, the relation $G(n) = 2F(n)$ is rather difficult to prove.

7. For $n = 3, 4, 5, 6, 7, 8$, compute $S(n)$ and use Theorem 5 to find the corresponding $r_3(n)$.

*8. Given a binary (or, more generally, an n-ary) quadratic form of discriminant d, and the number h of equivalence classes, find an upper bound for d as a function of h (and of n). This problem has been solved only in a few particular cases, e.g., that of positive definite binary forms for $h = 1$ and $h = 2$ (see [257], [258], [13], and [14]).

Chapter 5

Legendre's Theorem

§1. The Main Theorem and Early Results

In this chapter we consider a diagonal form more general than a simple sum of squares. We shall be concerned with the ternary quadratic forms $Q(x, y, z) = ax^2 + by^2 + cz^2$. If a, b, c are positive integers, then strictly speaking this form is also a sum of squares, because it can be written as

$$\underbrace{x^2 + x^2 + \cdots + x^2}_{a \text{ times}} + \underbrace{y^2 + y^2 + \cdots + y^2}_{b \text{ times}} + \underbrace{z^2 + z^2 + \cdots + z^2}_{c \text{ times}},$$

but the number of squares varies with the values of the coefficients a, b, c. In fact, however, Q is an arbitrary diagonal form, with a, b, c integers, but not necessarily positive.

Instead of asking, as in Chapters 2, 3, and 4, for the general set $\mathbf{N}_Q$ of representable integers, or the number $\mathbf{N}_Q(x)$ of such representable integers $n \leqslant x$, we shall attempt to answer only the much more modest question: What conditions on the set (a, b, c) will insure that zero is representable? As already pointed out in Chapter 1, we are not interested in the (always present) trivial solution $x = y = z = 0$. By *solution*, we understand only a nontrivial one with $x \neq 0$, $y \neq 0$, and $z \neq 0$.

Even this seemingly so simple question turns out to be nontrivial. The main result is contained in the following theorem, proved by Legendre in 1785, if not earlier (see [155, especially pp. 512–513]; also [154, pp. 49–50]).

Theorem 1. *Let a, b, c be nonzero, squarefree, rational integers such that $(a, b) = (b, c) = (c, a) = 1$. Then the Diophantine equation*

$$ax^2 + by^2 + cz^2 = 0 \tag{5.1}$$

has nontrivial solutions if and only if the following conditions hold:

(i) *a, b, c do not all have the same sign.*

(ii) *$-ab$, $-bc$, and $-ca$ are quadratic residues modulo c, a, and b, respectively.*

Diophantus (see [54, Chapter XIII]) considered a particular case of (5.1), namely

$$\frac{y^2 - x^2}{a_1} = \frac{z^2 - y^2}{b_1}, \tag{5.2}$$

which corresponds, essentially to (5.1) with $a > 0$, $c > 0$, and $b = -(a + c)$. Diophantus gave some particular solutions of (5.2), and so did Alkarkhi (ca. 1000). Leonardo da Pisa gave a systematic method for the solution of (5.2). Several later mathematicians, among them Genocchi and Woepke, commented on, interpreted, and extended Leonardo da Pisa's approach. Lagrange (in 1767) considered the particular case $c = -1$ of (5.1). Euler [74, §§181–187] made (in 1770) an important contribution to this problem, essentially by factoring $ax^2 + by^2$ in what we today call an algebraic number field. He stated some theorems, as well as some conjectures. Gauss [81, §§294–298] gave a new proof of Legendre's theorem by use of his theory of ternary quadratic forms. Dirichlet also gave such a proof [55].

Many other mathematicians made contributions to the solution of this problem, either by giving methods to find all solutions, by giving algorithms for the solution of particular cases of (5.1), or by streamlining earlier proofs. Among the names mentioned in [54, Chapter XIII] are those of Calzoleri, Réalis, Bachmann, Cantor, Werebrusow, Cunningham, Thue, Aubry, and Cahen.

§2. Some Remarks and a Proof That the Conditions Are Necessary

The requirement that a, b, c be squarefree is no real restriction. Indeed, if $a = a'm_1^2$, $b = b'm_2^2$, $c = c'm_3^2$, with a', b', c' squarefree, then $ax^2 + by^2 + cz^2 = a'x_1^2 + b'y_1^2 + c'z_1^2$ with $x_1 = m_1x$, $y_1 = m_2y$, $z_1 = m_3z$, and a', b', c' squarefree. Next, the fact that a, b, and c are nonzero follows automatically from $(a, b) = (b, c) = (c, a) = 1$ (except for some trivial cases, when the result is obvious).

We also observe that, along with x, y, z, the triple mx, my, mz is also a solution of (5.1); hence, it is sufficient to find the primitive solutions (i.e., solutions with $(x, y, z) = 1$; see Chapter 1). In any primitive solution, the variables are actually coprime in pairs. Indeed, if $p \mid x$, $p \mid y$, then $p^2 \mid ax^2 + by^2$, so that, by (5.1), $p^2 \mid cz^2$; but c is squarefree, so that $p \mid z$, contrary to the assumption that $(x, y, z) = 1$.

The necessity of condition (i) is obvious for the existence of *real* solutions. We now prove the necessity of (ii). Let x, y, z be a primitive solution of (5.1). By $c \neq 0$, (5.1) and $acx^2 + bcy^2 + c^2z^2 = 0$ are equivalent; hence, if $p \mid a$, then

$$-bcy^2 \equiv (cz)^2 \pmod{p}. \tag{5.3}$$

From $(a, c) = 1$ and $p \mid a$, it follows that $p \nmid c$; therefore, if $p \mid y$, then $p \mid z$, which is false for a primitive solution, and we conclude that $p \nmid y$. It follows that we can solve $yy_1 \equiv 1 \pmod{p}$ for y_1. We now multiply (5.3) by y_1^2 and set $cy_1z = u$; we obtain $-bc \equiv u^2 \pmod{p}$. As this holds for every prime divisor of the square-free integer a, it follows that $-bc$ is a quadratic residue modulo a. Permutation of the letters completes the proof of the necessity of (ii).

§3. The Hasse Principle

Condition (i) can be interpreted as the requirement that (5.1) be solvable in the field $\mathbb{R}$ of reals. Condition (ii) has a similar interpretation. As is made clear by the preceding proof, (ii) essentially means, that (5.1) is solvable as a congruence modulo each prime divisor of the product abc. Looked upon from this point of view, the necessity of the condition becomes obvious. To this, we may add two other remarks:

(a) If $p \nmid abc$, then (5.1) is always solvable as a congruence modulo p.
(b) If (5.1) is solvable modulo an odd prime p, then it also is solvable modulo p^m $(m = 1, 2, 3, \ldots)$. (The prime $p = 2$ requires a different consideration; see Problem 3.)

The proofs of both statements may be found, in [86]; see also Problem 3.

If a congruence holds modulo all powers of a given prime, then the corresponding equation has a solution in the p-adic field (see [135]; also Chapter 14). Consequently, while apparently weaker, condition (ii) is equivalent to the statement: "Equation (5.1) is solvable in all p-adic fields." This formulation is suggested by the theory presented by H. Hasse (1898–1979) in several papers published during the years 1923–1924 (see [100] to [104]). The so-called *Hasse principle* states that, under certain conditions, the obviously *necessary* conditions for the solvability of a diophantine equation, namely, that it should be solvable in the real field and in every p-adic field, are also sufficient. The proof of Legendre's Theorem 1 amounts to a proof that Hasse's principle is valid for equation (5.1). It may be worthwhile mentioning that Hasse's principle does not have universal validity; indeed, counterexamples are known. The exact extent of its validity is not yet known.

§4. Proof of Sufficiency of the Conditions of Theorem 1

The proof is based on two different ideas. The general scheme is as follows: It would be easy to prove the theorem if we could factor Q into two linear polynomials, because linear equations are easy to solve. This is, of course, not possible, in general. The next best thing to try is to obtain such a factorization as a congruence modulo an appropriate number, and this is the first idea of the proof. Indeed, we shall show that there exist integers $a_1, b_1, c_1, a_2, b_2, c_2$, such that

$$ax^2 + by^2 + cz^2 \equiv (a_1x + b_1y + c_1z)(a_2x + b_2y + c_2z) \pmod{abc}. \tag{5.4}$$

It is then easy to find values x_0, y_0, z_0, such that

$$ax_0^2 + by_0^2 + cz_0^2 \equiv a_1x_0 + b_1y_0 + c_1z_0 \equiv 0 \pmod{abc}.$$

This, of course, is not sufficient to insure that (5.1) holds. However—and this is the second idea of the proof—we shall show that the congruence

$$a_1x + b_1y + c_1z \equiv 0 \pmod{abc} \tag{5.5}$$

can be satisfied, in general, by "small" values x_0, y_0, z_0 such that

$$-2abc < ax_0^2 + by_0^2 + cz_0^2 < abc. \tag{5.6}$$

If we combine (5.5) and (5.6), it follows that $ax_0^2 + by_0^2 + cz_0^2 = 0$ or $-abc$. In the first case, we have obtained a solution of (5.1); in the second case, set $x_1 = -by_0 + x_0z_0$, $y_1 = ax_0 + y_0z_0$, $z_1 = z_0^2 + ab$ and verify that

$$\begin{aligned}
ax_1^2 + by_1^2 + cz_1^2 &= a(b^2y_0^2 - 2bx_0y_0z_0 + x_0^2z_0^2) \\
&\quad + b(a^2x_0^2 + 2ax_0y_0z_0 + y_0^2z_0^2) + c(z_0^4 + 2abz_0^2 + a^2b^2) \\
&= ab(ax_0^2 + by_0^2 + cz_0^2) + x_0y_0z_0(-2ab + 2ab) \\
&\quad + abc(z_0^2 + ab) + z_0^2(ax_0^2 + by_0^2 + cz_0^2) \\
&= ab(-abc) + abc(z_0^2 + ab) + z_0^2(-abc) = 0,
\end{aligned}$$

so that x_1, y_1, z_1 is a solution of (5.1). The omitted cases, not covered by (5.5), are quite trivial, with obvious solutions.

We pass now to the actual proof and first prove

Lemma 1. *If $(m, n) = 1$ and (5.4) is solvable modulo m and modulo n, then it is solvable also modulo mn.*

Proof. Let

$$\begin{aligned}
ax^2 + by^2 + cz^2 &\equiv (u_1x + v_1y + w_1z)(u_2x + v_2y + w_2z) \pmod{n} \\
&\equiv (r_1x + s_1y + t_1z)(r_2x + s_2y + t_2z) \pmod{m}.
\end{aligned}$$

By the Chinese remainder theorem (see, e.g., [86]), we may select a_1 so that $a_1 \equiv u_1 \pmod{n}$ and $a_1 \equiv r_1 \pmod{m}$, and similarly for the other five coefficients b_1, c_1, a_2, b_2, c_2. With these choices, $ax^2 + by^2 + cz^2 \equiv (a_1x + b_1y + c_1z)(a_2x + b_2y + c_2z) \pmod{mn}$, concluding the proof. $\square$

As $(a,b) = (b,c) = (c,a) = 1$, it is now sufficient to show that (5.4) holds, modulo a; analogous considerations will then hold for b and c, and (5.4) will follow by Lemma 1.

Since $(a,b) = 1$, there exists b' such that $bb' \equiv 1 \pmod{a}$. Next, by the assumption that $-bc$ is a quadratic residue modulo a, there exists an integer g such that $-bc \equiv g^2 \pmod{a}$. Consequently, modulo a, we have

$$\begin{aligned} ax^2 + by^2 + cz^2 &\equiv by^2 + cz^2 \equiv bb'(by^2 + cz^2) \\ &\equiv b'(b^2y^2 + bcz^2) \equiv b'(b^2y^2 - g^2z^2) \\ &= b'(by + gz)(by - gz) \equiv (y + b'gz)(by - gz) \\ &= (0 \cdot x + y + fz)(0 \cdot x + by - gz), \quad \text{with } f = b'g. \end{aligned}$$

This is the factorization modulo a and finishes the proof of the factorization (5.4).

Lemma 2. *Let α, β, γ be positive reals, such that $\alpha\beta\gamma = m \in \mathbb{Z}$. Then the congruence $a_1x + b_1y + c_1z \equiv 0 \pmod{m}$ has a nontrivial solution with $|x| \leqslant \alpha$, $|y| \leqslant \beta$, $|z| \leqslant \gamma$.*

Proof. We use Dirichlet's "drawer principle" (see §3.3). Let x, y, and z take on the values $0, 1, \ldots, [\alpha], 0, 1, \ldots, [\beta]$, and $0, 1, \ldots, [\gamma]$, respectively. This leads to $(1 + [\alpha])(1 + [\beta])(1 + [\gamma]) > \alpha\beta\gamma = m$ triplets. There are, however, only m residue classes modulo m; hence, there are two triplets, say (x_1, y_1, z_1) and (x_2, y_2, z_2), that lead to the same residue class for $a_1x + b_1y + c_1z$, i.e., $a_1x_1 + b_1y_1 + c_1z_1 \equiv a_1x_2 + b_1y_2 + c_1z_2 \pmod{m}$. This means, if we set $x = x_1 - x_2$, $y = y_1 - y_2$, $z = z_1 - z_2$, that $a_1x + b_1y + c_1z \equiv 0 \pmod{m}$, with $|x| = |x_1 - x_2| \leqslant \alpha$ and similarly, $|y| \leqslant \beta$, $|z| \leqslant \gamma$. $\square$

We now return to (5.5). Let $\alpha = \sqrt{|bc|}$, $\beta = \sqrt{|ca|}$, $\gamma = \sqrt{|ab|}$. Then $\alpha\beta\gamma = abc$ and (5.5) has solutions x_0, y_0, z_0, with

$$x_0^2 \leqslant |bc|, \qquad y_0^2 \leqslant |ca|, \qquad z_0^2 \leqslant |ab|. \tag{5.7}$$

Here we may first ignore the case $|a| = |b| = |c| = 1$, because $x^2 + y^2 - z^2 = 0$ has solutions, e.g., $x = z = 1$, $y = 0$. Next, recalling the pairwise coprimality of a, b, and c, the products in (5.7) cannot be perfect squares unless both factors equal 1 in absolute value. Except for this case, the inequalities in (5.7) are, in fact, strict. Without loss of generality, recalling condition (i), we now assume $a > 0$, $b < 0$, $c < 0$. Then, if we ignore for a moment the possibility $b = c = -1$, (5.7) becomes $x_0^2 < bc$, $y_0^2 \leqslant -ac$, $z_0^2 \leqslant -ab$. It then follows that $ax_0^2 + by_0^2 + cz_0^2 < abc +$ (nonpositive terms), and also that $ax_0^2 + by_0^2 + cz_0^2 \geqslant$ (positive terms) $+ b(-ac) + c(-ab) > -2abc$, and this finishes the proof of (5.6).

We already saw how to use (5.6) in order to complete the proof of Theorem 1. It remains to consider two "loose ends". First, in passing from x_0, y_0, z_0 to x_1, y_1, z_1, it may happen that $x_1 = y_1 = z_1 = 0$. Secondly, we have to look at the omitted case $b = c = -1$.

Concerning the first problem, $z_1 = z^2 + ab = 0$ means $z^2 = -ab$, which, since $(a, b) = 1$, $a > 0$, $b < 0$, is possible only for $a = 1$, $b = -1$. In that case, however, $x = 1$, $y = -1$, $z = 0$ is a nontrivial solution.

As for the case $b = c = -1$, we observe that $-bc = -1$ is, by (ii), a quadratic residue modulo a. Hence, if $p \mid a$, then $(-1/p) = +1$, so that $p \equiv 1 \pmod 4$. By Theorem 2.2 we then know that $a = y_0^2 + z_0^2$ for some integers y_0, z_0, and we conclude that $x = 1$, $y = y_0$, and $z = z_0$ is a solution of (5.1), because $a \cdot 1 - y_0^2 - z_0^2 = 0$. Theorem 1 is proved.

§5. Problems

1. Consider the equations

$$2x^2 + 3y^2 + 7z^2 = 0,$$

$$2x^2 - 3y^2 + 7z^2 = 0,$$

$$2x^2 - 3y^2 + 19z^2 = 0.$$

   By use of Theorem 1, decide which equations do and which do not have nontrivial solutions. Find the nontrivial solutions.

2. Find solutions of $3x + 4y + 7z \equiv 0 \pmod 8$ with $|x| \leqslant 2$, $|y| \leqslant 2$, $|z| \leqslant 2$ (observe: $2^3 = 8$), preferably by following the steps of the proof of Lemma 2.

*3. Show that if the conditions of Theorem 1 are satisfied, then (5.1) is solvable as a congruence modulo all prime powers. (This is not an easy problem. You may need the following lemma: modulo every odd prime p, there exist quadratic residues whose sum is a nonresidue, and quadratic nonresidues whose sum is a quadratic residue. Also, for $p = 2$ you need the full assumptions of Theorem 1.)

Chapter 6

Representations of Integers as Sums of Nonvanishing Squares

§1. Representations by $k \geqslant 4$ Squares

It is clear that, for $k \geqslant 4$, every nonnegative integer is representable as a sum of k squares. Indeed, one can always write $n = \sum_{i=1}^{4} x_i^2 + 0^2 + \cdots + 0^2$, with an arbitrary number of zeros. On the other hand, the number of representations $r_k(n)$ increases very rapidly with n (see Chapter 12), and besides the representations with $k - 4$ zeros, one usually finds others with fewer zeros or none at all. For example, if $k = 5$, then $5 = 2^2 + 1^2 + 0^2 + 0^2 + 0^2 = 1^2 + 1^2 + 1^2 + 1^2 + 1^2$.

§2. Representations by k Nonvanishing Squares

The more interesting problem is then to consider the representation of integers n by *exactly k nonvanishing squares*, a phrase that we shall abbreviate henceforth by "k nv. sq."

We observe that, for $k \leqslant 4$, there are infinitely many integers that *do not* have representations by exactly k nv. sq. Indeed, we recall (see Theorem 3.5) that, for n odd, there exists a 1–1 correspondence between the representations of $2n$ and of $2 \cdot 4^a n$ as sums of four squares. It follows that if, for n odd, $2n$ has no representation by four nv. sq., the same will be the case for all integers $m = 4^a \cdot 2n$. Now, $2 \cdot 1 = 1^2 + 1^2 + 0^2 + 0^2$, so that $2 \cdot 4^a = (2^a)^2 + (2^a)^2 + 0 + 0$, and these integers have no other representations as sums of four squares, in particular no representations as sums of 4 nv. sq. For instance, $8 = 2^2 + 2^2 + 0^2 + 0^2$, $32 = 4^2 + 4^2 + 0^2 + 0^2$, etc. Similarly, if $4 \mid n$ and $n = \sum_{i=1}^{3} x_i^2$, then all x_i are even and $n/4 = \sum_{i=1}^{3} (x_i/2)^2$. Hence, if there exists a representation as sum of three nv. sq. for $n = 4^a n_1$, $4 \nmid n_1$, then such a representation exists also for n_1. By observing, for instance, that $25 = 3^2 + 4^2 + 0^2$, but that 25 has no representation as a sum of three nv. sq., it follows that the infinite set of integers $n = 4^a \cdot 25$ also has only the representations $n = (2^a \cdot 3)^2 + (2^a \cdot 4)^2 + 0^2$. Finally, the set of integers 4^a consists of squares, but not of sums of two nv. sq.

With these facts in mind, the following theorem should not appear devoid of interest.

Theorem 1. *For each $k \geqslant 5$, all but a finite set of integers are sums of exactly k nonvanishing squares.*

Proof. Let us consider first the case $k = 5$. We observe that the integer 169 has the marvelous property that it can be represented as a sum of 1, 2, 3, 4, 5 (in fact also 6, 7, 8, and many more) nv. sq. Indeed, $169 = 13^2 = 12^2 + 5^2 = 12^2 + 3^2 + 4^2 = 11^2 + 4^2 + 4^2 + 4^2 = 12^2 + 4^2 + 2^2 + 2^2 + 1^2$. Consider now any integer $n \geqslant 170$. Then $n - 169 = \sum_{i=1}^{4} x_i^2$, with, say $x_1 \geqslant x_2 \geqslant x_3 \geqslant x_4$. If all $x_i^{\prime s}$ are different from zero, then $n = 13^2 + \sum_{i=1}^{4} x_i^2$ is the sum of 5 nv. sq.; if only three of the x_i are different from zero, then $n = 12^2 + 5^2 + \sum_{i=1}^{3} x_i^2$; if only two of the x_i are different from zero, then $n = 12^2 + 3^2 + 4^2 + x_1^2 + x_2^2$; finally, if $n - 169 = x_1^2\ (>0)$, then $n = 11^2 + 4^2 + 4^2 + 4^2 + x_1^2$ and n always appears as the sum of five nv. sq. As for $n < 170$, we can test this finite set of integers and find that all are sums of five nv. sq., except for $n = 1, 2, 3, 4$ (which are obvious) and also 6, 7, 9, 10, 12, 15, 18, 33.

If $k = 6$, we may proceed in two ways. We may again use $n - 169$, where now we use also the decomposition $169 = 12^2 + 4^2 + 2^2 + 2^2 + 1^2$ into five nv. sq. Alternatively, we may consider $n - 1$, which, for $n \geqslant 171$, can be represented as a sum of 5 nv. sq., so that $n = \sum_{i=1}^{5} x_i^2 + 1^2$ is, indeed, the sum of six nv. sq. By checking $n \leqslant 170$, we find as exceptions the obvious 1, 2, 3, 4, 5, and then only 7, 8, 10, 11, 13, 16, and 19. For $k = 7$, once more, we represent $n - 1 = \sum_{i=1}^{6} x_i^2$, with all $x_i \neq 0$, and then $n = 1^2 + \sum_{i=1}^{6} x_i^2$. The only possible exceptional integers are $n = 1$ and the previous exceptional integers plus 1. Indeed, $1, 2, \ldots, 6$ are obviously not sums of 7 nv. sq., so we look at 8, 9, 11, 12, 14, 17, 20. If any of these were the sum of 7 nv. sq., then the smallest square would be 1^2 (because $7 \cdot 2^2 = 28 > 20$). Hence, if any of these integers were the sum of seven nv. sq., then $7 = 8 - 1, 8 = 9 - 1, \ldots$ or $19 = 20 - 1$ would be sums of six nv. sq., contrary to what has been shown.

In general, let us assume that we have shown already that, for a given $k \geqslant 6$, all integers are sums of k nv. sq., except for $1, 2, \ldots, k - 1$ and all $k + b$, where $b \in \mathbf{B} = \{1, 2, 4, 5, 7, 10, 13\}$. Then we claim that this result holds also for $k + 1$. Indeed, *a fortiori*, $1, 2, \ldots, k$ cannot be represented as sums of $k + 1$ nv. sq. If $n - 1 = \sum_{i=1}^{k} x_i^2$, $\prod_i x_i \neq 0$, then $n = \sum_{i=1}^{k+1} x_i^2$ holds with $x_{k+1} = 1$. It only remains to consider the integers $b + k + 1$, $b \in \mathbf{B}$. However, for $k \geqslant 6$, in any such representation, the smallest square is always 1^2, because $(k + 1)2^2 > 13 + k + 1$. Consequently, if, for any $b \in \mathbf{B}$, $b + k + 1 = \sum_{i=1}^{k+1} x_i^2$, $\prod_i x_i \neq 0$, then $b + k = \sum_{i=1}^{k} x_i^2$, $\prod_i x_i \neq 0$, contrary to the induction assumption. This finishes the proof of the following refinement of Theorem 1. □

Theorem 2. *With the previous definition of the set* $\mathbf{B}$, *every integer n is the sum of k nonvanishing squares, provided that $k \geqslant 6$ and that $n \neq 1, 2, \ldots, k - 1$, $n \neq k + b$, with $b \in \mathbf{B}$. For $k = 5$ the same statement holds with $b \in \mathbf{B} \cup \{28\}$.*

This theorem seems to have been proved first by Dubouis [60] in 1911, and was discovered independently by G. Pall [204]; see also [88], [198], and [199].

§3. Representations as Sums of Four Nonvanishing Squares

The situation is very different for $k \leqslant 4$. If $k = 4$, it is sufficient to consider only the integers $n \not\equiv 0 \pmod 8$, because if $n_1 = \sum_{i=1}^{4} x_i^2$, $\prod_{i=1}^{4} x_i \neq 0$, then also $n = 4^a n_1 = \sum_{i=1}^{4} y_i^2$, $y_i = 2^a x_i \neq 0$ $(i = 1, 2, 3, 4)$ and, by Theorem 3.5, the converse also holds, provided that n_1 is even. If n_1 is odd, then we have to write $n = 4^{a-1}(4n_1)$.

We start as before, with $n - 169$ if $n \equiv 2, 3, 4, 6$, or $7 \pmod 8$, $n \geqslant 170$; and $n - 4 \cdot 169 = n - 676$, $n \geqslant 677$, if $n \equiv 1$ or $5 \pmod 8$. In the first case, $n - 169 \equiv 1, 2, 3, 5$, or $6 \pmod 8$; in the second case, $n - 676 \equiv 1$ or $5 \pmod 8$. In either case, the difference is equal to the sum of one, two, or three nv. sq., so that in all cases n can be written as the sum of exactly four nv. sq., by using $169 = 13^2 = 5^2 + 12^2 = 3^2 + 4^2 + 12^2$ or $676 = 26^2 = 10^2 + 24^2 = 6^2 + 8^2 + 24^2$, respectively. This shows that, if $n \not\equiv 0 \pmod 8$, and if $n \geqslant 677$ for $n \equiv 1 \pmod 4$, $n \geqslant 170$ otherwise, then every integer is the sum of four nv. sq. Let us call such integers *ordinary*, and those that are not sums of four nv. sq. *exceptional*. It remains to investigate (i) the integers $n \leqslant 169$ ($n \leqslant 676$, if $n \equiv 1 \pmod 4$); and (ii) the integers of the form $n = 4^a n_1$, $4 \nmid n_1$, $a \geqslant 1$, $n_1 \leqslant 169$ ($n_1 \leqslant 676$, if $n_1 \equiv 1 \pmod 4$). If n_1 is ordinary, then so is $n = 4^a n_1$; hence in (ii) it is sufficient to restrict ourselves to the exceptional integers found in (i). However, as we want to conclude, so to speak, in the wrong direction, we have to be careful to write n as $4^{a-1}(4n_1)$ if n_1 is odd.

A tedious but easy verification shows that among the integers $n \leqslant 169$ ($n \leqslant 676$, if $n \equiv 1 \pmod 4$) such that $n \not\equiv 0 \pmod 8$, only the elements of the set $\mathbf{A} = \{1, 2, 3, 5, 6, 9, 11, 14, 17, 29, 41\}$ are exceptional. For each $n_1 \in \mathbf{A}$, we now examine the set $4^a n_1$ as follows: If $n_1 = 1$, then $4^a n_1 = 4^{a-1} \cdot 4 = 4^a(1^2 + 1^2 + 1^2 + 1^2)$ and these integers are all ordinary. For $n_1 = 2$, $4^a \cdot n_1 = 4^a \cdot 2$, and these integers *cannot* be represented as sums of 4 nv. sq.; otherwise, 2 itself could be so represented. Hence the integers $4^a \cdot 2$ are all exceptional. Similarly, $4^a \cdot 3 = 4^{a-1} \cdot 12 = 4^{a-1}(3^2 + 1^2 + 1^2 + 1^2)$, and these are ordinary integers. Also $4^a \cdot 5 = 4^{a-1} \cdot 20 = 4^{a-1}(3^2 + 3^2 + 1^2 + 1^2)$ are ordinary. However, by Theorem 3.5, the integers $4^a \cdot 6$ and $4^a \cdot 14$ are exceptional, because 6 and 14 are exceptional and even. All other $n_1 \in \mathbf{A}$, lead to sets $4^a \cdot n_1$ of integers, ordinary for $a \geqslant 1$, or back to one of the three sets $4^a \cdot n_1$ ($n_1 = 2, 6$, or 14) of exceptional integers already found. So, e.g., $4^a \cdot 41 = 4^{a-1} \cdot 164 = 4^{a-1}(11^2 + 5^2 + 3^2 + 3^2)$, and these are ordinary integers, while $4^a \cdot 56 = 4^{a+1} \cdot 14$, and these belong to one of the three mentioned sets of exceptional integers.

We summarize the results found in the following theorem.

Theorem 3. *With the set* **B** *as defined in Section 2, all integers are sums of four nonvanishing squares, except for the finite set consisting of* 1, 2, 3, *and* $n =$

$4 + b$, where $b \in \mathbf{B} \cup \{25, 37\}$, and the three infinite sets $4^a n_1$, with $n_1 = 2, 6$, or 14.

This theorem had been conjectured already by Descartes [53] and has been proved by Dubouis [60]. The present proof appears to be new.

§4. Representations as Sums of Two Nonvanishing Squares

By Theorem 2.2, we know that $n = x_1^2 + x_2^2$ if and only if $n = 2^a n_1 n_2^2$, with $a \geqslant 0$, $n_1 = \prod_{p_i \equiv 1 \pmod 4} p_i^{a_i}$, $n_2 = \prod_{q_j \equiv 3 \pmod 4} q_j^{b_j}$. Since all representations of n as a sum of two squares are of the form $n = (n_2 x_1)^2 + (n_2 x_2)^2$, where $2^a n_1 = x_1^2 + x_2^2$, it is sufficient to consider the case where $n_2 = 1$, i.e., $n = 2^a n_1$.

Let $x_i = 2^{b_i} y_i$, y_i odd; if $b_1 = b_2 = b$, say, then $n = 2^a n_1 = 2^{2b}(y_1^2 + y_2^2)$, so that $y_1^2 \equiv y_2^2 \equiv 1 \pmod 8$, $y_1^2 + y_2^2 \equiv 2 \pmod 8$, $y_1 y_2 \neq 0$, $x_1 x_2 \neq 0$. In this case, $a = 2b + 1$, and $n_1 = \frac{1}{2}(y_1^2 + y_2^2) \equiv 1 \pmod 4$, as expected. If $b_1 < b_2$, set $b = b_1$, $c = b_2 - b_1 > 0$, so that $n = 2^a n_1 = 2^{2b}(y_1^2 + 4^c y_2^2)$, and $a = 2b$, $n_1 = y_1^2 + 4^c y_2^2 \equiv 1 \pmod 4$. It follows that $y_1 \neq 0$, so that $x_1 \neq 0$, but $x_2 = y_2 = 0$ is not ruled out (this corresponds to $b_2 = \infty$). However, unless this happens, $y_1 y_2 \neq 0$, so that $x_1 x_2 \neq 0$. The only case when n_1 (and, hence, n itself) is not a sum of two nv. sq. is therefore if $n = 2^a n_1 = 2^{2b} y_1^2 = (2^b y_1)^2 = m^2$, say, where m^2 is a perfect square, whose only representations as a sum of two squares are the four representations $(\pm m)^2 + 0^2 = 0^2 + (\pm m)^2$. Here $n_1 = y_1^2 = p_1^{2a_1} p_2^{2a_2} \cdots p_r^{2a_r}$, $p_i \equiv 1 \pmod 4$, $i = 1, 2, \ldots, r$. In case $n_1 > 1$, then, by Theorem 2.3, $r_2(n_1) = 4 \prod (2a_i + 1) \geqslant 4 \cdot 3 = 12 > 4$; hence, $r_2(n) = r_2(m^2) = r_2(n_1) \geqslant 12$, and m^2 has at least one essentially distinct representation from $m^2 + 0^2$. In this case, $y_1 y_2 \neq 0$, so that also $x_1 x_2 \neq 0$. Consequently, the only possibility of failing to have $n = 2^a n_1 n_2^2$ representable as sum of two nv. sq. is the case $a = 2b$, $n_1 = 1$, i.e., $n = 4^b n_2^2$. We formulate this result as

Theorem 4.* *With a, n_1, and n_2 defined as above, let $n = 2^a n_1 n_2^2$; then n is the sum of two nonvanishing squares, unless $n_1 = 1$ and a is even.*

§5. Representations as Sums of Three Nonvanishing Squares

We have answered completely the problem of representations of a natural integer by exactly k nv. sq., except for $k = 3$. In fact, in this case, the complete answer is still not known and depends on the difficult, and still unsolved, problem of the determination of all discriminants of binary, positive definite quadratic forms with exactly one class in each genus. It is known that the

* Observe the omission of $2^a n_1 n_2^2$ with a odd and $n_1 = 1$ in [88].

number of such discriminants is finite (in fact, $\lim_{n\to\infty}(h(n)/g(n)) = \infty$; see Chowla [42]), and it is believed that the known list of 101 such discriminants (Euler's *numeri idonei*) is complete, the largest being $4 \cdot 1848$, but so far we have no proof of this conjecture.

Let us denote by $\mathbf{M}$ the set of integers n of the form $n = 4^a(8m + 7)$. An integer n can be represented as a sum of at most three nv. sq. if and only if $n \notin \mathbf{M}$. If that is the case, it may still happen that in any one of the representations of n as a sum of three squares, at least one of them is always zero, i.e., that $n = x_1^2$ or $n = x_1^2 + x_2^2$, but $n \neq x_1^2 + x_2^2 + x_3^2$, with all x_i nonzero. Nevertheless, $r_3(n) > 0$, because a representation $n = x_1^2 + x_2^2$ is counted by $r_3(n)$; in fact, it is counted, in general, triply, namely as $x_1^2 + x_2^2 + 0^2$, as $x_1^2 + 0^2 + x_2^2$, and as $0^2 + x_1^2 + x_2^2$. Similarly, if $n = m^2$, then a representation $m^2 + 0^2$, counted by $r_2(n)$, is counted by $r_3(n)$ twice, namely, as $(0^2) + m^2 + 0^2$ and as $m^2 + 0^2 + (0^2)$ (the latter is, naturally, not counted as distinct from $m^2 + (0^2) + 0^2$). It follows that, if $n = m^2$ has no representations by two nv. sq., then the four representations counted by $r_2(n)$ count as six representations in $r_3(n)$, namely $(\pm m)^2 + 0^2 + 0^2$, $0^2 + (\pm m)^2 + 0^2$, and $0^2 + 0^2 + (\pm m)^2$. Hence, for $n = m^2$

$$r_3(n) \geqslant \tfrac{3}{2} r_2(n). \tag{6.1}$$

We know from Theorem 4 that equality can occur here only if $m = 2^b n_2$, where all prime factors q of n_2 satisfy $q \equiv 3 \pmod 4$.

If $n \neq m^2$, then all representations counted by $r_2(n)$ are counted triply by $r_3(n)$, so that $r_3(n) \geqslant 3r_2(n)$. If we have equality here, then all representations counted by $r_3(n)$ are actually representations by two nv. sq. and one zero, and in this case, n has no representation as a sum of three nv. sq. It follows that, for $n \neq m^2$, n is the sum of three nv. sq. precisely when

$$r_3(n) > 3r_2(n). \tag{6.2}$$

Even if $n = m^2$, (6.2) is a *sufficient* condition for n to be a sum of three nv. sq.

Let us first consider squarefree integers. Then, by Theorem 2.2, $r_2(n) = 0$ unless $n = 2^a p_1 p_2 \cdots p_t$, with $a = 0$ or 1 and $p_j \equiv 1 \pmod 4$, $j = 1, 2, \ldots, t$, in which case, by Theorem 2.3, $r_2(n) = 4 \cdot 2^t$. Also,* by Theorem 4.2, if h is the class number then $r_3(n) = 12h$ if $n \equiv 1, 2, 5, 6 \pmod 8$, and $r_3(n) = 24h$ if $n \equiv 3 \pmod 8$. However, if $n \equiv 3$ or $6 \pmod 8$, then $r_2(n) = 0$ and (6.2) obviously holds, and so it is sufficient to consider the cases $n \equiv 1, 2,$ or $5 \pmod 8$, with $r_3(n) = 12h$. The discriminant corresponding to n is $D = -4n$, so that (regardless of the value 0 or 1 of a) D has $t + 1$ prime factors and $g = 2^t$ genera. Consequently, if there are k classes in each genus, then $h = gk = 2^t k$ and $r_3(n) = 12 \cdot 2^t k$. The inequality (6.2) becomes $12 \cdot 2^t k > 12 \cdot 2^t$, or $k > 1$. We

*For simplicity of exposition, we exclude here the trivial cases $1 = 1^2 + 0^2 + 0^2$ and $3 = 1^2 + 1^2 + 1^2$, for which $\delta_n \neq 1$.

conclude that precisely those squarefree integers $n \notin \mathbf{M}$ that correspond to discriminants with exactly one class per genus have no representations as sums of three nv. sq. Assuming the stated conjecture, that $D = -4 \cdot 1848$ is the (absolutely) largest discriminant with one class per genus, we find that the corresponding squarefree integers $n \notin \mathbf{M}$ are (including $n = 1$) $\mathbf{S} = \{1, 2, 5, 10, 13, 37, 58, 85, 130\}$. We formulate the result obtained as

Lemma 1. *There exists a finite set* $\mathbf{T}$ *of squarefree integers,* $\mathbf{T} \supset \mathbf{S}$, *such that every squarefree integer* $n \notin \mathbf{M} \cup \mathbf{T}$ *is a sum of three nonvanishing squares.*

If n_0 is not squarefree, then $n_0 = nm^2$ with n squarefree, and if n is a sum of three nv. sq., then so is n_0. The converse, however, does not hold. For that reason, we have to investigate the conditions under which nr^2, for some prime r, can be a sum of three nv. sq. when n is not. First, assume that $n \in \mathbf{M}$. Then, with the earlier notation, also 2^2n, np^2, and nq^2 belong to $\mathbf{M}$, because $p^2 \equiv q^2 \equiv 1 \pmod 8$; hence, also $r_3(n_0) = 0$. Next, for $n \notin \mathbf{M}$, we know that $r_3(4n) = r_3(n)$ and $r_2(4n) = r_2(n)$, so that, if n is not a sum of three nv. sq., then $2^2 n$ cannot be either. It is therefore sufficient to consider only the cases nq^2 and np^2, for $n \in \mathbf{T}$.

By the corollary to Theorem 2.3, $r_2(n) = r_2(nq^2)$. Hence, recalling the way in which representations counted by $r_2(n) = r_2(nq^2)$ are counted as representations by $r_3(n)$ and $r_3(nq^2)$, respectively, we see that in order to show that nq^2 has a representation as a sum of three nv. sq., it is sufficient to show that $r_3(nq^2) > r_3(n)$. This is indeed the case and follows from (4.8). Indeed, by (4.8) and (4.10),

$$r_3(nq^2) = \sum_{d^2 \mid nq^2} R_3\left(\frac{nq^2}{d^2}\right) \geqslant R_3(nq^2) + R_3(n) > R_3(n) = r_3(n),$$

because n is squarefree, $n \equiv 1$ or $2 \pmod 4$, and $nq^2 \equiv n \pmod 8$. In conclusion: If $n_0 \notin \mathbf{M}$ and $q^2 \mid n_0$, where $q \equiv 3 \pmod 4$, then n_0 is a sum of three nv. sq.

Next, let

$$n = 2^a p_1 p_2 \cdots p_t, \qquad a = 0 \text{ or } 1, \tag{6.3}$$

and let us consider the effect of multiplying n by p^2. First, let $p \nmid n$ (i.e., $p \neq p_i$, $i = 1, 2, \ldots, t$). Then, by Theorem 2.3, $r_2(p^2n) = 3r_2(n)$. From Theorem 4.4 and (4.12), it follows that, if $\chi_1(m) = (\frac{-4np^2}{m})$, $\chi_2(m) = (\frac{-4n}{m})$, then $L(1, \chi_1) = L(1, \chi_2)(1 - (\frac{-4n}{p})1/p) = L(1, \chi_2)(1 - (\frac{n}{p})1/p)$, because $p \equiv 1 \pmod 4$. It follows further that with $p(n)$ defined as in Theorem 4.4,

$$P(np^2) = 1 + \frac{1}{p - (\frac{-n}{p})} = 1 + \frac{1}{p - (\frac{n}{p})},$$

and, by (4.9),

$$\begin{aligned} r_3(np^2) &= \frac{24}{\pi}\sqrt{np^2}\,L(1,\chi_1)\left(1+\frac{1}{p-\left(\frac{n}{p}\right)}\right) \\ &= \frac{24}{\pi}p\sqrt{n}\,L(1,\chi_2)\left(1-\left(\frac{n}{p}\right)\frac{1}{p}\right)\left(1+\frac{1}{p-\left(\frac{n}{p}\right)}\right) \\ &= \frac{24}{\pi}\sqrt{n}\,L(1,\chi_2)\left(p-\left(\frac{n}{p}\right)+1\right) \\ &= r_3(n)\left(p-\left(\frac{n}{p}\right)+1\right) \geqslant pr_3(n). \end{aligned}$$

It nows follows that (6.2) holds for np^2 provided that $r_3(n) \geqslant 3r_2(n)$ is valid, and in particular for $n \neq m^2$. Indeed, if that is the case, we obtain $r_3(np^2) \geqslant pr_3(n) \geqslant 3pr_2(n) \geqslant 3 \cdot 5r_2(n) = 5r_2(np^2) > 3r_2(np^2)$. If $n = m^2$, we use only (6.1), and then $pr_3(n) \geqslant p \cdot \frac{3}{2}r_2(n) = (p/2)r_2(p^2n) > 3r_2(p^2n)$ holds only if $p > 6$; in particular, the proof goes through for all $p \equiv 1 \pmod 4$, *except* for $p = 5$. However, $n = m^2 \in \mathbf{T}$ means $n = 1$, and $n_0 = 1 \cdot 5^2 = 25$ is the only integer np^2, $p \nmid n$, $n \not\equiv 0 \pmod 4$, that has not been shown to be a sum of three nv. sq.

We now consider the same question when $p \mid n$, and remark that in this case $p \mid n$ and $n \in \mathbf{T}$ imply that $n \neq m^2$. Proceeding as before, we observe that now the left hand side of (6.1) is multiplied by $p(1 + 1/p) = p + 1$, while for n as in (6.3) and $p \mid n$,

$$r_2(p^2n) = \frac{3+1}{1+1}r_2(n) = 2r_2(n);$$

hence,

$$\begin{aligned} r_3(np^2) &\geqslant (p+1)r_3(n) \geqslant 3(p+1)r_2(n) \\ &= 3(p+1)\frac{r_2(p^2n)}{2} = \frac{p+1}{2}3r_2(p^2n) > 3r_2(p^2n), \end{aligned}$$

and the sufficient condition (6.2) holds for all integers np^2 with $n \in \mathbf{T}$. The verification that indeed every representation of 25 as a sum of three squares contains a zero summand now completes the proof of

Lemma 2. *The only odd squares that are not sums of three nonvanishing squares are* 1 *and* 25.

By Lemma 2 and the remark that $4^a n$ has a representation as a sum of three nv. sq. if and only if n has one, we obtain the following theorem, due to Hurwitz (1859–1909) (see [119]):

Theorem 5. *The only squares that are not sums of three nonvanishing squares are the integers* 4^a *and* $25 \cdot 4^a$.

As is seen, for any integer n and prime p, if $n \notin \mathbf{M}$, then np^2 has a representation as a sum of three nv. sq., except for $n = 4^a$, if also $p = 5$. This, with Theorem 5, implies the following theorem of G. Pall [204]:

Theorem 6. *Every integer* $n \notin \mathbf{M}$ *and containing an odd square factor larger than one is the sum of three nonvanishing squares, except for* $n = 4^a \cdot 25$.

Finally, by using also Lemma 1, we obtain [88]

Theorem 7. *There exists a finite set* $\mathbf{T}$ *of integers, having* t *elements, that contains the set* $\mathbf{S}_1 = \mathbf{S} \cup \{25\}$, *and such that every integer* $n \notin \mathbf{M}$ *and not of the form* $n = 4^a n_1$, *with* $n_1 \in \mathbf{T}$, *is a sum of three nonvanishing squares.*

Conjecture. $\mathbf{T} = \{1, 2, 5, 10, 13, 25, 37, 58, 85, 130\}$, $t = 10$.

Arguments in favor of the conjecture may be found in [88].

In case the conjecture fails, $\mathbf{T}$ may contain at most one more integer and that must exceed $5 \cdot 10^{10}$. Its existence is, of course, highly unlikely. The proof uses the fact, proved by P. J. Weinberger [279], that there can exist at most one fundamental discriminant $-D$ with one class per genus, and $D > 2 \cdot 10^{11}$. The conclusion for $\mathbf{T}$ is formulated in a recent paper by F. Halter-Koch [93]. In [93] are discussed, in addition to the present problem, such questions as the representation of integers by k squares under a variety of conditions, for instance, that the squares should be distinct, coprime, etc. Similar representation problems are considered for a few quadratic forms more general than sums of squares. The results are that, for $k \geqslant 5$, all positive integers have such representations, with the exception of a finite, explicitly determined set of integers. In the case $k = 4$, the exceptional set consists of the union of a finite set and an infinite set of integers of form $n = 4^h a$ with $a \in \mathbf{A}$, $\mathbf{A}$ a finite set. Analogous, somewhat more complicated results hold for $k = 3$.

§6. On the Number of Integers $n \leqslant x$ That Are Sums of k Nonvanishing Squares

We recall that $\mathbf{N}_k(x)$ stands for the number of integers $n \leqslant x$ that are sums of k squares. In counting those integers, no distinction is made between integers that have and those that do not have representations by k nv. sq. So, e.g., $n = 1$

is counted by all $\mathbf{N}_k(x)$, $k \geqslant 1$, because $1 = 1^2 + 0^2 + \cdots + 0^2$, with $k-1$ zeros. Let us denote by $\bar{\mathbf{N}}_k(x)$ the number of integers $n \leqslant x$ that have representations as sums of k nv. sq. In order to avoid exceptions that are trivial, but complicated to state, let us assume to the end of this section that $x \geqslant k + 13$.

For $k = 1$, the problem is, once more, trivial, because $\bar{\mathbf{N}}_1(x) = \mathbf{N}_1(x) = [\sqrt{x}]$.

For $k \geqslant 5$, Theorem 2 solves the problem completely, as it shows that $\bar{\mathbf{N}}_5(x) = [x] - 12$, and $\bar{\mathbf{N}}_k(x) = [x] - (k + 6)$ for $k \geqslant 6$.

It remains to consider the cases $k = 2$, 3, and 4.

By §8 of Chapter 2, we know that $\mathbf{N}_2(x) = bx/\sqrt{\log x} + o(x/\sqrt{\log x})$, with $b \simeq 0.764\ldots$. To obtain $\bar{\mathbf{N}}_2(x)$, we have to eliminate only those integers $n \leqslant x$ that are squares $n = m^2$, but not sums of 2 nv. sq. Obviously, the number of these squares is less than the total number of squares $m^2 \leqslant x$; hence it is less than $\sqrt{x}$, and the difference between $\mathbf{N}_2(x)$ and $\bar{\mathbf{N}}_2(x)$ is absorbed by the error term. This reasoning justifies

Theorem 8. $\bar{\mathbf{N}}_2(x) = \mathbf{N}_2(x) + O(\sqrt{x}) = bx/\sqrt{\log x} + o(x/\sqrt{\log x})$.

For $k = 4$, by Theorem 3, there are four sets of integers that are not sums of four nv. sq.; one of them is the finite set of eight elements $\{1, 3, 5, 9, 11, 17, 29, 41\}$, and the other three are $\mathbf{S}_i = 4^a n_i$, with $n_1 = 2$, $n_2 = 6$, $n_3 = 14$. The number of elements $4^a n_i \leqslant x$, $a \geqslant 0$, equals $[f(\log x - \log n_i)] + 1$, where, as before, f stands for $(\log 4)^{-1}$. It follows that $\mathbf{S}_i(x) = f(\log x - \log n_i) + \theta_i$. Here and in what follows, $0 \leqslant \theta_i \leqslant 1$. Consequently, the number of elements in $\mathbf{S}_1 \cup \mathbf{S}_2 \cup \mathbf{S}_3$ with $n \leqslant x$ equals $f(3\log x - \log n_1 n_2 n_3) + \sum_{i=1}^{3} \theta_i = f(3\log x - \log 42) - 1 + \theta$. By using $[x] = x - \theta_4$, $f\log 42 = 2.696158\ldots$, and $0 \leqslant \theta = \sum_{j=1}^{3} \theta_i < 3$, we obtain

$$\bar{\mathbf{N}}_4(x) = [x] - \frac{3}{\log 4}\log x + \frac{\log 42}{\log 4} + 1 - \theta - 8,$$

and an easy computation yields

Theorem 9. *With the above notation,* $\bar{\mathbf{N}}_4(x) = x - (3/\log 4)\log x - C + R(x)$, *where* $C = 9 - (\log 42/\log 4) \simeq 6.303\ldots$ *and* $|R(x)| \leqslant 2$.

For $k = 3$, let us denote by $\mathbf{M}_i$ $(i = 1, 2)$ the sets of integers that are sums of i, but not of three, nv. sq., and set $\mathbf{M}_0 = \mathbf{M}_1 \cap \mathbf{M}_2$. Then it is clear that

$$\begin{aligned}\bar{\mathbf{N}}_3(x) &= [x] - \mathbf{M}(x) - \mathbf{M}_1(x) - \mathbf{M}_2(x) + \mathbf{M}_0(x) \\ &= \mathbf{N}_3(x) - \mathbf{M}_1(x) - \mathbf{M}_2(x) + \mathbf{M}_0(x).\end{aligned} \tag{6.4}$$

Here $\mathbf{M}(x)$ is the number of integers $n \leqslant x$ that are not even sums of three squares, and $\mathbf{M}_j(x)$ are the numbers of integers $n \in \mathbf{M}_j$, $n \leqslant x$.

We recall from Theorem 4.6 that

$$\mathbf{N}_3(x) = [x] - \mathbf{M}(x) = x - \theta_1 - \left(\frac{x}{6} - \frac{7}{8}f\log x + C + R(x)\right),$$

where

$$C = \frac{7\log 7}{8}f - \frac{7}{6} \quad \text{and} \quad R(x) = \frac{7}{6} + \theta_2\left(f\log\frac{x}{7} + \theta_3\right) - \frac{7}{8}\theta_3 - \frac{7}{6}\cdot 4^{-\theta_3},$$

with $f = (\log 4)^{-1}$. Next, by Theorem 5, $\mathbf{M}_1(x)$ is the number of squares $n = m^2 \leqslant x$ of the form 4^a or $25\cdot 4^a$. In order to be counted also by $\mathbf{M}_0(x)$, n has to be a sum of two nonvanishing squares; the integers $25\cdot 4^a$ are of this form, while the integers 4^a are not. Hence, $\mathbf{M}_1(x) - \mathbf{M}_0(x)$ counts precisely the integers $n = 4^a \leqslant x$, and their number equals $1 + [(\log x/(\log 4)] = f\log x - \theta' + 1 = f\log x + \theta_4$.

Finally, let n_1 be counted by $\mathbf{M}_2(x)$, $4 \nmid n_1$; then the integers counted by $\mathbf{M}_2(x)$ are precisely the integers of the form $n = 4^a n_1 \leqslant x$, $a = 0, 1, \ldots$. Here $n_1 \not\equiv 7 \pmod 8$, because otherwise n_1 would require four nv. sq. and then $n_1 \notin \mathbf{M}_2$. Furthermore, by Theorem 7 and the remark that $n_1 \neq 1, n_1 \in \mathbf{T} \subset \mathbf{N}_2$, it follows that, if we set $\mathbf{T}_1 = \{n \mid n \in \mathbf{T}, n \neq 1\}$, $t_1 = t - 1$, then

$$\mathbf{M}_2(x) = \sum_{n_1\in\mathbf{T}_1}\left(\left[\frac{\log(x/n_1)}{\log 4}\right] + 1\right) = \sum_{n_1\in\mathbf{T}_1}\{f(\log x - \log n_1) + 1 - \theta_5\}$$

$$= t_1 f\log x - f\sum_{n_1\in\mathbf{T}_1}\log n_1 + t_1\theta_6$$

$$(0 < \theta_6 = 1 - \theta_5 \leqslant 1).$$

By replacing $\mathbf{N}_3(x)$, $\mathbf{M}_1(x) - \mathbf{M}_0(x)$, and $\mathbf{M}_2(x)$ with these values in (6.4), we obtain

$$\begin{aligned}\bar{\mathbf{N}}_3(x) = {} & \frac{5x}{6} + \frac{7}{8}f\log x - \frac{7\log 7}{8}f + \frac{7}{6} \\ & - \left(\frac{7}{6} + \theta_2\left(f\log\frac{x}{7} + \theta_3\right) - \frac{7}{8}\theta_3 - \frac{7}{6}\cdot 4^{-\theta_3} + \theta_1\right) \\ & - f\log x - \theta_4 - t_1 f\log x + f\sum_{n_1\in\mathbf{T}_1}\log n_1 - t_1\theta_6,\end{aligned}$$

or

$$\bar{\mathbf{N}}_3(x) = \frac{5x}{6} - f\log x\left(\frac{1}{8} + t_1\right) + f\left(\sum_{n_1\in\mathbf{T}_1}\log n_1 - \frac{7\log 7}{8}\right) + R, \tag{6.5}$$

where $R = R(x;\ \theta_1, \theta_2, \theta_3, \theta_4, \theta_6) = -(\theta_2 f \log(x/7) + \theta_3(\theta_2 - \frac{7}{8}) - \frac{7}{6} \cdot 4^{-\theta_3} + \theta_1 + \theta_4 + \theta_6 t_1)$. One easily finds that

$$-\left(f \log\frac{x}{7} + \frac{11}{6} + t_1\right) = R(x;\, 1,1,1,1,1) \leqslant R(x;\, \theta_1, \theta_2, \theta_3, \theta_4, \theta_6)$$

$$\leqslant R(x;\, 0,0,0,0,0) = \tfrac{7}{6}.$$

An elementary computation now completes the proof of

Theorem 10.

$$\bar{\mathbf{N}}_3(x) = \frac{5x}{6} - \left(t - \frac{3}{8}\right)\frac{\log x}{\log 4}$$

$$+ \frac{1}{\log 4}\left(\sum_{n \in \mathbf{T}} \log n - \frac{3}{8}\log 7\right) - \frac{t}{2} + \frac{1}{6} + R(x),$$

where

$$|R(x)| \leqslant \frac{1}{2\log 4}\log\frac{x}{7} + \frac{t}{2} + 1.$$

Proof. Replace $t_1 + 1$ by t in (6.5) and in R, and observe that $\sum_{n_1 \in \mathbf{T}_1} \log n_1 = \sum_{n \in \mathbf{T}} \log n$, because $\mathbf{T} = \mathbf{T}_1 \cup \{1\}$. Next, set $R_1(x) = R + (f/2)\log(x/7) + t/2 - \frac{1}{6}$, and in (6.5) subtract $(f/2)\log(x/7) + t/2 - \frac{1}{6} = ((f/2)\log x) - ((f/2)\log 7 - t/2 + \frac{1}{6})$ from the term in $\log x$ and from the constant term, respectively. Finally, drop the useless subscript from R_1 and observe that now $-(f/2)\log(x/7) - t/2 - 1 \leqslant R \leqslant (f/2)\log(x/7) + t/2 + 1$. □

Corollary 1. $\lim_{x\to\infty} \bar{\mathbf{N}}_3(x)/x = \lim_{x\to\infty} \mathbf{N}_3(x)/x = \frac{5}{6}$.

Corollary 2. $\bar{\mathbf{N}}_3(x) - \mathbf{N}_3(x) = O(\log x)$.

Corollary 3. *By assuming the conjecture that* $\mathbf{T}$ *has* $t = 10$ *elements (see* §5), *Theorem* 10 *implies*

$$\bar{\mathbf{N}}_3(x) = \frac{5x}{6} - \frac{77}{8\log 4}\log x + a + R,$$

where

$$a = \frac{1}{\log 4}\left(\sum_{n \in \mathbf{T}} \log n - \frac{3\log 7}{8}\right) - \frac{29}{6} \simeq 14.38\ldots \quad \textit{and} \quad |R| \leqslant \frac{\log(x/7)}{\log 16} + 6.$$

If the conjecture fails, then the coefficient of $\log x$ *is* $-85/(8 \log 4)$ *and the constant a is increased by* $(\log X/(\log 4)$, *where X is the hypothetical eleventh element of* **T**.

§7. Problems

1. Find representations of 169 as sum of k nonvanishing squares, for as many of $k = 1, 2, 3, 4, \ldots, 50, 51, 52, \ldots, 169$ as you have the patience for (it starts by being amusing and may become boring). What are the values of k for which 169 *is not* a sum of k nonvanishing squares?
2. Explain why 169 and no smaller integer could be used in the proofs of Theorems 1, 2, and 3, and determine the next smallest integer that could have been used instead.
3. For what values of k is 126 a sum of k nonvanishing squares? For what values is it not?

*4. It is known that the Riemann hypothesis permits one to prove the conjecture at the end of §5; in fact, much weaker assumptions suffice (see [87]). Try to prove the conjecture under the weakest possible unproven assumption (ideally, of course, without any unproven assumption).

5. Verify Theorem 8 for $m = 9, 11, 13$, and 15.
6. Verify Theorem 9 for $m = 11$ and $m = 22$.
7. Verify Theorem 10 for $m = 22$.
8. Verify Theorem 11 for $m = 2 \cdot 3 \cdot 5 = 30$.
9. Verify Theorem 12 for $m = 30$.
10. Show that there is no contradication between Corollary 3 and the result in [88].

Chapter 7

The Problem of the Uniqueness of Essentially Distinct Representations

§1. The Problem

In 1948 D. H. Lehmer raised an interesting question [156] and answered it almost completely. The remaining gap has only been filled partially during the more than 30 years that have since passed [20].

The question is the following: In what precedes, we have made the distinction between representations by k squares that are essentially distinct and those that are counted as different by $r_k(n)$, but are not essentially distinct, in that they differ only by the order of the summands, or by the sign of a number to be squared. So, e.g., 9 has only the representation $3^2 + 0^2$ as a sum of two squares; however, $r_2(9) = 4(d_1(9) - d_3(9)) = 4(2 - 1) = 4$, corresponding to the four representations $3^2 + 0^2 = (-3)^2 + 0^2 = 0^2 + 3^2 = 0^2 + (-3)^2$. As the number k of squares increases, the number of representations counted by $r_k(n)$, for each essentially distinct one, increases very rapidly. In the "general" case, i.e., when all squares are different from each other and different from zero, each essentially distinct representation (what, from here on, following Lehmer, we shall call a *partition* of n) corresponds to $c_k = k!\,2^k$ representations counted by $r_k(n)$. Already $c_4 = 4!\,2^4 = 384$, $c_5 = 5!\,2^5 = 3840$, etc. For example, $55 = 7^2 + 2^2 + 1^2 + 1^2 + 0^2 = 6^2 + 3^2 + 1^2 + 3^2 + 0^2 = 6^2 + 4^2 + 1^2 + 1^2 + 1^2 = 5^2 + 5^2 + 2^2 + 1^2 + 0^2 = 5^2 + 4^2 + 3^2 + 2^2 + 1^2$ has only five partitions into five squares. However, $r_5(55) = 7104$, because the first, second, and fourth partitions count each for $5 \cdot 6 \cdot 2 \cdot 16 = 960$ representations, the last one for $5!\,2^5 = 3840$, and the third one for $6 \cdot 2 \cdot 2^5 = 384$ representations. This value of 7104 certainly looks "inflated", as Lehmer says, although it does have a perfectly valid geometric meaning; indeed, it is the number of lattice points on the 4-dimensional sphere $\sum_{i=1}^{5} x_i^2 = 55$ in 5-dimensional Euclidean space. If however, we are, interested in the number of partitions, the dicrepancy between 5 and 7104 may be shocking.

The situation would not be so bad if, at least for a fixed k, one could use c_k as defined above: as the ratio between $r_k(n)$ and the number of partitions of n into k squares, say $p_k(n)$. However, as seen above, c_k works only for the "general" case. This is, indeed, the most frequent one when n is very large with respect to k. Yet, even for the modest $k = 5$, with $n = 55$ by no means exceedingly small, we found only one out of five partitions that fits the "general" model.

The general problem of finding a formula that should give the number of partitions of an arbitrary integer n into k squares (where we don't distinguish between partitions that contain zeros and those that do not) is unsolved.*

We shall restrict ourselves to the much easier problem of characterizing, for each k, the set $\mathbf{L}_k$ of integers that have exactly one partition into k squares.

§2. Some Preliminary Remarks

We observe that, for every k, if $n = \sum_{i=1}^{k} x_i^2$, then $4n = \sum_{i=1}^{k} (2x_i)^2$, whence $p_k(4n) \geqslant p_k(n)$. Consequently, if $p_k(n) > 1$, then $p_k(4n) > 1$. Next, $p_{k+1}(n) \geqslant p_k(n)$ follows, by remembering that 0^2 is accepted as a summand in any partition.

As in previous chapters, it will turn out that the answer to our present problem is quite different in the two cases $k \geqslant 5$ and $k \leqslant 4$, with $k = 4$ some kind of a watershed, and $k = 3$ by far the most difficult case. In all cases, Lehmer's results of [156] are quite complete, except for $k = 3$.

§3. The Case $k = 4$

A simple way to treat this case is based on the fact that $c_4 = 4!\,2^4 = 384$, whence $p_4(n) \geqslant r_4(n)/384$. Consequently, by Theorem 3.4, if n_1 is odd, then

$$p_4(n_1) \geqslant \frac{r_4(n_1)}{384} = \frac{\sum_{d \mid n_1} d}{48}, \qquad p_4(2n_1) \geqslant \frac{r_4(2n_1)}{384} = \frac{\sum_{d \mid 2n_1} d}{48},$$

and

$$p_4(4n_1) \geqslant \frac{r_4(4n_1)}{384} = \frac{r_4(2n_1)}{384} = \frac{\sum_{d \mid 2n_1} d}{48} = \frac{3\sum_{d \mid n_1} d}{48} = \frac{\sum_{d \mid n_1} d}{16}.$$

Hence, if $n \not\equiv 0 \pmod 4$, the first two inequalities imply that $p_4(n) \geqslant (n+1)/48$, so that $p_4(n) > 1$ for $n \geqslant 48$, while the last inequality shows that, for $n \equiv 4 \pmod 8$,

$$p_4(n) \geqslant \frac{\sum_{d \mid (n/4)} d}{16} \geqslant \frac{n/4 + 1}{16} = \frac{n+4}{64};$$

in this case, $p_4(n) > 1$ if $n > 60$. It follows that it is sufficient to check only the

* Since the manuscript of this book has been written, this problem has been (essentially) solved by the author; see [1] of the Addenda to the Bibliography.

integers $n \not\equiv 0 \pmod 4$ for $n < 48$ and $n \not\equiv 4 \pmod 8$ up to $n \leqslant 60$. Even this relatively easy verification is superfluous, as the partitions of all $n \leqslant 100$ are tabulated [173]. It turns out that none of the integers $n \equiv 4 \pmod 8$, $n \leqslant 60$, leads to $p_4(n) = 1$. For odd integers, we find the set $\mathbf{A} = \{1, 3, 5, 7, 11, 15, 23\}$. Of these, all $n_1 \in \mathbf{A}$, $n_1 \leqslant 15$, have $4n_1 \leqslant 60$; hence for them $p_4(4^a n_1) > 1$. By direct verification (or from [173]), also $p_4(4 \cdot 23) = p_4(92) = 3 > 1$, so that also $n = 4^a \cdot 23$, $a \geqslant 1$, lead to $p_4(n) > 1$. Finally, among the integers $n \equiv 2 \pmod 4$, $n < 48$, only $p_4(2) = p_4(6) = p_4(14) = 1$. For these integers, however, also $p_4(4^a n) = p_4(4n) = p_4(n) = 1$ because n is even, and this leads to the three infinite sets $\mathbf{S}_i = \{4^a n_i\}$, $n_1 = 2$, $n_2 = 6$, $n_3 = 14$, $a \geqslant 0$, each with only one partition for each one of its elements, namely $(2^a)^2 + (2^a)^2 + 0^2 + 0^2$, $(2^{a+1})^2 + (2^a)^2 + (2^a)^2 + 0^2$, and $(3.2^a)^2 + (2^{a+1})^2 + (2^a)^2 + 0^2$, respectively. This result should be compared with Theorem 6.3; we formulate it here as

Theorem 1. *The only integers, with a single partition into four squares are those that belong to the set* $\mathbf{L}_4 = \mathbf{A} \cup \mathbf{S}_1 \cup \mathbf{S}_2 \cup \mathbf{S}_3$.

§4. The Case $k \geqslant 5$

Since $p_{k+1}(n) \geqslant p_k(n)$, it is sufficient to restrict our attention to the set $\mathbf{L}_4$. We observe that $2 \cdot 4 = 8$, $6 \cdot 4 = 24$, and $14 \cdot 4 = 96$ have $p_5(96) > p_5(24) = p_5(8) = 2$; hence, for all elements $n \in \mathbf{S}_1 \cup \mathbf{S}_2 \cup \mathbf{S}_3$, with $n \equiv 0 \pmod 4$, $p_5(n) \geqslant 2$, and it is sufficient to consider only the finite set $\mathbf{A} \cup \{2, 6, 14\}$. However, also $p_5(5) = p_5(11) = p_5(14) = p_5(23) = 2$, so that $\mathbf{L}_5$ reduces to the set $\{1, 2, 3, 6, 7, 15\}$.

For $k = 6$, we observe that $6 = 2^2 + 1^2 + 1^2 + 0^2 + 0^2 + 0^2 = 1^2 + 1^2 + 1^2 + 1^2 + 1^2$ and $15 = 3^2 + 2^2 + 1^2 + 1^2 + 0^2 + 0^2 = 2^2 + 2^2 + 2^2 + 1^2 + 1^2 + 1^2$, so that $p_6(6) = p_6(15) = 2$ and $\mathbf{L}_6 = \{1, 2, 3, 7\}$. Finally, $7 = 2^2 + 1^2 + 1^2 + 1^2 + 3 \cdot 0^2 = 7 \cdot 1^2$, so that $p_7(7) \geqslant 2$ and $\mathbf{L}_7 = \{1, 2, 3\}$. It should be obvious that for $k > 7$ we have $L_k = L_7$, and this finishes the proof of the following theorem.

Theorem 2. *The only integers that satisfy* $p_k(n) = 1$ *for* $k \geqslant 5$ *are those belonging to the finite sets* $\mathbf{L}_5 = \{1, 2, 3, 6, 7, 15\}$, $\mathbf{L}_6 = \{1, 2, 3, 7\}$, *and* $\mathbf{L}_k = \{1, 2, 3\}$ *for* $k \geqslant 7$.

While the proof given for Theorem 2 is simple and leaves nothing to be desired, we present also a second proof for $\mathbf{L}_5$, because of its esthetic appeal; this proof is due to D. H. Lehmer. The way to obtain $\mathbf{L}_6$ and $\mathbf{L}_k$, $k \geqslant 7$, from $\mathbf{L}_5$ need not be repeated.

Assume that $p_5(n) = 1$ for some $n \geqslant 16$. Then each of the five integers $n - r^2$ $(r = 0, 1, 2, 3, 4)$ is nonnegative and hence is of the form $\sum_{i=1}^{4} x_{r,i}^2$, so that $n =$

$r^2 + \sum_{i=1}^4 x_{r,i}^2$ are five representations of n as a sum of five squares. By our assumption that $p_5(n) = 1$, these five representations must coincide; hence, recalling that each of the five values of r must occur in this unique representation, it follows that $n = 0^2 + 1^2 + 2^2 + 3^2 + 4^2 = 30$. However, we also have $30 = 3^2 + 3^2 + 2^2 + 2^2 + 2^2$, so that $p_5(30) \geqslant 2$, and hence every n with $p_5(n) = 1$ must satisfy $n \leqslant 15$. By Theorem 1, the only candidates are $n = 1, 2, 3, 5, 6, 7, 8, 11, 14$, or 15. If we eliminate from this set the integers 5, 8, 11, and 14, for which $p_5(n) > 1$, we end up with $\mathbf{L}_5 = \{1, 2, 3, 6, 7, 15\}$, as claimed.

§5. The Cases $k = 1$ and $k = 2$

As usual, $k = 1$ is trivial, with $p_1(n) = 1$ if and only if $n = m^2$, $m \in \mathbb{Z}$.

If $k = 2$, then $n = 2^a n_1 n_2$, $n_1 = \prod_{p_i \equiv 1 \pmod 4} p_i^{a_i}$, $n_2 = \prod_{q_j \equiv 3 \pmod 4} q_j^{b_j}$ leads to $p_2(n) = r_2(n) = 0$, unless all b_j''s are even; if that is the case, we may write n_2^2 rather than n_2 and set $n = 2^a n_1 n_2^2$. Next, if $a = 2b + c$, $c = 0$ or 1, then $n = 2^c n_1 (2^b n_2)^2$, and it follows from the Corollary to Theorem 2.3 that there exists a 1–1 correspondence between the representations $x_1^2 + x_2^2$ of $2^c n_1$ and the representations $(2^b n_2 x_1)^2 + (2^b n_2 x_2)^2$ of n. Hence, it is sufficient to consider only integers of the form $n = 2^c n_1$, $c = 0$ or 1, $n_1 = \prod_{p_i \equiv 1 \pmod 4} p_i^{a_i}$, and select those integers $m = 2^c n_1$ for which $p_2(m) = 1$. If we denote by $\mathbf{M}_1$ the set of those integers m, then $\mathbf{L}_2 = \{n \mid n = 4^a m n_2^2, m \in \mathbf{M}_1\}$.

By Theorem 2.3, it follows that for $p \equiv 1 \pmod 4$, $r_2(p) = 4 \cdot 2 = 8$, so that a prime $p \equiv 1 \pmod 4$ has exactly one partition into two squares, $p = x_1^2 + x_2^2$. Indeed, with two permutations of the terms and four choices for the signs of x_1 and x_2, this partition accounts for those eight representations counted by $r_2(p)$.

It now follows from (2.2) that if $p_1 p_2 \mid n_1$, where $p_1 = a^2 + b^2$, $p_2 = c^2 + d^2$, then we are led to at least two distinct partitions for $p_1 p_2$. Indeed, with the notation of Chapter 2, we can have neither $ac + bd = |ac - bd|$, as this would imply $abcd = 0$, nor $ac + bd = ad + bc$, which is equivalent to $(a - b)(c - d) = 0$, whence $1 \equiv p_1 = a^2 + b^2 \equiv 0$ or 2 (mod 4) or $1 \equiv p_2 = c^2 + d^2 \equiv 0$ or 2 (mod 4), which are both absurd. By induction, each new prime factor $p \equiv 1 \pmod 4$ of n_1 multiplies the number of partitions of n_1 by 2 (observe that this leads to another proof of Theorem 2.3). Hence, $p_2(n) = 1$ is possible only if $n_1 = p \equiv 1 \pmod 4$. On the other hand, $2 = 1^2 + 1^2$, so that (2.2) becomes $2p = (1^2 + 1^2)(a^2 + b^2) = (a + b)^2 + (a - b)^2 = (a - b)^2 + (a + b)^2$, and $2p$, like p itself, has a single partition into two squares, as also follows, of course, directly from Theorem 2.3, with $r_2(2p) = 4d_1(2p) = 4 \cdot 2 = 8$. We formulate this result in the form of a theorem that has been stated already by Fermat, but was first proved by Euler.

Theorem 3. *The set $\mathbf{L}_2$ of integers n such that $p_2(n) = 1$ consists of the integers of the form $n = 2^a p n_2^2$, $a \geqslant 0$, $p \equiv 1 \pmod 4$, $n_2 = \prod_{q_j \equiv 3 \pmod 4} q_j^{b_j}$.*

§6. The Case $k = 3$

As already observed (see §4.2), there is a 1–1 correspondence between the representations—and, similarly, the partitions—of n and of $4^a n$. Hence, it is sufficient to consider only integers $n \not\equiv 0 \pmod 4$. We also recall that for $n \equiv 7 \pmod 8$, $r_3(n) = p_3(n) = 0$, and it remains to consider only the cases $n \equiv 1, 2, 3, 5$, or $6 \pmod 8$. Once we determine the set $\mathbf{L}'_3$ of integers $n \not\equiv 0, 4, 7 \pmod 8$, with $p_3(n) = 1$, we have $\mathbf{L}_3 = \{n = 4^a n' \mid n' \in \mathbf{L}'_3\}$. The problem can be simplified further by reducing it to the case of squarefree n. Indeed, if $n = m^2 f$ with f squarefree, then, with the notation of Chapter 4, $r_3(n) = \sum_{d \mid m} R_3(m^2 f/d^2) \geqslant R(m^2 f) + R_3(f)$. By (4.7) with $\delta_n = 1$, it is clear that $R_3(m^2 f) R_3(f) > 0$, because m is odd, $m^2 \equiv 1 \pmod 8$, so that $f \equiv n \equiv 1, 2, 3, 5$, or $6 \pmod 8$. It follows that $n = m^2 f = x_1^2 + x_2^2 + x_3^2$ has a solution with $(x_1, x_2, x_3) = 1$. Similarly, also $f = y_1^2 + y_2^2 + y_3^2$, with $(y_1, y_2, y_3) = 1$. Consequently, $n = m^2 f = (my_1)^2 + (my_2)^2 + (my_3)^2 = z_1^2 + z_2^2 + z_3^2$, say, with $(z_1, z_2, z_3) = m > 1$. It follows that n has at least two different partitions into three squares. From here on, we proceed, essentially, as in the case $k = 4$. From $c_3 = 3!\,2^3 = 48$, it follows that $p_3(n) \geqslant r_3(n)/48$; hence, $p_3(n) > 1$, provided that $r_3(n) > 48$.

First, let $n \equiv 3 \pmod 8$. Then, recalling that for squarefree n we have $r_3(n) = R_3(n)$, it follows from (4.7) of Chapter 4 that $r_3(n) = 24h > 48$ unless $h = 1$ or 2. However, all negative discriminants with class numbers $h = 1$ or $h = 2$ are known; specifically (see [257], [258], [13], and [14]), for $h = 1$, these are the discriminants $-d = 3, 4, 7, 8, 11, 19, 43, 67$, and 163. For $n \equiv 3 \pmod 8$ and $-d = n$, we find in this list $n = 3, 11, 19, 43, 67$, and 163. We verify that indeed, for all six, $p_3(n) = 1$. The 18 negative discriminants corresponding to $h = 2$ are (see [258]) $-d = 15, 20, 24, 35, 40, 51, 52, 88, 91, 115, 123, 148, 187, 232, 235, 267, 403$, and 427. In this list, there are ten entries with $n \equiv 3 \pmod 8$, namely 35, 51, 91, 115, 123, 187, 235, 267, 403, and 427; but only six of them, namely 35, 91, 115, 235, 403, and 427, lead to $p_3(n) = 1$. This finishes the proof of

Lemma 1. *The only integers $n \equiv 3 \pmod 8$ for which $p_3(n) = 1$ are the 12 entries of the set* $\{3, 11, 19, 35, 43, 67, 91, 115, 163, 235, 403, 427\}$.

Let now $n \equiv 1, 2, 5$, or $6 \pmod 8$; then equation (4.7) yields $r_3(n) = 12h$, and this exceeds 48 only for $h > 4$. Among the discriminants $d = -n$ with $n \equiv 1, 2, 5$, or $6 \pmod 8$ and $h = 1$, we find only $d = -4$ and $d = -8$, corresponding to $n = 1$ and $n = 2$, respectively and indeed, $p_3(1) = p_3(2) = 1$. As a parenthetical remark, we observe that all discriminants with class number $h = 1$ have been accounted for and lead to $p_3(n) = 1$, except for $-d = n = 7$, for which we know that $p_3(7) = r_3(7) = 0$.

The discriminants $-d = 4n$, *with* $h = 2$ are 20, 24, 40, 52, 88, 148, and 232, corresponding to $n = 5, 6, 10, 13, 22, 37$, and 58, and all these lead indeed to $p_3(n) = 1$.

The case of class number $h = 3$ is rather curious. We do not need to know the discriminants with $h = 3$. Indeed, none of them can correspond to an n with $p_3(n) = 1$. To see this, observe that, by (4.7), if $h = 3$ then $r_3(n) = 12h = 36$. However, no single partition can lead to $r_3(n) = 36$. Indeed, if all summands are nonvanishing and distinct, the partition corresponds to $c_3 = 48$ representations. If the partition has all summands positive, but two of them equal, then it corresponds to $3 \cdot 2^3 = 24$ representations. If it has one summand zero and the others distinct, it corresponds to $3! \cdot 2^2 = 24$ representations. If it consists of three equal summands, it corresponds to $2^3 = 8$ representations. If it has two equal, positive summands and one zero, it corresponds to $3 \cdot 2^2 = 12$ summands. Finally, if it has two summands zero, it corresponds to $3 \cdot 2 = 6$ representations. It follows that $r_3(n) = 36$ is possible only for integers that have exactly two partitions into three squares, of which one is of the form $a^2 + a^2 + 0^2$ and the other one is either of the form $a^2 + a^2 + b^2$ or $a^2 + b^2 + 0^2$, with $0 \neq a \neq b \neq 0$.

It remains to consider only the integers $n \equiv 1, 2, 5$, or $6 \pmod 8$ and such that the discriminant $d = -4n$ has class number $h = 4$. As of this writing, we know that the number of these discriminants is finite, and it is believed that there are no discriminants with $h = 4$ besides the known ones [31]; however, we have no proof that there may not be some (absolutely) very large discriminant ($-d > 100{,}000$) with $h = 4$. Among the known discriminants, one finds $p_3(n) = 1$ precisely for the values of $n = -d/4$ equal to an element of the set $\{14, 21, 30, 42, 46, 70, 78, 93, 133, 142, 190, 253\}$. A search was made for further discriminants with $h = 4$, but up to 100,000 none was found. Nevertheless, as we do not possess a proof that our list of discriminants with $h = 4$ is complete, the best that we can assert is

Theorem 4. *The only integers for which* $p_3(n) = 1$ *are those of the set* $\mathbf{L}_3 = \{n \mid n = 4^a m,\ 0 \leqslant a \in \mathbb{Z},\ m \in \mathbf{M}_1\}$, *where* $\mathbf{M}_1 \supset \mathbf{M}_2 = \{1, 2, 3, 5, 6, 10, 11, 13, 14, 19, 21, 22, 30, 35, 37, 42, 43, 46, 58, 67, 70, 78, 91, 93, 115, 133, 142, 163, 190, 235, 253, 403, 427\}$; $\mathbf{M}_1$ *is a finite set, and any* $n \in \mathbf{M}_1$, $n \notin \mathbf{M}_2$ *satisfies* $n > 100{,}000$, *with* $n \equiv 1, 2, 5$, or $6 \pmod 8$), *and if* $d = -4n$ *then* $h(\mathrm{d}) = 4$.

Conjecture. $\mathbf{M}_2 = \mathbf{M}_1$.

§7. Problems

1. Determine the partitions of 10 into four and into five squares.

2. Prove $\lim_{n\to\infty} r_k(n)/p_k(n) = c_k$ for all k.

3. Does Theorem 1 imply Theorem 6.3? Is it implied by it? Can one prove one of the theorems from the other, perhaps with the help of some additional considerations or information?

4. Prove formally the two inequalities $p_{k+1}(n) \geqslant p_k(n)$ and $p_k(4n) \geqslant p_k(n)$. Can you generalize one or the other of them?

*5. Prove the conjecture that $\mathbf{M}_2 = \mathbf{M}_1$.

*6. Try to find a formula for $p_k(n)$.

*7. Determine all discriminants of positive definite binary quadratic forms over $\mathbb{Z}$, with $h = 4$. The solution of this problem would greatly simplify and improve Theorem 4.

Chapter 8

Theta Functions

§1. Introduction

We recall from Chapter 1 that the determination of $r_k(n)$ can be reduced to that of the coefficient $a_n^{(k)}$ in the Taylor series expansion of the function $(\sum_{m=-\infty}^{\infty} x^{m^2})^k = (1 + 2\sum_{m=1}^{\infty} x^{m^2})^k$, because this series, denoted traditionally by $\{\theta_3(x)\}^k$, equals $\sum_{-\infty<m_i<\infty} x^{m_1^2+m_2^2+\cdots+m_k^2} = 1 + \sum_{n=1}^{\infty} r_k(n)x^n$. It follows that if $\{\theta_3(x)\}^k = \sum_{n=0}^{\infty} a_n^{(k)}x^n$, then $a_0^{(k)} = 1$ for all k, and for $n \geqslant 1$, $r_k(n) = a_n^{(k)}$.

In order to prepare for the computation of the coefficients of $\{\theta_3(x)\}^k$, we start by studying functions of the general type of $\theta_3(x) = \sum_{-\infty}^{\infty} x^{m^2}$. (We often suppress the mention of the summation variable when there is no danger of confusion.) These are the *theta functions*. In this chapter, only that part of their theory will be sketched that will be needed in Chapter 9 in connection with the computation of $r_k(n)$. For a more extensive treatment, the reader may consult the classical work of Jacobi [121], or the standard treatises by Tannery and Molk [262] and Krazer [141]. The presentation in this chapter will follow, in the main, that of Rademacher [221], and an effort has been made to keep much of the notation of [221]. For a different approach, by hypergeometric series, see [6].

The theory of theta functions is replete with an abundance of often very elegant-looking formulae—perhaps more formulae per page than in any comparable chapter of mathematics, as a glance at any treatise will show. With so many relations at one's disposal, it is not surprising that one should be able to solve them for the successive powers of $\theta_3(x)$, especially if the relations turn out to be linear. And that is ultimately our aim. It turns out, however, as we shall see, that this program is only partially successful. Indeed, we obtain the desired result only for even powers of $\theta_3(x)$, and thus only for the $r_{2k}(n)$.

§2. Preliminaries

Before we introduce the theta functions, let us recall, without proofs (these may be found, e.g., in [2]) some results concerning simply and doubly periodic functions.

Any periodic, continuously differentiable function $f(x)$ has a convergent Fourier series. If $f(z+\omega)=f(z)$ for every z and some constant ω, then (see, e.g., [120] or [265]) $f(z)=\sum_{-\infty}^{\infty} a_n e^{(2\pi/\omega)inz}$, where $a_n=(1/\omega)\int_0^{\omega} f(z)\times e^{-2\pi inz/\omega}\,dz$. If $f(z+\omega_1)=f(z+\omega_2)=f(z)$ for every z, and ω_1/ω_2 is rational, then $f(z)$ is a simply periodic function. The function $f(z)$ reduces to a constant if ω_1/ω_2 is real but irrational. Only if the ratio $\tau=\omega_1/\omega_2$ is not real, say $\operatorname{Im}\tau>0$, is it the case that $f(z)$ is properly speaking *doubly periodic*. A meromorphic, doubly periodic function is called an *elliptic function*. If ω_1 and ω_2 are periods, so are $m\omega_1+n\omega_2$ $(m,n\in\mathbb{Z})$. The set $\Omega=\{\omega\mid\omega=m\omega_1+n\omega_2\}$ of all periods forms a lattice in the z-plane. Let us take the smallest period, say ω_1, as unity and take its direction as the real axis, by replacing, if necessary, z by $z'=ze^{i\alpha}$, $\alpha=\arg\omega_1$. Next, let ω_2 be the smallest period with $\arg\omega_2>0$ such that $\tau=\omega_2/\omega_1=\omega_2$ and such that 0, 1, τ, and $1+\tau$ are the vertices of a *period parallelogram* P that contains no other lattice points. The whole plane is tiled by such parallelograms, all translates of P. An elliptic function that is regular in a closed parallelogram of periods is bounded on that compact set, and hence in the whole plane, so that it is a bounded, entire function and reduces to a constant. The same conclusion follows if we know that the elliptic function has only a single pole of order at most one in P.

The *Weierstrass $\wp$-function*, defined by the series

$$\wp(z)=\frac{1}{z^2}+\sum_{\substack{\omega\in\Omega\\ \omega\neq 0}}\left\{\frac{1}{(z-\omega)^2}-\frac{1}{\omega^2}\right\},$$

convergent for every $z\notin\Omega$, is therefore in some sense the simplest elliptic function. It has a pole of order 2 at the origin and hence at every lattice point $\omega\in\Omega$, with principal part z^{-2} at the origin and $(z-\omega)^{-2}$ at the lattice point $z=\omega$.

The theta functions may be thought of as entire functions that come as close to being doubly periodic as it is possible to make them without reducing them to constants. They are simply periodic, and for all of them $\theta(v+1)=\pm\theta(v)$; they have also a "quasiperiod" τ, $\operatorname{Im}\tau>0$, for which $\theta(v+\tau)=A\theta(v)$, where A is a *nonconstant* factor (see §4). The elliptic functions themselves can be built up as ratios of theta functions (see §7).

§3. Poisson Summation and Lipschitz's Formula

Poisson summation plays a fundamental role in the general theory of the theta functions. However, we shall make very few applications of it, and only two in the present chapter, one of which is the proof of Lipschitz's formula. This, in turn, will be used only in some two or three instances, but its use will be crucial. For those reasons, we shall give only somewhat sketchy proofs of these important theorems.

Poisson's summation formula states that, under proper restrictions on the function $f(x)$, the sum $S = \sum_{-\infty}^{\infty} f(n)$ satisfies $S = \sum_{-\infty}^{\infty} A_k$, where $A_k = \int_{-\infty}^{\infty} f(u)e^{-2\pi iku}\,du$. For a proof, observe that the function $S(u) = \sum_{-\infty}^{\infty} f(n+u)$ is periodic, because $S(u+1) = S(u)$. It follows that, having the period 1, $S(u) = \sum_{-\infty}^{\infty} A_k e^{2\pi iku}$, with $A_k = \int_0^1 f(u)e^{-2\pi iku}\,du$. In particular, $S = S(0) = \sum_{-\infty}^{\infty} A_k$, as claimed. Several sets of conditions on $f(x)$ insure the validity of those manipulations. In particular, all infinite sums and integrals that occur have to converge. In the following theorem, two such sets of sufficient conditions are stated.

Theorem 1. *Let us assume that, for a function $f(x)$, one of the following two sets of conditions is satisfied:*

(A) *$f(x)$ is twice differentiable, and the integrals $\int_{-\infty}^{\infty} f(x)\,dx$ and $\int_{-\infty}^{\infty} |f''(x)|\,dx$ both converge.*

(B) *There exists a constant $M > 0$ such that $f(x)$ is of bounded variation for $|x| < M$ and is twice differentiable for $|x| > M$, and the three integrals $\int_{-\infty}^{\infty} f(x)\,dx$, $\int_{-\infty}^{-M} |f''(x)|\,dx$, and $\int_M^{\infty} |f''(x)|\,dx$ all converge.*

Then $S = \sum_{n=-\infty}^{\infty} f(n)$ converges and equals $\sum_{k=-\infty}^{\infty} A_k$, where $A_k = \int_{-\infty}^{\infty} f(u)e^{-2\pi iku}\,du$.

Complete proofs of Theorem 1 may be found in [221] and [22]; see also [70].

Corollary 1. *Let $\psi(t,\alpha) = \sum_{n=-\infty}^{\infty} e^{-\pi(n+\alpha)^2 t}$, where $\operatorname{Re} t > 0$ and $\alpha \in \mathbb{C}$. Then $\psi(t,\alpha) = t^{-1/2}e^{-\pi t\alpha^2}\psi(t^{-1}, -it\alpha)$. In particular,*

$$\psi(t,0) = \sum_{-\infty}^{\infty} e^{-\pi t n^2} = \frac{1}{\sqrt{t}}\psi(t^{-1},0) = \frac{1}{\sqrt{t}}\sum_{k=-\infty}^{\infty} e^{-\pi k^2/t}. \tag{8.1}$$

Proof. Here $f(x) = e^{-\pi(x+\alpha)^2 t}$ and it is easy to verify that $f(x)$ satisfies conditions (A) of Theorem 1. Hence, $A_k = \int_{-\infty}^{\infty} e^{-\pi t(u+\alpha)^2}e^{-2\pi iku}\,du = \int_{-\infty}^{\infty} e^{-\pi t(u+\alpha+(ik/t))^2}\,e^{2\pi ik\alpha-\pi k^2/t}\,du = e^{2\pi ik\alpha-\pi k^2/t}\int_L e^{-\pi tz^2}\,dz$, where $z = u + \alpha + (ik/t)$ and the line L of integration is parallel to the real axis and may be defined by $x + iy$, with $-\infty < x < \infty$, $y = (k/t) + \operatorname{Im}\alpha$. By considering the rectangle $\mathscr{R}$ with vertices $\pm M$, $\pm M + iy$, we verify that the contributions of the vertical sides to the integral $\int_{\mathscr{R}}$ approach zero as $M \to \infty$; hence, by Cauchy's theorem, $\int_{\mathscr{R}} = \int_{-\infty}^{\infty}$ and $A_k = Ie^{2\pi ik\alpha-\pi k^2/t}$, where $I = \int_{-\infty}^{\infty} e^{-\pi tz^2}\,dz = (1/\sqrt{\pi t})\int_{-\infty}^{\infty} e^{-u^2}\,du = (2/\sqrt{\pi t})\int_0^{\infty} e^{-u^2}\,du = (2/\sqrt{\pi t})\int_0^{\infty} e^{-v}\,dv/2\sqrt{v} = (1/\sqrt{\pi t})\int_0^{\infty} v^{-1/2}e^{-v}\,dv = (\pi t)^{-1/2}\Gamma(\frac{1}{2}) = t^{-1/2}$. Substituting $t^{-1/2}$ for I, we conclude that $\sum_{n=-\infty}^{\infty} e^{-\pi(n+u+\alpha)^2 t} = \sum_{k=-\infty}^{\infty} A_k e^{2\pi iku} = t^{-1/2}\sum_{k=-\infty}^{\infty} e^{2\pi ik\alpha-\pi(k^2/t)+2\pi iku}$ and for $u = 0$,

$$\psi(t,\alpha)=t^{-1/2}\sum_{k=-\infty}^{\infty}e^{2\pi ik\alpha-\pi k^2/t}$$

$$=t^{-1/2}\sum_{k=-\infty}^{\infty}e^{-(\pi/t)(k^2-2ikt\alpha-t^2\alpha^2)}e^{-\pi t\alpha^2}$$

$$=t^{-1/2}e^{-\pi t\alpha^2}\sum_{k=-\infty}^{\infty}e^{-(\pi/t)(k-it\alpha)^2}$$

$$=t^{-1/2}e^{-\pi t\alpha^2}\psi(t^{-1},-it\alpha),$$

as claimed. □

Comment. A remarkably simply proof of Corollary 1 that does not use Theorem 1 is worth mentioning. It is due to G. Pólya, and the interested reader may wish to consult his original paper [217], or Bellman's book [21] for the details. An outline of this proof follows.

Let $\omega = e^{2\pi i/k}$ be a kth root of unity, $k \neq 1$; then $\omega^k = 1$ and $\sum_{v=0}^{k-1}\omega^v = \sum_{-k/2<v\leqslant k/2}\omega^v = 0$. Now $(z^{1/2}+z^{-1/2})^{2m} = \sum_{v=-m}^{m}\binom{2m}{m+v}z^v$; and if we replace z by $\omega^v z$ and sum over v, by using the above identities, we obtain

$$\sum_{-k/2<v\leqslant k/2}\{(\omega^v z)^{1/2}+(\omega^v z)^{-1/2}\}^{2m} = k\sum_{|v|\leqslant[m/k]}\binom{2m}{m+kv}z^{kv}.$$

Here the square brackets stand, as on earlier occasions, for the greatest integer function.

Now, for any $t>0$, take $k=[\sqrt{mt}]$, and for arbitrary $s\in\mathbb{C}$, set $z=e^{s/k}$. Then, after 2^{2m} is factored out, the last identity becomes

$$\sum_{-k/2<v\leqslant k/2}\left\{\frac{e^{(s+2\pi iv)/2k}+e^{-(s+2\pi iv)/2k}}{2}\right\}^{2m} = \sum_{|v|\leqslant[m/k]}\frac{r}{v}2^{-2m}\binom{2m}{m+r}e^{sv},$$

where $r=v[\sqrt{tm}]$. We now let $k\to\infty$, use the well-known fact that if $x_n\to x$ then $\lim_{n\to\infty}(1+x_n/n)^n = e^x$ and the easily verified one (use Stirling's formula) that if $\lim_{n\to\infty} rn^{-1/2} = x$ then $\lim_{n\to\infty}2^{-2n}\sqrt{n}\binom{2n}{n+r} = e^{-x^2}\pi^{-1/2}$, and obtain

$$\sum_{v=-\infty}^{\infty}e^{(s+2\pi iv)^2/4t} = \sqrt{\frac{t}{\pi}}\sum_{v=-\infty}^{\infty}e^{-tv^2+sv}.$$

If we write here α for $-s/2t$ and then replace t by πt, we obtain Corollary 1.

Perhaps even more interesting than the proof itself is the idea behind it, which comes from physics. The heat equation (8.3), mentioned already in the Introduction is satisfied (as we shall presently see) by all theta functions—and $\psi(t,\alpha)$ is essentially one of them. This equation governs the process of continuous diffusion, which however may be considered as the limit of the random

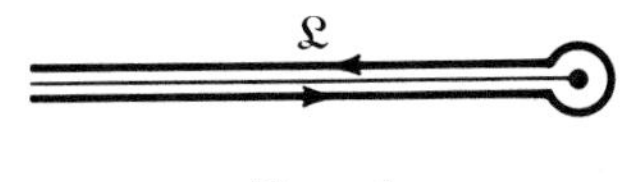

Figure 1

movement of individual molecules, a process governed by the binomial expansion. Hence, a function that is solution of (8.3) will satisfy a limiting form of an identity based on the binomial expansion.

Theorem 2 (Lipschitz's formula; see [221] or [167]). *Let* $\operatorname{Re} z > 0, 0 \leqslant \alpha < 1$, *and* $\sigma = \operatorname{Re}(s)$. If $\sigma > 1$, *then*

$$\frac{(2\pi)^s}{\Gamma(s)} \sum_{m=0}^{\infty} (m + \alpha)^{s-1} e^{-2\pi z(m+\alpha)} = \sum_{n=-\infty}^{\infty} (z + ni)^{-s} e^{2\pi i n \alpha},$$

where we define $(z + ni)^s$ *by* $|z + ni|^s e^{is \arg(z+ni)}$, *with* $|\arg(z + ni)| < \pi/2$. *If* $0 < \alpha < 1$, *the equality holds for* $\sigma > 0$.

Proof. For $\sigma > 1$, $S = \sum_{-\infty}^{\infty} (z + ni)^{-s} e^{2\pi i n\alpha}$ converges absolutely and one verifies that conditions (A) of Theorem 1 hold. Hence, $S = \sum_{-\infty}^{\infty} A_k$, where $A_k = \int_{-\infty}^{\infty} (z + vi)^{-s} e^{2\pi i v\alpha} e^{-2\pi i k v}\, dv$. To compute the integral, set $z + vi = w$; then $A_k = i^{-1} \int_{z-i\infty}^{z+i\infty} w^{-s} e^{2\pi(w-z)(\alpha-k)}\, dw = i^{-1} e^{2\pi z(k-\alpha)} \int_{z-i\infty}^{z+i\infty} w^{-s} e^{-2\pi w(k-\alpha)}\, dw$, so $A_{-k} = i^{-1} e^{-2\pi z(k+\alpha)} \int_{z-i\infty}^{z+i\infty} w^{-s} e^{2\pi w(k+\alpha)}\, dw$. If $k + \alpha \leqslant 0$, then for $M > 0$, $\int_{z-iM}^{z+iM} w^{-s} e^{2\pi w(k+\alpha)}\, dw = \int_{\mathscr{C}} w^{-s} e^{2\pi w(k+\alpha)}\, dw$, where $\mathscr{C}$ is a semicircle in the right half plane, with diameter from $z - iM$ to $z + iM$. When we let $M \to \infty$, the integral vanishes in the limit and $A_{-k} = 0$ for $k + \alpha < 0$, i.e. (recall that $0 \leqslant \alpha < 1$) for $k < 0$. If $k + \alpha > 0$, then $\int_{z-i\infty}^{z+i\infty} w^{-s} e^{2\pi w(k+\alpha)}\, dw = (2\pi)^{s-1} \times (k + \alpha)^{s-1} \int_{a-i\infty}^{a+i\infty} \zeta^{-s} e^{\zeta}\, d\zeta$, where $\zeta = 2\pi w(k + \alpha)$ and $a = 2\pi(k + \alpha) \operatorname{Re} z$. For $\sigma > 1$, the path of integration can be deformed into the path $\mathscr{L}$ from $-\infty$ to 0, with a turn of 2π in the positive direction around the origin and back again to $-\infty$ (see Fig. 1). It is well known (see [221], [22], or [114]) that $(2\pi i)^{-1} \int_{\mathscr{L}} \zeta^{-s} e^{\zeta}\, d\zeta = 1/\Gamma(s)$. We now replace the integral by this value in the above formula and obtain $A_{-k} = e^{-2\pi z(k+\alpha)}(2\pi)^s(k + \alpha)^{s-1}/\Gamma(s)$, valid for $k \geqslant 0$, and the result follows for $\sigma > 1$. However, if $\alpha > 0$, the left hand side converges for $\sigma > 0$ (in fact, for all σ), so that Theorem 2 holds, by analytic continuation, wherever the right hand side converges. For $\alpha > 0$, this is the case for all $\sigma > 0$. To prove this, one has to verify that $\lim_{N\to\infty} \sum_{n=N+1}^{N+M} |(z + ni)^{-s} e^{2\pi i n\alpha}| = 0$. We leave this rather tedious task to the interested reader. □

§4. The Theta Functions

The function of most immediate interest to us and that we have denoted by $\theta_3(x)$ is the value for $v = 0$ and $q = x$ of the function traditionally defined by $\theta_3(v \mid \tau) = \sum_{-\infty}^{\infty} q^{n^2} e^{2\pi i n v}$, where $q = e^{i\pi\tau}$. The other three theta functions are similarly defined, as follows:

$$
\begin{aligned}
\theta_1(v\,|\,\tau) &= -i\sum_{-\infty}^{\infty}(-1)^n q^{(n+1/2)^2} e^{(2n+1)\pi i v} = 2\sum_{n=0}^{\infty}(-1)^n q^{(n+1/2)^2}\sin(2n+1)\pi v\\
\theta_2(v\,|\,\tau) &= \sum_{-\infty}^{\infty} q^{(n+1/2)^2} e^{(2n+1)\pi i v} = 2\sum_{n=0}^{\infty} q^{(n+1/2)^2}\cos(2n+1)\pi v\\
\theta_3(v\,|\,\tau) &= \sum_{-\infty}^{\infty} q^{n^2} e^{2\pi i n v} = 1 + 2\sum_{n=1}^{\infty} q^{n^2}\cos 2\pi n v\\
\theta_4(v\,|\,\tau) &= \sum_{-\infty}^{\infty}(-1)^n q^{n^2} e^{2\pi i n v} = 1 + 2\sum_{n=1}^{\infty}(-1)^n q^{n^2}\cos 2\pi n v.
\end{aligned}
\tag{8.2}
$$

Convergence of these series requires $|q| < 1$, or, equivalently, $\operatorname{Im}\tau > 0$. In what follows, we shall assume, without further mention, that these inequalities are satisfied. Then the series in (8.2) converge absolutely and, for fixed q, represent entire functions of v.

Several authors have adopted a different notation for these functions which has, at least, a mnemonic advantage. Let $\theta_{00} = \theta_3$, $\theta_{01} = \theta_4$, $\theta_{10} = \theta_2$, and $\theta_{11} = i\theta_1$; then all four formulae (8.2) can be condensed into a single one. For μ and ν either 0 or 1, $\theta_{\mu\nu}(v\,|\,\tau) = \sum_{-\infty}^{\infty}(-1)^{\nu n} e^{(n+\mu/2)^2\pi i\tau} e^{2\pi i(n+\mu/2)v}$. One immediately verifies that, for all four values of (μ, ν), the function $\theta_{\mu\nu}(v\,|\,\tau)$ satisfies the parabolic (essentially the heat) equation

$$
\frac{\partial^2\theta_{\mu\nu}(v\,|\,\tau)}{\partial v^2} = 4\pi i\frac{\partial\theta_{\mu\nu}(v\,|\,\tau)}{\partial\tau}. \tag{8.3}
$$

This justifies a claim made without proof in the Introduction. In order to simplify the writing, in what follows, whenever τ is kept fixed, we shall suppress it in the notation, if this is possible without creating confusion, and we shall write, e.g., $\theta_2(v)$, rather than $\theta_2(v\,|\,\tau)$.

From the first set of equalities (8.2), we easily verify the following identities, in which the factor $q^{-1}e^{-2\pi i v}$ is abbreviated by A:

$$
\begin{aligned}
\theta_1(v+1) &= -\theta_1(v), & \theta_1(v+\tau) &= -A\theta_1(v),\\
\theta_2(v+1) &= -\theta_2(v), & \theta_2(v+\tau) &= A\theta_2(v),\\
\theta_3(v+1) &= \theta_3(v), & \theta_3(v+\tau) &= A\theta_3(v),\\
\theta_4(v+1) &= \theta_4(v), & \theta_4(v+\tau) &= -A\theta_4(v).
\end{aligned}
\tag{8.4}
$$

We also observe from the second set of identities (8.2) that $\theta_1(v)$ is an odd function, while the other three are even functions of v. Finally, (8.2) shows that, e.g., $\theta_3((v+\frac{1}{2})\,|\,\tau) = \sum_{-\infty}^{\infty} q^{n^2} e^{2\pi inv} e^{\pi in} = \sum_{-\infty}^{\infty} (-1)^n q^{n^2} e^{2\pi inv} = \theta_4(v\,|\,\tau)$ and similarly, each theta function can be expressed with the help of any of the others, appropriately shifted. The complete result may be given in tabular form, with B an abbreviation for the factor $q^{-1/4}e^{-\pi iv}$, as follows:

	v	$v+\frac{1}{2}$	$v+\tau/2$	$v+(1+\tau)/2$
θ_1	$\theta_1(v)$	$\theta_2(v)$	$iB\theta_4(v)$	$B\theta_3(v)$
θ_2	$\theta_2(v)$	$-\theta_1(v)$	$B\theta_3(v)$	$-iB\theta_4(v)$
θ_3	$\theta_3(v)$	$\theta_4(v)$	$B\theta_2(v)$	$iB\theta_1(v)$
θ_4	$\theta_4(v)$	$\theta_3(v)$	$iB\theta_1(v)$	$B\theta_2(v)$

(8.5)

§5. The Zeros of the Theta Functions

From the first identities in (8.2) and from (8.4) it follows that $\theta_1(0\,|\,\tau) = \theta_1(1\,|\,\tau) = \theta_1(\tau\,|\,\tau) = 0$. By induction on m and n it follows that $\theta_1(v\,|\,\tau) = 0$ for all v belonging to the lattice points of $L_1 = \{v\,|\,v = m + n\tau\}$. In fact, $\theta_1(v\,|\,\tau)$, considered as a function of v, vanishes nowhere else, and L_1 is therefore called the *lattice of zeros* of θ_1. To prove our last assertion, let us observe that the v-plane is tiled by the parallelograms like RSTU (see Fig. 2) of vertices $\pm\frac{1}{2} \pm \tau/2$ that are the *centers* of the four parallelograms of periods around the origin. The closed parallelogram contains a single point of L_1, namely the origin. Our claim will be proved if we show that $\theta_1(v)$ vanishes exactly once in RSTU. This, however, follows from the argument principle. If N is the number of zeros inside RSTU, then

$$N = \frac{1}{2\pi i}\int_{\mathscr{C}} \frac{\theta_1'(v)}{\theta_1(v)}\,dv,$$

where $\mathscr{C}$ is the contour RSTU, deformed appropriately, if necessary, to avoid passing through any zeros of $\theta_1(v)$. By (8.4), we may pair off integrals along parallel sides in $\mathscr{C}$ and obtain

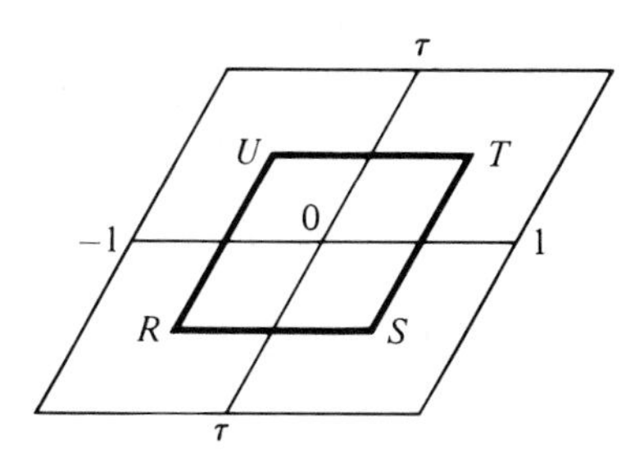

Figure 2

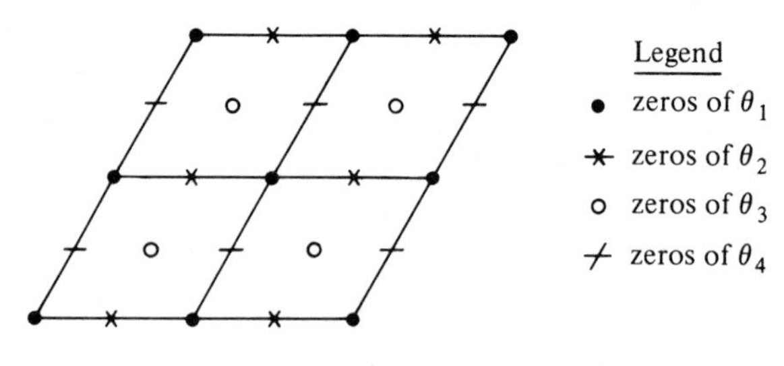

Figure 3

$$N = \frac{1}{2\pi i}\left\{\int_R^S \left(\frac{d}{dv}\log\theta_1(v) - \frac{d}{dv}\log\theta_1(v+\tau)\right)dv\right.$$

$$\left.+\int_U^R \left(\frac{d}{dv}\log\theta_1(v) - \frac{d}{dv}\log\theta_1(v+1)\right)dv\right\}$$

$$= \frac{1}{2\pi i}\left\{\int_R^S \frac{d}{dv}\log\frac{\theta_1(v)}{\theta_1(v+\tau)}dv + \int_U^R \frac{d}{dv}\log\frac{\theta_1(v)}{\theta_1(v+1)}dv\right\}.$$

By (8.4),

$$\log\frac{\theta_1(v)}{\theta_1(v+\tau)} = \log(-1) - \log A$$

$$= \pi i + \log q + 2\pi i k + 2\pi i v, \qquad k \in \mathbb{Z},$$

where the summand in k is due to the fact that the logarithm is not single-valued. However, the derivative needed equals $2\pi i$ and does not depend on the unknown constant integer k. Similarly,

$$\frac{d}{dv}\log\frac{\theta_1(v)}{\theta_1(v+1)} = \frac{d}{dv}(\log(-1) + 2\pi i k) = 0,$$

whence $N = (1/2\pi i)\{2\pi i + 0\} = 1$, which proves our claim.

By using the table (8.5), we obtain the lattices of zeros for all four theta functions, as follows (see also Fig. 3):

the zero set of θ_1 is $L_1 = \{m + n\tau\}$;
the zero set of θ_2 is $L_2 = \{m + \frac{1}{2} + n\tau\}$;
the zero set of θ_3 is $L_3 = \{m + \frac{1}{2} + (n + \frac{1}{2})\tau\}$;
the zero set of θ_4 is $L_4 = \{m + (n + \frac{1}{2})\tau\}$.

§6. Product Formulae

Like all entire functions, the theta functions can be represented by infinite series (see (8.2)), as well as by infinite Weierstrass type products that exhibit their zeros. We are able to write down the latter with relative ease, because, from §5, we know those zeros. In many formulae, it is convenient to replace the somewhat complicated expression $e^{2\pi i v}$ simply by z; accordingly, we shall use either the notation $\theta_j(v \mid \tau)$ or the notation $\theta_j(z; q)$, but even in the first case, $e^{\pi i \tau}$ will usually be replaced by q.

Since $|q| < 1$, the infinite product $F(z) = \prod_{n=1}^{\infty} (1 + q^{2n-1}z)(1 + q^{2n-1}z^{-1})$ converges for $z \neq 0$ and has simple zeros precisely at $z = e^{2\pi i(m+1/2+(n+1/2)\tau)} = e^{\pi i(2n+1)\tau}e^{\pi i} = -q^{2n+1}$, which are the zeros of $\theta_3(z; q)$. Also, considered as a function of v, $F(e^{2\pi i(v+1)}) = F(e^{2\pi i v})$, while

$$
\begin{aligned}
F(e^{2\pi i(v+\tau)}) &= \prod_{n=1}^{\infty} (1 + q^{2n-1}e^{2\pi i(v+\tau)})(1 + q^{2n-1}e^{-2\pi i(v+\tau)}) \\
&= \prod_{n=1}^{\infty} (1 + q^{2n-1}e^{2\pi i\tau}z)(1 + q^{2n-1}e^{-2\pi i\tau}z^{-1}) \\
&= \prod_{n=1}^{\infty} (1 + q^{2n+1}z)(1 + q^{2n-3}z^{-1}) = \frac{1 + q^{-1}z^{-1}}{1 + qz} F(z),
\end{aligned}
$$

with $(1 + q^{-1}z^{-1})/(1 + qz) = 1/qz = q^{-1}e^{-2\pi i v} = A$. Consequently, $\theta_3(v)/F(e^{2\pi i v})$ is invariant under $v \to v + 1$ and under $v \to v + \tau$, so that this ratio is doubly periodic. It is also a meromorphic function, and the zeros of the denominator are canceled by those of the numerator; hence, it is an entire, doubly periodic function and reduces to a constant, independent of v. The value of that constant may well depend, in general, on q.

So far, we have shown that $\theta_3(z; q) = T(q) \prod_{n=1}^{\infty} (1 + q^{2n-1}z)(1 + q^{2n-1}z^{-1})$, and it remains to determine $T(q)$. Following Gauss, set $z = -1$ in (8.2) and obtain $\theta_3(-1; q) = \sum_{-\infty}^{\infty} (-1)^n q^{n^2} = T(q) \prod_{m=1}^{\infty} (1 - q^{2m-1})^2$, so that $T(q) \prod_{m=1}^{\infty} (1 - q^{2m})^{-1} = (\sum_{-\infty}^{\infty} (-1)^n q^{n^2}) \{\prod_{m=1}^{\infty} (1 - q^{2m})(1 - q^{2m-1})^2\}^{-1} = (\sum_{-\infty}^{\infty} (-1)^n q^{n^2}) \{\prod_{m=1}^{\infty} (1 - q^m)(1 - q^{2m-1})\}^{-1} = G(q)$, say. Similarly, if we substitute $z = i$, we obtain, observing that the odd powers cancel, $\theta_3(i; q) = \sum_{-\infty}^{\infty} i^n q^{n^2} = \sum_{-\infty}^{\infty} (-1)^k q^{4k^2} = T(q) \prod_{m=1}^{\infty} (1 + iq^{2m-1})(1 - iq^{2m-1}) = T(q) \prod_{m=1}^{\infty} (1 + q^{4m-2})$ and $T(q) \prod_{m=1}^{\infty} (1 - q^{2m})^{-1} = (\sum_{-\infty}^{\infty} (-1)^k q^{4k^2}) \{\prod_{m=1}^{\infty} (1 - q^{2m})(1 + q^{4m-2})\}^{-1} = (\sum_{-\infty}^{\infty} (-1)^n q^{4n^2}) \{\prod_{m=1}^{\infty} (1 - q^{4m}) \times (1 - q^{4m-2})(1 + q^{4m-2})\}^{-1} = (\sum_{-\infty}^{\infty} (-1)^n q^{4n^2}) \{\prod_{m=1}^{\infty} (1 - q^{4m})(1 - q^{8m-4})\}^{-1}$. The last product is precisely $G(q^4)$, and by equating the two values found for $T(q) \prod_{m=1}^{\infty} (1 - q^{2m})^{-1}$, we infer that $G(q) = G(q^4)$, so that, by induction, $G(q) = G(q^{4^r})$, for all integers $r \geqslant 0$. However, since $|q| < 1$, we have $\lim_{r \to \infty} G(q^{4^r}) = G(0) = 1$, and this finishes the proof (for another proof see

[98]) that $T(q) = \prod_{m=1}^{\infty}(1 - q^{2m})$. Recalling the definition of $T(q)$, we observe that we have also completed the proof of

Lemma 1 (Jacobi's triple product identity). *For* $|q| < 1$ *and* $z \neq 0$,

$$\theta_3(z; q) = \sum_{-\infty}^{\infty} q^{n^2} z^n = \prod_{m=1}^{\infty} (1 - q^{2m})(1 + q^{2m-1}z)(1 + q^{2m-1}z^{-1}).$$

By use of the table (8.5) we now easily obtain the product representations of the other three theta functions, all valid for $|q| < 1$, namely

$$\begin{aligned}
\theta_1(v \mid \tau) &= 2q^{1/4} \sin \pi v \prod_{m=1}^{\infty} (1 - q^{2m})(1 - q^{2m}e^{2\pi i v})(1 - q^{2m}e^{-2\pi i v}),\\
\theta_2(v \mid \tau) &= 2q^{1/4} \cos \pi v \prod_{m=1}^{\infty} (1 - q^{2m})(1 + q^{2m}e^{2\pi i v})(1 + q^{2m}e^{-2\pi i v}),\\
\theta_3(v \mid \tau) &= \prod_{m=1}^{\infty} (1 - q^{2m})(1 + q^{2m-1}e^{2\pi i v})(1 + q^{2m-1}e^{-2\pi i v}),\\
\theta_4(v \mid \tau) &= \prod_{m=1}^{\infty} (1 - q^{2m})(1 - q^{2m-1}e^{2\pi i v})(1 - q^{2m-1}e^{-2\pi i v}).
\end{aligned} \tag{8.6}$$

By setting $v = 0$ in (8.6), we obtain useful identities for $\theta_j(0 \mid \tau)$ $(j = 2, 3, 4)$. For $j = 1$, both sides vanish; however, $\lim_{v \to \infty} \theta_1(v \mid \tau)/v = \theta_1'(0 \mid \tau) = \theta_1'$ is meaningful. As is customary, we put $\theta_j(0 \mid \tau) = \theta_j$. With this notation, we can state these results as follows.

Theorem 3. *For* $|q| < 1$,

$$\begin{aligned}
\theta_1' &= 2\pi q^{1/4} \prod_{m=1}^{\infty} (1 - q^{2m})^3,\\
\theta_2 &= 2q^{1/4} \prod_{m=1}^{\infty} (1 - q^{2m})(1 + q^{2m})^2,\\
\theta_3 &= \prod_{m=1}^{\infty} (1 - q^{2m})(1 + q^{2m-1})^2,\\
\theta_4 &= \prod_{m=1}^{\infty} (1 - q^{2m})(1 - q^{2m-1})^2.
\end{aligned}$$

If we differentiate $\theta_1(v \mid \tau)$ with respect to v and then set $v = 0$, we obtain, by comparison with Theorem 3,

Corollary 2 (Jacobi). *For* $|q| < 1$, $\prod_{m=1}^{\infty}(1 - q^{2m})^3 = \sum_{-\infty}^{\infty}(-1)^n(2n-1)q^{n(n+1)}$.

By setting $q^2 = x$ in Corollary 2, we obtain

Corollary 3. *For* $|x| < 1$, $\prod_{m=1}^{\infty}(1 - x^m)^3 = \sum_{n=0}^{\infty}(-1)^n(2n+1)x^{n(n+1)/2}$.

We observe that here the exponents on the right are precisely the triangular numbers $t_n = n(n+1)/2$.

Corollary 4. $\theta_1' = \pi\theta_2\theta_3\theta_4$.

Proof. From Theorem 3, it follows that $\pi\theta_2\theta_3\theta_4 = \theta_1' H^2(q)$, where $H(q) = \prod_{m=1}^{\infty}(1 + q^{2m})(1 + q^{2m-1})(1 - q^{2m-1}) = \prod_{m=1}^{\infty}(1 + q^{2m})(1 - q^{4m-2}) = \prod_{m=1}^{\infty}(1 + q^{4m})(1 + q^{4m-2})(1 - q^{4m-2}) = \prod_{m=1}^{\infty}(1 + q^{4m})(1 - q^{8m-4}) = H(q^2)$ and, by induction, $H(q) = H(q^{2^r}) = \lim_{r\to\infty} H(q^{2^r}) = H(0) = 1$. □

The proofs given here are in the spirit of those presented by H. Rademacher in his lectures (see also [221]). As one would expect, such beautifully simple results as Jacobi's triple product identity (Lemma 1) and the corollaries induced many mathematicians to look for proofs that are either simpler or more revealing than the original ones. Among them, some that differ from the ones presented here are those mentioned by Andrews [5] and also those in Hardy and Wright [98].

§7. Some Elliptic Functions

As already mentioned at the end of §2, we can build up elliptic functions as ratios of theta functions. Indeed, for $j = 2$, 3, and 4, let us define the functions

$$f_j(v \mid \tau) = \frac{\theta_1'}{\theta_j}\frac{\theta_j(v \mid \tau)}{\theta_1(v \mid \tau)}. \tag{8.7}$$

We verify that for $j = 2$, 3, or 4, $f_j(v \mid \tau)$ has a single pole of first order at $v = 0$, with residue 1. Because of this, we know that it cannot be an elliptic function with periods 1, τ, but by (8.4),

$$\begin{aligned} f_2(v+1) &= f_2(v), & f_2(v+\tau) &= -f_2(v), \\ f_3(v+1) &= -f_3(v), & f_3(v+\tau) &= -f_3(v), \\ f_4(v+1) &= -f_4(v), & f_4(v+\tau) &= f_4(v). \end{aligned} \tag{8.8}$$

From (8.7) and the results of §6 we immediately find that $f_2(\tfrac{1}{2}) = f_3((1+\tau)/2) =$

$f_4(\tau/2) = 0$. By combining this with (8.5) and Corollary 4, we obtain the values of all $f_j(v)$'s for $v = \frac{1}{2}$, $\tau/2$, and $(1 + \tau)/2$. For example,

$$f_4(\tfrac{1}{2}) = \frac{\theta'_1}{\theta_4}\frac{\theta_4(\frac{1}{2})}{\theta_1(\frac{1}{2})} = \frac{\theta'_1}{\theta_4}\frac{\theta_3}{\theta_2} = \frac{\pi\theta_2\theta_3\theta_4}{\theta_4}\frac{\theta_3}{\theta_2} = \pi\theta_3^2.$$

The other computations are similar, and their result may be summarized as follows:

$f_j \backslash v$	$\frac{1}{2}$	$\tau/2$	$(1 + \tau)/2$
f_2	0	$-i\pi\theta_3^2$	$-i\pi\theta_4^2$
f_3	$\pi\theta_4^2$	$-i\pi\theta_2^2$	0
f_4	$\pi\theta_3^2$	0	$\pi\theta_2^2$

(8.9)

It follows from (8.8) that the squares $f_j^2(v)$ *are* elliptic functions of periods 1, τ; also, the $f_j(v)$ themselves are elliptic, but with larger periods, so that each of their parallelograms of periods contains two poles of first order, with residues of opposite signs. Specifically, f_2 is periodic with periods $1, 2\tau$; f_3 with periods $2, 2\tau$; and f_4 with periods $2, \tau$. We verify that the $f_j^2(v \mid \tau)$ are all even functions; they have a pole of order 2 at the origin, with principal part v^{-2}, so that the differences $f_j^2(v \mid \tau) - f_k^2(v \mid \tau)$ are doubly periodic, without singularities, and hence are constants, which we denote by c_{jk}. For the same reason, the differences $f_j^2(v \mid \tau) - \wp(v \mid \tau)$ are also constant. By using the Taylor expansions

$$f_j^2(v \mid \tau) = \left(\frac{\theta'_1}{\theta_j}\right)^2 \frac{(\theta_j + \theta''_j v^2/2 + \cdots)^2}{(\theta'_1 v + \theta'''_1 v^3/3! + \cdots)^2}$$

$$= v^{-2}\frac{1 + \dfrac{\theta''_j}{\theta_j}v^2 + \cdots}{1 + \dfrac{\theta'''_1}{\theta'_1}\dfrac{v^2}{3} + \cdots} = v^{-2}\left(1 + \left(\frac{\theta''_j}{\theta_j} - \frac{\theta'''_1}{3\theta'_1}\right)v^2 + \cdots\right),$$

we compute

$$c_{jk} = \frac{\theta''_j}{\theta_j} - \frac{\theta''_k}{\theta_k}$$

and, by (8.3),

$$c_{jk} = 4\pi i\left(\theta_j^{-1}\frac{\partial}{\partial\tau}\theta_j - \theta_k^{-1}\frac{\partial}{\partial\tau}\theta_k\right) = 4\pi i\frac{\partial}{\partial\tau}\log\frac{\theta_j(0 \mid \tau)}{\theta_k(0 \mid \tau)}.$$

We now compute c_{42} in two different ways: by substituting $\frac{1}{2}$, and then $(1+\tau)/2$ into $f_4^2(v \mid \tau) - f_2^2(v \mid \tau)$. By use of (8.7), (8.9), and Corollary 4, we obtain

$$c_{42} = f_4^2(\tfrac{1}{2}) - f_2^2(\tfrac{1}{2}) = f_4^2(\tfrac{1}{2}) = \left(\frac{\theta_1'}{\theta_4}\frac{\theta_4(\frac{1}{2} \mid \tau)}{\theta_1(\frac{1}{2} \mid \tau)}\right)^2$$

$$= \left(\frac{\theta_1'\theta_3}{\theta_4\theta_2}\right)^2 = \left(\frac{\theta_1'\theta_3}{\theta_2\theta_3\theta_4}\right)^2 = \pi^2\theta_3^4.$$

Also,

$$c_{42} = f_4^2\left(\frac{1+\tau}{2}\middle|\tau\right) - f_2^2\left(\frac{1+\tau}{2}\middle|\tau\right) = \left(\frac{\theta_1'\theta_2}{\theta_4\theta_3}\right)^2 + \left(\frac{\theta_1'\theta_4}{\theta_2\theta_3}\right)^2 = \pi^2(\theta_2^4 + \theta_4^4).$$

By comparing these two representations of c_{42}, we obtain

$$\theta_3^4 = \theta_2^4 + \theta_4^4. \tag{8.10}$$

Also the function $\theta_1'(v \mid \tau)/\theta_1(v \mid \tau) = (d/dv)\log\theta_1(v \mid \tau) = \Phi(v \mid \tau)$, say, is of interest, but presents the following difficulty. It has poles at all the zeros of $\theta_1(v \mid \tau)$, and all with the same residue $+1$. Hence, it cannot be elliptic. We observe that $\Phi(v+1 \mid \tau) = \Phi(v \mid \tau)$ and, by (8.4), that $\Phi(v+\tau \mid \tau) = \Phi(v \mid \tau) - 2\pi i$. With the benefit of hindsight, one may now consider the combination

$$\psi(v \mid \tau) = \frac{1}{2}\left\{\Phi\left(\frac{v}{2}\middle|\frac{\tau}{2}\right) - \Phi\left(\frac{v+1}{2}\middle|\frac{\tau}{2}\right)\right\}; \tag{8.11}$$

this function has the periods $2, \tau$, and the poles of first order at $v = 2m + n\tau$ (with residues $+1$) and at $v = 2m + 1 + n\tau$ (with residues -1). These are precisely the periods, poles, and residues of $f_4(v \mid \tau)$; hence, the difference $f_4(v \mid \tau) - \psi(v \mid \tau)$ is an entire elliptic function, i.e., a constant. In fact, this constant is zero. To see that, observe that $f_4(v \mid \tau) - \frac{1}{2}\Phi(v/2 \mid \tau/2)$ is holomorphic at 0 and odd, so that it vanishes for $v = 0$ and

$$\lim_{v\to 0}(f_4(v \mid \tau) - \psi(v \mid \tau)) = \tfrac{1}{2}\Phi(\tfrac{1}{2} \mid \tau/2) = \frac{1}{2}\frac{\theta_1'(\frac{1}{2} \mid \tau/2)}{\theta_1(\frac{1}{2} \mid \tau/2)} = \frac{1}{2}\frac{\theta_2'(0 \mid \tau/2)}{\theta_2(0 \mid \tau/2)} = 0,$$

by use of (8.5) and the remark that $\theta_2(v \mid \tau)$ is an even function. We record the result in the form

$$f_4(v \mid \tau) = \frac{\theta_1'}{\theta_4}\frac{\theta_4(v \mid \tau)}{\theta_1(v \mid \tau)} = \psi(v \mid \tau). \tag{8.12}$$

§8. Addition Formulae

The theta functions have addition formulae that may be considered as analogues or generalizations of those for trigonometric or elliptic functions (for the latter, see, e.g., [2]). We conclude this chapter with the proof of the following addition theorem.

Theorem 4. *Let i, j, k stand for an arbitrary, but fixed permutation of the indices 2, 3, 4. Then the following identity holds:*

$$\theta_j\theta_k\theta_1(u+v)\theta_i(u-v) = \theta_j(v)\theta_k(v)\theta_1(u)\theta_i(u) + \theta_1(v)\theta_i(v)\theta_j(u)\theta_k(u). \tag{8.13}$$

Proof. Set $G_i(v) = \theta_1(v+c)\theta_i(v-c)$ where c is an arbitrary constant; then it follows from (8.4) that

$$G_i(v+1) = \varepsilon G_i(v), \qquad \varepsilon = \begin{cases} 1 & \text{for } i = 2, \\ -1 & \text{for } i = 3, 4; \end{cases}$$

$$G_i(v+\tau) = \eta q^{-2}e^{-4\pi i v}G_i(v) \qquad \eta = \begin{cases} 1 & \text{for } i = 4, \\ -1 & \text{for } i = 2, 3. \end{cases}$$

Similarly, set $H_i(v) = \theta_j(v)\theta_k(v)$; then we find that

$$H_i(v+1) = \varepsilon H_i(v),$$

$$H_i(v+\tau) = \eta q^{-2}e^{-4\pi i v}H_i(v),$$

with the same values for ε and η as before. It follows that $H_i(v)/G_i(v)$ is an elliptic function with the periods 1 and τ. The same holds also for

$$\frac{\theta_1(v)\theta_i(v)}{G_i(v)} = \frac{\theta_1(v)\theta_i(v)}{\theta_1(v+c)\theta_i(v-c)},$$

because

$$\frac{\theta_1(v+1)}{\theta_1(v+c+1)} = \frac{\theta_1(v)}{\theta_1(v+c)}, \quad \frac{\theta_i(v+1)}{\theta_i(v-c+1)} = \frac{\theta_i(v)}{\theta_i(v-c)} \qquad (i = 2, 3, \text{ or } 4)$$

and similarly for the period τ. Consequently,

$$C(v) = \frac{\theta_1(v)\theta_i(v)}{\theta_1(-c)\theta_i(-c)\theta_1(v+c)\theta_i(v-c)} \quad \text{and}$$

$$D(v) = \frac{\theta_j(v)\theta_k(v)}{\theta_j(-c)\theta_k(-c)\theta_1(v+c)\theta_i(v-c)}$$

are both elliptic functions, with the same periods 1 and τ. At $v = -c$, both have a pole of first order, due to the zero of $\theta_1(v + c)$ in the denominator. The corresponding residues are easily computed, and both are equal to $1/\theta'_1\theta_i(-2c)$. The functions $C(v)$ and $D(v)$ have, of course, also another pole, but this also is common and of first order, namely the zero of $\theta_i(v - c)$. This shows that the difference $C(v) - D(v)$ is an elliptic function with at most one pole of first order in its parallelogram of periods; hence, it reduces to a constant. (This proves, incidentally, that $C(v)$ and $D(v)$ have the same residue also at the zero of $\theta_i(v - c)$.)

By remembering that $\theta_1(v)$ is odd and that all other $\theta_i(v)$ are even, we obtain

$$\frac{\theta_1(v)\theta_i(v)}{\theta_1(c)\theta_i(c)} + \frac{\theta_j(v)\theta_k(v)}{\theta_j(c)\theta_k(c)} = C\theta_1(v + c)\theta_i(v - c). \tag{8.14}$$

To find the value of the constant C, set $v = 0$ and use $\theta_1(0) = 0$ to obtain $\theta_j\theta_k/\theta_j(c)\theta_k(c) = C\theta_1(c)\theta_i(c)$, whence $C = \theta_j\theta_k/\theta_1(c)\theta_2(c)\theta_3(c)\theta_4(c)$. If we substitute this value of C in (8.14), clear of fractions and, to improve the symmetry, rename the arguments, we obtain (8.13), as claimed. □

§9. Problems

1. Prove that, if $f(z + \omega_1) = f(z + \omega_2) = f(z)$ holds for all $z \in \mathbb{C}$ and $\omega_2/\omega_1 = a/b$, $a, b \in \mathbb{Z}$, $(a, b) = 1$, then $f(z)$ is simply periodic. Find its smallest period.

2. Prove that, if $f(z)$ is a continuous function with $f(z + \omega_1) = f(z + \omega_2) = f(z)$ for all $z \in \mathbb{C}$ and ω_2/ω_1 is real and irrational, then $f(z)$ reduces to a constant.

3. Prove that $S(z) = \sum_{\omega \in \Omega, \omega \neq 0} \{1/(z - \omega)^2 - 1/\omega^2\}$ converges and that the sum $\wp(z) = z^{-2} + S(z)$ is a nonconstant, even, doubly periodic function. (Hint: Consider first $\wp'(z) = -2\sum_{\omega \in \Omega}(z - \omega)^{-3}$.

4. Use Theorem 1 in order to show that $\theta_3(v/\tau \mid -1/\tau) = \sqrt{\tau/i}\, e^{\pi i v^2/\tau}\theta_3(v \mid \tau)$.

5. Complete the proof of Theorem 2 by verifying that, for $\sigma > 0$, $\alpha > 0$, and $M > 0$, $\lim_{N\to\infty} \sum_{n=N+1}^{N+M} (z + ni)^{-s} e^{2\pi i n\alpha} = 0$. (Hint: Set $s_n = \sum_{r=0}^{n} e^{2\pi i r\alpha}$; then $e^{2\pi i n\alpha} = s_n - s_{n-1}$; use partial summation and observe that the sum of the geometric series s_n leads to $|s_n| < \csc \pi\alpha$, etc. If you have difficulties and are still curious, see the complete proof in [221].)

6. Verify the statements concerning the lattices L_i.

7. Compute the constants c_{43} and c_{23}.

8.

   (a) Determine the periods, the poles, and their residues for the function $\psi_2(v \mid \tau) = \Phi(v \mid 2\tau) - \Phi(v + \tau \mid 2\tau)$. Observe that these coincide with the periods, poles, and residues of $f_2(v \mid \tau)$, and evaluate the constant $f_2(v \mid \tau) - \psi_2(v \mid \tau)$.

(b) Consider the same problem with

$$\psi_3(v \mid \tau) = \frac{1}{2}\left\{\Phi\left(\frac{v}{2}\middle|\frac{1+\tau}{2}\right) - \Phi\left(\frac{v+1}{2}\middle|\frac{1+\tau}{2}\right)\right\} \quad \text{and} \quad f_3(v \mid \tau).$$

9. Compute the residues of the poles of $C(v)$ and $D(v)$ at the zero of $\theta_2(v - c)$. (Assume $(i, j, k) = (2, 3, 4.)$)

Chapter 9

Representations of Integers as Sums of an Even Number of Squares

§1. A Sketch of the Method

In the previous chapter, we have presented those fundamental properties of the theta functions which will be used in the present one to obtain the number $r_k(n)$ of representations of a natural number n as a sum of an even number k of squares.

Before we proceed, it appears worthwhile to sketch what we intend to do; otherwise, it may seem to the reader that we perform a series of unmotivated manipulations that, somewhat miraculously, yield the desired result. Here then is our program.

In §7 of Chapter 8 we have studied the elliptic functions $f_j(v \mid \tau)$, $j = 2, 3, 4$. Now we shall see that these, as well as their squares, can be expanded into *Lambert series*, of which the following are typical:

$$f_4(v \mid \tau) = \pi \csc \pi v + 4\pi \sum_{m=1}^{\infty} \frac{q^{2m-1}}{1 - q^{2m-1}} \sin (2m - 1)\pi v,$$

$$f_4^2(v \mid \tau) = \pi^2 \csc^2 \pi v + C_4(q) - 8\pi^2 \sum_{m=1}^{\infty} \frac{mq^{2m}}{1 - q^{2m}} \cos 2\pi m v, \tag{9.1}$$

$$f_4^2\left(v + \frac{\tau}{2} \middle| \tau\right) = -8\pi^2 \sum_{m=1}^{\infty} \frac{mq^m}{1 - q^{2m}} \cos 2\pi m v + C_4(q),$$

with $C_4(q) = 8\pi^2 \sum_{m=1}^{\infty} q^{2m-1}/(1 - q^{2m-1})^2$.

For a fixed value of v, (9.1) and similar formulae represent ordinary Lambert series in q. These Lambert series can be expanded into power series in q. On the other hand, by differentiating the addition formula (8.13) of Theorem 8.4, we obtain identities that connect the elliptic functions $f_i(v)$, for specific values of v, with polynomials in $\theta_2, \theta_3, \theta_4$. These identities permit us to solve for any *even* power of any $\theta_i (i = 2, 3, 4)$, and we shall do so for the even powers of θ_3. In this way, we shall succeed in expressing $\theta_3^2, \theta_3^4, \theta_3^6$, and θ_3^8 as polynomials—in fact, as linear forms—in $f_2(v)$, $f_4(v)$, $f_2^2(v)$, $f_4^2(v)$, and some of their derivatives, all computed at $v = \frac{1}{2}$, or at $v = \tau/2$. If we now replace in these identities the elliptic functions by their Lambert series and expand the latter into power series, we shall have obtained the desired representation of θ_3^k, for k even, by power series in q. The coefficients of q^n turn out to be divisor functions—and, as we know, they will be precisely the wanted $r_k(n)$.

For even $k \geqslant 10$, this program does not succeed completely. Indeed, in those cases θ_3^k contains in its representation, besides the mentioned linear form in the elliptic functions, additional terms in θ_1', θ_2, θ_3, θ_4 that cannot be eliminated. Nevertheless, the approach is still useful, because if the additional terms are expanded into power series, their coefficients of q^n turn out to be of a lower order than the one obtained from the linear form and may be considered as error terms in the representation of $r_k(n)$ as a sum of divisor functions.

§2. Lambert Series

From (8.6) it follows that

$$\Phi\left(v\,\middle|\,\frac{\tau}{2}\right) = \frac{d}{dv}\log\theta_1\left(v\,\middle|\,\frac{\tau}{2}\right)$$

$$= \pi\cot\pi v + 2\pi i \sum_{m=1}^{\infty}\left\{\frac{-q^m e^{2\pi i v}}{1 - q^m e^{2\pi i v}} + \frac{q^m e^{-2\pi i v}}{1 - q^m e^{-2\pi i v}}\right\}.$$

From the convergence of the product in (8.6), the convergence for $|q| < 1$ of the sum of the terms in braces follows, but not necessarily that of the two infinite sums taken separately. In any case, neither fraction can be expanded into a power series, unless the stronger conditions $|qe^{2\pi i v}| < 1$ and $|qe^{-2\pi i v}| < 1$ hold. By $q = e^{\pi i v}$, these conditions mean $|e^{2\pi i(\tau/2 \pm v)}| < 1$, i.e., $\mathrm{Im}(\tau/2 \pm v) > 0$, or

$$|\mathrm{Im}\, v| < \mathrm{Im}\frac{\tau}{2}. \tag{9.2}$$

For v inside this horizontal strip of width $\mathrm{Im}\,\tau$ along the real axis, one has $-q^m e^{2\pi i v}/(1 - q^m e^{2\pi i v}) = -\sum_{k=1}^{\infty} q^{km} e^{2\pi i k v}$, $(q^m e^{-2\pi i v})/(1 - q^m e^{-2\pi i v}) = \sum_{k=1}^{\infty} q^{km} e^{-2\pi i k v}$, so that

$$\Phi\left(v\,\middle|\,\frac{\tau}{2}\right) = \pi\cot\pi v - 2\pi i \sum_{m=1}^{\infty}\sum_{k=1}^{\infty}\left(q^{km}e^{2\pi i k v} - q^{km}e^{-2\pi i k v}\right)$$

$$= \pi\cot\pi v + 4\pi \sum_{m=1}^{\infty}\sum_{k=1}^{\infty} q^{km}\sin 2\pi k v$$

$$= \pi\cot\pi v + 4\pi \sum_{k=1}^{\infty}\frac{q^k}{1 - q^k}\sin 2\pi k v.$$

We now replace v successively by $v/2$ and by $(v+1)/2$ and obtain

$$\Phi\left(\frac{v}{2}\,\middle|\,\frac{\tau}{2}\right) = \pi\cot\frac{\pi v}{2} + 4\pi\sum_{k=1}^{\infty}\frac{q^k}{1-q^k}\sin\pi k v$$

and

$$\Phi\left(\frac{v+1}{2}\,\middle|\,\frac{\tau}{2}\right) = -\pi\tan\frac{\pi v}{2} + 4\pi\sum_{k=1}^{\infty}(-1)^k\frac{q^k}{1-q^k}\sin\pi k v,$$

so that, by (8.11) and (8.12),

$$f_4(v\,|\,\tau) = \psi(v\,|\,\tau) = \frac{\pi}{2}\left\{\cot\frac{\pi v}{2} + \tan\frac{\pi v}{2}\right\} + 2\pi\sum_{k=1}^{\infty}\frac{q^k(1-(-1)^k)}{1-q^k}\sin\pi k v,$$

and finally,

$$f_4(v\,|\,\tau) = \frac{\pi}{\sin\pi v} + 4\pi\sum_{k=1}^{\infty}\frac{q^{2k-1}}{1-q^{2k-1}}\sin(2k-1)\pi v,$$

thus proving the first identity in (9.1).

If, instead of starting from $\psi(v\,|\,\tau)$, we start from $\Phi(v\,|\,2\tau) - \Phi(v+\tau\,|\,2\tau)$, we may show in the same way that $f_2(v\,|\,\tau) - \Phi(v\,|\,2\tau) + \Phi(v+\tau\,|\,2\tau) = C_2(q)$ is a constant in v. Moreover, $f_2(v\,|\,\tau)$ and $\Phi(v\,|\,2\tau) = \theta_1'(v\,|\,2\tau)/\theta_1(v\,|\,2\tau)$ are odd functions, holomorphic for $v = 0$; hence, they vanish for $v = 0$. Next, if we set $\tau' = 2\tau$ and use (8.5),

$$\Phi(v+\tau\,|\,2\tau) = \frac{\theta_1'(v+\tau\,|\,2\tau)}{\theta_1(v+\tau\,|\,2\tau)} = \frac{\theta_1'(v+\tau'/2\,|\,\tau')}{\theta_1(v+\tau'/2\,|\,\tau')} = \frac{\theta_4'(v\,|\,\tau')}{\theta_4(v\,|\,\tau')} = \frac{\theta_4'(v\,|\,2\tau)}{\theta_4(v\,|\,2\tau)},$$

and this is also an odd function of v and vanishes at $v = 0$ (a fact obvious also from (8.2)), so that $C_2(q) = 0$ and $f_2(v\,|\,\tau) = \Phi(v\,|\,2\tau) - \Phi(v+\tau\,|\,2\tau)$. If we replace here $\Phi(v\,|\,2\tau)$ and $\Phi(v+\tau\,|\,2\tau)$ by their values $(d/dv)\log\theta_1(v\,|\,2\tau)$ and $(d/dv)\log\theta_4(v\,|\,2\tau)$, respectively, and use (8.6), a simple computation yields

$$f_2(v\,|\,\tau) = \pi\cot\pi v - 4\pi\sum_{n=1}^{\infty}\frac{q^{2n}}{1+q^{2n}}\sin 2\pi n v = \pi\cot\pi v - 4\pi S(v\,|\,\tau),$$

where we have put

$$S(v\,|\,\tau) = \sum_{n=1}^{\infty}\frac{q^n}{1+q^{2n}}\sin 2\pi n v.$$

It is somewhat less simple to compute the value of $f_2(v+\tau/2\,|\,\tau)$, which will be needed later. We have

$$\cot\left(\pi v + \frac{\pi\tau}{2}\right) = i\frac{e^{i\pi v}e^{i\tau\pi/2} + e^{-i\pi v}e^{-i\pi\tau/2}}{e^{i\pi v}e^{i\tau\pi/2} - e^{-i\pi v}e^{-i\pi\tau/2}} = i\frac{e^{2\pi iv}q + 1}{e^{2\pi iv}q - 1}.$$

Next,

$$\sin 2\pi n\left(v + \frac{\tau}{2}\right) = \sin 2\pi nv \cos \pi n\tau + \cos 2\pi nv \sin \pi\tau n$$

$$= \frac{1}{2}(q^n + q^{-n}) \sin 2\pi nv - \frac{i}{2}(q^n - q^{-n}) \cos 2\pi nv.$$

We substitute this in the sum $S(v + \tau/2)$, which becomes

$$\frac{1}{2}\sum_{n=1}^{\infty} \frac{q^{2n}}{1 + q^{2n}}\{q^n(\sin 2\pi nv - i\cos 2\pi nv) + q^{-n}(\sin 2\pi nv + i\cos 2\pi nv)\}$$

$$= -\frac{i}{2}\sum_{n=1}^{\infty} \frac{q^{3n}e^{2\pi inv}}{1 + q^{2n}} + \frac{i}{2}\sum_{n=1}^{\infty} \frac{q^n e^{-2\pi inv}}{1 + q^{2n}}$$

$$= -\frac{i}{2}\sum_{n=1}^{\infty} \frac{q^{3n}e^{2\pi inv}}{1 + q^{2n}} + \frac{i}{2}\sum_{n=1}^{\infty} \frac{q^n}{1 + q^{2n}} \cdot 2\cos 2\pi nv - \frac{i}{2}\sum_{n=1}^{\infty} \frac{q^n}{1 + q^{2n}} e^{2\pi inv}$$

$$= i\sum_{n=1}^{\infty} \frac{q^n}{1 + q^{2n}} \cos 2\pi nv - \frac{i}{2}\sum_{n=1}^{\infty} \frac{q^n(q^{2n} + 1)}{1 + q^{2n}} e^{2\pi inv}.$$

The last sum equals $\sum_{n=1}^{\infty} (qe^{2\pi iv})^n = qe^{2\pi iv}/(1 - qe^{2\pi iv})$; hence,

$$f_2\left(v + \frac{\tau}{2}\middle|\tau\right) = \pi i\frac{e^{2\pi iv}q + 1}{e^{2\pi iv}q - 1} - 4\pi i\sum_{n=1}^{\infty} \frac{q^n}{1 + q^{2n}} \cos 2\pi nv + 2\pi i\frac{qe^{2\pi iv}}{1 - qe^{2\pi iv}}.$$

Adding the first and last terms, we obtain

$$\pi i\frac{2qe^{2\pi iv} - (qe^{2\pi iv} + 1)}{1 - qe^{2\pi iv}} = -\pi i,$$

so that, finally,

$$f_2\left(v + \frac{\tau}{2}\middle|\tau\right) = -\pi i - 4\pi i\sum_{n=1}^{\infty} \frac{q^n}{1 + q^{2n}} \cos 2\pi nv. \tag{9.3}$$

Similar expansions hold for the other f_j-functions, as well as for the squares of the $f_j(v \mid \tau)$. In fact, as already mentioned, all $f_j^2(v \mid \tau)$ have the same principal part v^{-2} at the origin. Hence, not only do we have $f_j^2(v \mid \tau) - f_k^2(v \mid \tau) = c_{jk}$, but furthermore $f_j^2(v \mid \tau) - \wp(v; 1, \tau)$ are all elliptic functions without singularities and consequently reduce to constants.

We have already computed the c_{jk} in Chapter 8. As for $f_4(v \mid \tau) - \wp(v; 1, \tau)$, if we set $v = \tau/2$, we obtain the constant $0 - \wp(\tau/2; 1, \tau)$. It follows that

$$\begin{aligned}
f_4^2(v \mid \tau) &= \wp(v; 1, \tau) - \wp(\tfrac{\tau}{2}; 1, \tau) \\
&= \sum_{m_1, m_2 = -\infty}^{\infty} \left\{ \frac{1}{(v + m_1 + m_2\tau)^2} - \frac{1}{(\frac{\tau}{2} + m_1 + m_2\tau)^2} \right\} \\
&= \sum_{m_1 = -\infty}^{\infty} \frac{1}{(v + m_1)^2} + \sum_{m_1 = -\infty}^{\infty} \sum_{m_2 > 0}^{\infty} \left\{ \frac{1}{(v + m_1 + m_2\tau)^2} \right. \\
&\qquad \left. - \frac{1}{(m_1 + (m_2 - \frac{1}{2})\tau)^2} \right\} \\
&\qquad + \sum_{m_1 = -\infty}^{\infty} \sum_{m_2 < 0}^{\infty} \left\{ \frac{1}{(v + m_1 + m_2\tau)^2} - \frac{1}{(m_1 + (m_2 + \frac{1}{2})\tau)^2} \right\} \\
&= \frac{\pi^2}{\sin^2 \pi v} + \sum_{m_2 > 0}^{\infty} \left\{ \sum_{m_1 = -\infty}^{\infty} \frac{1}{(v + m_1 + m_2\tau)^2} \right. \\
&\qquad \left. - \sum_{m_1 = -\infty}^{\infty} \frac{1}{(m_1 + (m_2 - \frac{1}{2})\tau)^2} \right\} \\
&\qquad + \sum_{m_2 > 0}^{\infty} \left\{ \sum_{m_1 = -\infty}^{\infty} \frac{1}{(v + m_1 - m_2\tau)^2} - \sum_{m_1 = -\infty}^{\infty} \frac{1}{(m_1 - (m_2 - \frac{1}{2})\tau)^2} \right\} \\
&= \frac{\pi^2}{\sin^2 \pi v} + \sum_{m_2 > 0}^{\infty} \left\{ \sum_{m_1 = -\infty}^{\infty} \frac{-1}{((v + m_2\tau)/i + m_1 i)^2} \right. \\
&\qquad \left. + \sum_{m_1 = -\infty}^{\infty} \frac{1}{((m_2 - \frac{1}{2})\tau/i + m_1 i)^2} \right\} \\
&\qquad + \sum_{m_2 > 0}^{\infty} \left\{ \sum_{m_1 = -\infty}^{\infty} \frac{-1}{((-v + m_2\tau)/i + m_1 i)^2} \right. \\
&\qquad \left. + \sum_{m_1 = -\infty}^{\infty} \frac{1}{((m_2 - \frac{1}{2})\tau/i + m_1 i)^2} \right\}.
\end{aligned}$$

We make again the assumption that (9.2) holds; then the conditions for the validity of Theorem 8.2 are satisfied, and when we transform these four sums, we obtain

$$f_4^2(v\,|\,\tau) = \frac{\pi^2}{\sin^2 \pi v} + \frac{(2\pi)^2}{\Gamma(2)} \sum_{m_2>0}^{\infty} \left\{ -\sum_{m=1}^{\infty} m e^{2\pi i(v+m_2\tau)m} + \sum_{m=1}^{\infty} m e^{2\pi i(m_2-1/2)\tau m} \right\}$$

$$+ \frac{(2\pi)^2}{\Gamma(2)} \sum_{m_2>0}^{\infty} \left\{ -\sum_{m=1}^{\infty} m e^{2\pi i(-v+m_2\tau)m} + \sum_{m=1}^{\infty} m e^{2\pi i(m_2-1/2)\tau m} \right\}.$$

Each of the four sums converges separately (even absolutely); hence, we may group them differently. The first and third sum combine, including the outside factor $4\pi^2$, into $-8\pi^2 \sum_{m_2=1}^{\infty} \sum_{m=1}^{\infty} m q^{2m_2 m} \cos 2\pi m v$, while the second and fourth sums are identical and, with $e^{\pi i \tau}$ replaced by q, add up to $8\pi^2 \sum_{m_2=1}^{\infty} \sum_{m=1}^{\infty} m q^{(2m_2-1)m}$. On summing over m, this sum becomes $C_4(q) = 8\pi^2 \sum_{m=1}^{\infty} q^{2m-1}/(1-q^{2m-1})^2$, while if we first sum over m_2, it becomes $C_4(q) = 8\pi^2 \sum_{m=1}^{\infty} m q^{-m} \sum_{m_2=1}^{\infty} q^{2mm_2} = 8\pi^2 \sum_{m=1}^{\infty} m q^m/(1-q^{2m})$. Finally, summing the other double sum with respect to m_2, we obtain

$$f_4^2(v\,|\,\tau) = \frac{\pi^2}{\sin^2 \pi v} - 8\pi^2 \sum_{m=1}^{\infty} \frac{m q^{2m}}{1-q^{2m}} \cos 2\pi m v + C_4(q),$$

the second identity in (9.1). If we replace here v by $v + \tau/2$, computations similar to those used to prove (9.3) lead to

$$f_4^2\left(v + \frac{\tau}{2}\,\middle|\,\tau\right) = -8\pi^2 \sum_{m=1}^{\infty} \frac{m q^m}{1-q^{2m}} \cos 2\pi m v + C_4(q),$$

the last identity in (9.1).

From (9.1) one can also compute $f_j^2(v\,|\,\tau)$ and $f_j^2(v+\tau/2\,|\,\tau)$, for $j = 2$ and 3, by estimating for some numerical value the constant differences $f_4^2(v\,|\tau) - f_j^2(v\,|\,\tau)$ or $f_4^2(v+\tau/2\,|\,\tau) - f_j^2(v+\tau/2\,|\,\tau)$.

§3. The Computation of the Powers θ_3^{2k}

By differentiating (8.13) with respect to v and then setting $v = 0$, we obtain

$$\theta_j\theta_k(\theta_1'(u)\theta_i(u) - \theta_1(u)\theta_i'(u)) = \theta_1'\theta_i\theta_j(u)\theta_k(u), \tag{9.4}$$

because $\theta_j'(u)$ are odd functions, so that $\theta_2' = \theta_3' = \theta_4' = 0$. If we multiply both sides of (9.4) by $-\theta_1'/\theta_2\theta_3\theta_4\theta_1^2(u)$, we obtain

$$\frac{\theta_1'}{\theta_i}\frac{d}{du}\frac{\theta_i(u)}{\theta_1(u)} = -\frac{(\theta_1')^2}{\theta_j\theta_k}\frac{\theta_j(u)\theta_k(u)}{\theta_1^2(u)}.$$

Taking (8.7) into account, we recognize this as

$$f_i'(u) = -f_j(u)f_k(u). \tag{9.5}$$

By simple iterations, we obtain the successive derivatives. For example, from $f_2'(u) = -f_3(u)f_4(u)$ we obtain $f_2''(u) = -(f_3'(u)f_4(u) + f_3(u)f_4'(u)) = f_2(u)(f_4^2(u) + f_3^2(u))$, $f_2'''(u) = -f_3(u)f_4(u)(4f_2^2(u) + f_3^2(u) + f_4^2(u))$, etc. The corresponding derivatives for $f_3(u)$ and $f_4(u)$ may be obtained from those of $f_2(u)$ by circular permutations of the subscripts.

The identity (9.5) also permits us also to compute the derivatives of $f_i^2(u)$. Indeed, $(d/du)f_i^2(u) = 2f_i(u)f_i'(u) = -2f_2(u)f_3(u)f_4(u)$, the same for any of the subscripts $i = 2$, 3, or 4. The result was to be expected, because we already observed that $f_i^2(u) - f_j^2(u) = c_{ij}$ are constants, being elliptic functions without singularities. By recalling also that $f_i^2(u) - \wp(u; 1, \tau)$ are constants, we may record the result as

$$\{f_i^2(u)\}' = -2f_2(u)f_3(u)f_4(u) = \wp'(u; 1, \tau). \tag{9.6}$$

By successive differentiations, we find, by use of (9.5) that, e.g.,

$$\frac{d^2}{du^2} f_i^2(u) = -2\{f_2(u)f_3(u)f_4'(u) + f_2(u)f_3'(u)f_4(u) + f_2'(u)f_3(u)f_4(u)\}$$

$$= 2\{(f_2(u)f_3(u))^2 + (f_3(u)f_4(u))^2 + (f_4(u)f_2(u))^2\}$$

and after a short computation,

$$\frac{d^4}{du^4} f_i^2(u) = 8\{(f_2(u)f_3(u))^2 + (f_3(u)f_4(u))^2 + (f_4(u)f_2(u))^2\}$$

$$\times (f_2^2(u) + f_3^2(u) + f_4^2(u)) + 48(f_2(u)f_3(u)f_4(u))^2.$$

We can compute the derivatives for the numerical values $u = \frac{1}{2}$, $u = \tau/2$, and $u = (1 + \tau)/2$ by substituting these values in the right hand members and using (8.9). In this way, we obtain

$$\begin{aligned}
f_2''(\tau/2) &= i\pi^3\theta_3^2\theta_2^4, & \{f_i^2(u)\}''_{u=1/2} &= 2\pi^4\theta_3^4\theta_4^4,\\
f_2^{(4)}(\tau/2) &= -i\pi^5\theta_3^2\theta_2^4(4\theta_3^4 + \theta_2^4), & \{f_i^2(u)\}^{(4)}_{u=1/2} &= 8\pi^6\theta_3^4\theta_4^4(\theta_3^4 + \theta_4^4),\\
f_4''(\tfrac{1}{2}) &= \pi^3\theta_3^2\theta_4^4, & \{f_i^2(u)\}''_{u=\tau/2} &= 2\pi^4\theta_2^4\theta_3^4,\\
f_4^{(4)}(\tfrac{1}{2}) &= \pi^5\theta_3^2\theta_4^4(4\theta_3^4 + \theta_4^4), & \{f_i^2(u)\}^{(4)}_{u=\tau/2} &= -8\pi^6\theta_2^4\theta_3^4(\theta_3^4 + \theta_2^4).
\end{aligned} \tag{9.7}$$

For example, to prove the first, we set $u = \tau/2$ in previous representation of $f_2''(u)$ and use (8.9); the result is $f_2(u)(f_4^2(u) + f_3^2(u)) = -i\pi\theta_3^2(0 + (-i\pi\theta_2^2)^2) = i\pi^3\theta_3^2\theta_2^4$. From (8.9), we also record that $\pi\theta_3^2 = f_4(\tfrac{1}{2})$ and, by squaring, $\pi^2\theta_3^4 = f_4^2(\tfrac{1}{2})$. Next, from (9.7), $f_4''(\tfrac{1}{2}) - if_2''(\tau/2) = \pi^3\theta_3^2\theta_4^4 + \pi^3\theta_3^2\theta_2^4 = \pi^3\theta_3^2(\theta_4^4 + \theta_2^4)$, so that, by (8.10), $\pi^3\theta_3^6 = f_4''(\tfrac{1}{2}) - if_2''(\tau/2)$.

Proceeding in the same way, we obtain the following set of formulae, which could be continued indefinitely:

$$\begin{aligned}
\pi\theta_3^2 &= f_4(\tfrac{1}{2}),\\
\pi^2\theta_3^4 &= f_4^2(\tfrac{1}{2}),\\
\pi^3\theta_3^6 &= f_4''(\tfrac{1}{2}) - if_2''(\tau/2),\\
2\pi^4\theta_3^8 &= \{f_2^2(u)\}''_{u=1/2} + \{f_2^2(u)\}''_{u=\tau/2},\\
5\pi^5\theta_3^{10} &= f_4^{(4)}(\tfrac{1}{2}) + if_2^{(4)}(\tau/2) + 2\pi^5\theta_3^2\theta_2^4\theta_4^4,\\
16\pi^6\theta_3^{12} &= \{f_2^2(u)\}^{(4)}_{u=1/2} - \{f_2^2(u)\}^{(4)}_{u=\tau/2} + 16\pi^2(\theta_1')^4.
\end{aligned} \tag{9.8}$$

§4. Representation of Powers of θ_3 by Lambert Series

We recall that, by (9.1) and (9.3),

$$f_4(v \mid \tau) = \frac{\pi}{\sin \pi v} + 4\pi \sum_{m=1}^{\infty} \frac{q^{2m-1}}{1 - q^{2m-1}} \sin(2m-1)\pi v$$

and

$$f_2\left(v + \frac{\tau}{2} \,\middle|\, \tau\right) = \pi i - 4\pi i \sum_{m=1}^{\infty} \frac{q^m}{1 + q^{2m}} \cos 2m\pi v.$$

Hence, by differentiation,

$$\begin{aligned}
f_4''(v \mid \tau) = {} & \pi^3 \csc \pi v\,(2\cot^2 \pi v + 1)\\
& - 4\pi^3 \sum_{m=1}^{\infty} \frac{(2m-1)^2 q^{2m-1}}{1 - q^{2m-1}} \sin(2m-1)\pi v,
\end{aligned}$$

$$\begin{aligned}
f_4^{(4)}(v \mid \tau) = {} & \pi^5 \csc \pi v\,(24\cot^4 \pi v + 28\cot^2 \pi v + 5)\\
& + 4\pi^5 \sum_{m=1}^{\infty} \frac{(2m-1)^4 q^{2m-1}}{1 - q^{2m-1}} \sin(2m-1)\pi v,
\end{aligned}$$

$$f_2''\left(v + \frac{\tau}{2}\middle|\tau\right) = 16i\pi^3 \sum_{m=1}^{\infty} \frac{m^2 q^m}{1 + q^{2m}} \cos 2m\pi v,$$

$$f_2^{(4)}\left(v + \frac{\tau}{2}\middle|\tau\right) = -64i\pi^5 \sum_{m=1}^{\infty} \frac{m^4 q^m}{1 + q^{2m}} \cos 2m\pi v.$$

Next, by (9.1), and recalling that $f_i^2(v) - f_j^2(v)$ is a constant, we have for all three subscripts $j = 2$, 3, or 4,

$$\frac{d^2}{dv^2} f_j^2(v\,|\,\tau) = \frac{\pi^4}{\sin^2 \pi v}(6\cot^2 \pi v + 2) + 32\pi^4 \sum_{m=1}^{\infty} \frac{m^3 q^{2m}}{1 - q^{2m}} \cos 2\pi m v$$

$$\frac{d^2}{dv^4} f_j^2(v\,|\,\tau) = \frac{\pi^6}{\sin^2 \pi v}(120\cot^4 \pi v + 120\cot^2 \pi v + 16)$$

$$- 128\pi^6 \sum_{m=1}^{\infty} \frac{m^5 q^{2m}}{1 - q^{2m}} \cos 2\pi m v$$

$$\frac{d^2}{dv^2} f_j^2\left(v + \frac{\tau}{2}\middle|\tau\right) = 32\pi^4 \sum_{m=1}^{\infty} \frac{m^3 q^m}{1 - q^{2m}} \cos 2\pi m v$$

$$\frac{d^4}{dv^4} f_j^2\left(v + \frac{\tau}{2}\middle|\tau\right) = -128\pi^6 \sum_{m=1}^{\infty} \frac{m^5 q^m}{1 - q^{2m}} \cos 2\pi m v.$$

By substituting for v either 0 or $\frac{1}{2}$, we obtain from (9.1) and from those derivatives the following Lambert series in q:

$$f_4(\tfrac{1}{2}) = \pi + 4\pi \sum_{m=1}^{\infty} (-1)^{m-1} \frac{q^{2m-1}}{1 - q^{2m-1}},$$

$$f_4^2(\tfrac{1}{2}) = \pi^2 + 8\pi^2 \sum_{m=1}^{\infty} \frac{m q^m}{1 - q^{2m}} - 8\pi^2 \sum_{m=1}^{\infty} (-1)^m \frac{m q^{2m}}{1 - q^{2m}},$$

$$f_4''(\tfrac{1}{2}) = \pi^3 + 4\pi^3 \sum_{m=1}^{\infty} (-1)^m \frac{(2m-1)^2 q^{2m-1}}{1 - q^{2m-1}},$$

$$f_4^{(4)}(\tfrac{1}{2}) = 5\pi^5 + 4\pi^5 \sum_{m=1}^{\infty} (-1)^{m-1} \frac{(2m-1)^4 q^{2m-1}}{1 - q^{2m-1}},$$

$$f_2''(\tau/2) = 16i\pi^3 \sum_{m=1}^{\infty} \frac{m^2 q^m}{1 + q^{2m}}, \tag{9.9}$$

$$f_2^{(4)}(\tau/2) = -64i\pi^5 \sum_{m=1}^{\infty} \frac{m^4 q^m}{1+q^{2m}},$$

$$\left\{\frac{d^2}{dv^2} f_j^2(v\,|\,\tau)\right\}_{v=1/2} = 2\pi^4 + 32\pi^4 \sum_{m=1}^{\infty} (-1)^m \frac{m^3 q^{2m}}{1-q^{2m}},$$

$$\left\{\frac{d^2}{dv^2} f_j^2(v\,|\,\tau)\right\}_{v=\tau/2} = 32\pi^4 \sum_{m=1}^{\infty} \frac{m^3 q^{2m}}{1-q^{2m}},$$

$$\left\{\frac{d^4}{dv^4} f_j^2(v\,|\,\tau)\right\}_{v=1/2} = 16\pi^6 - 128\pi^6 \sum_{m=1}^{\infty} (-1)^m \frac{m^5 q^{2m}}{1-q^{2m}},$$

$$\left\{\frac{d^4}{dv^4} f_j^2(v\,|\,\tau)\right\}_{v=\tau/2} = -128\pi^6 \sum_{m=1}^{\infty} \frac{m^5 q^m}{1-q^{2m}}.$$

We now substitute these values in (9.8), divide by the proper power of π, and obtain

$$\begin{aligned}
\theta_3^2 &= 1 + 4\sum_{m=1}^{\infty} (-1)^{m-1} \frac{q^{2m-1}}{1-q^{2m-1}},\\
\theta_3^4 &= 1 + 8\sum_{m=1}^{\infty} \frac{mq^m}{1-q^{2m}} - 8\sum_{m=1}^{\infty} (-1)^m \frac{mq^{2m}}{1-q^{2m}},\\
\theta_3^6 &= \pi^{-3}\left\{\pi^3 + 4\pi^3 \sum_{m=1}^{\infty} (-1)^m \frac{(2m-1)^2 q^{2m-1}}{1-q^{2m-1}} - i\left(16i\pi^3 \sum_{m=1}^{\infty} \frac{m^2 q^m}{1+q^{2m}}\right)\right\}\\
&= 1 + 4\sum_{m=1}^{\infty} (-1)^m \frac{(2m-1)^2 q^{2m-1}}{1-q^{2m-1}} + 16\sum_{m=1}^{\infty} \frac{m^2 q^m}{1+q^{2m}},\\
\theta_3^8 &= (2\pi^4)^{-1}\left\{2\pi^4 + 32\pi^4 \sum_{m=1}^{\infty} (-1)^m \frac{m^3 q^{2m}}{1-q^{2m}} + 32\pi^4 \sum_{m=1}^{\infty} \frac{m^3 q^m}{1-q^{2m}}\right\}\\
&= 1 + 16\sum_{m=1}^{\infty} (-1)^m \frac{m^3 q^{2m}}{1-q^{2m}} + 16\sum_{m=1}^{\infty} \frac{m^3 q^m}{1-q^{2m}},\\
\theta_3^{10} &= (5\pi^5)^{-1}\left\{5\pi^5 + 4\pi^5 \sum_{m=1}^{\infty} (-1)^{m-1} \frac{(2m-1)^4 q^{2m-1}}{1-q^{2m-1}}\right.\\
&\quad \left. + i\left(-64i\pi^5 \sum_{m=1}^{\infty} \frac{m^4 q^m}{1+q^{2m}}\right)\right\} + \frac{2}{5}\theta_3^2\theta_2^4\theta_4^4
\end{aligned} \tag{9.10}$$

$$= 1 + \frac{4}{5}\sum_{m=1}^{\infty}(-1)^{m-1}\frac{(2m-1)^4 q^{2m-1}}{1-q^{2m-1}} + \frac{64}{5}\sum_{m=1}^{\infty}\frac{m^4 q^m}{1+q^{2m}} + \frac{2}{5}\theta_3^2\theta_2^4\theta_4^4,$$

$$\theta_3^{12} = (16\pi^6)^{-1}\left\{16\pi^6 - 128\pi^6\sum_{m=1}^{\infty}(-1)^m\frac{m^5 q^{2m}}{1-q^{2m}} + 128\pi^6\sum_{m=1}^{\infty}\frac{m^5 q^m}{1-q^{2m}}\right\}$$

$$+\left(\frac{\theta_1'}{\pi}\right)^4$$

$$= 1 - 8\sum_{m=1}^{\infty}(-1)^m\frac{m^5 q^{2m}}{1-q^{2m}} + 8\sum_{m=1}^{\infty}\frac{m^5 q^m}{1-q^{2m}} + \left(\frac{\theta_1'}{\pi}\right)^4.$$

By proving the first two formulae (9.10), we pay an old debt. Indeed, by the first, we finally justify the last equality in the Jacobi formula quoted in §2.4, while the second of (9.10) is equivalent to Jacobi's formula (3.14) (see Problem 5), also accepted without proof in Chapter 3.

§5. Expansions of Lambert Series into Divisor Functions

In this section, we shall use the symbols $d(n)$, $d_j(n)$, $\sigma(n)$, $\sigma'(n)$, etc., defined in Chapters 2 and 3. All congruences under a summation sign without mention of a modulus are understood modulo 4.

In the case of two squares,

$$\sum_{m=1}^{\infty}(-1)^{m-1}\frac{q^{2m-1}}{1-q^{2m-1}} = \sum_{m=1}^{\infty}(-1)^{m-1}\sum_{r=1}^{\infty}q^{r(2m-1)} = \sum_{n=1}^{\infty}q^n\sum_{2m-1\mid n}(-1)^{m-1}$$

$$= \sum_{n=1}^{\infty}q^n\left\{\sum_{\substack{d\mid n\\ d\equiv 1}}1 - \sum_{\substack{d\mid n\\ d\equiv 3}}1\right\} = \sum_{n=1}^{\infty}q^n\{d_1(n) - d_3(n)\}$$

and

$$\theta_3^2 = 1 + \sum_{n=1}^{\infty}r_2(n)q^n = 1 + 4\sum_{n=1}^{\infty}q^n\{d_1(n) - d_3(n)\}. \tag{9.11}$$

In the case of four squares, we observe that

$$\frac{mq^m}{1-q^{2m}} - \frac{(-1)^m mq^{2m}}{1-q^{2m}} = \frac{mq^m(1+q^m) - (1+(-1)^m)mq^{2m}}{1-q^{2m}}$$

$$= \frac{mq^m}{1-q^m} - \frac{1+(-1)^m}{2}\frac{2mq^{2m}}{1-q^{2m}}.$$

Also, $\sum_{m=1}^{\infty} mq^m/(1-q^m) = \sum_{m=1}^{\infty} m \sum_{r=1}^{\infty} q^{rm} = \sum_{n=1}^{\infty} q^n \sum_{m|n} m = \sum_{n=1}^{\infty} \sigma(n)q^n$, and

$$\sum_{m=1}^{\infty} \frac{1+(-1)^m}{2} \cdot 2m \frac{q^{2m}}{1-q^{2m}} = \sum_{m=1}^{\infty} \frac{1+(-1)^m}{2} \cdot 2m \sum_{r=1}^{\infty} q^{2rm}$$

$$= \sum_{n=1}^{\infty} q^n \sum_{2m|n} \frac{1+(-1)^m}{2}(2m) = \sum_{n=1}^{\infty} q^n \sum_{\substack{d|n \\ d\equiv 0}} d,$$

so that the difference of the two sums is $\sum_{n=1}^{\infty} \sigma'(n)q^n$, and

$$\theta_3^4 = 1 + \sum_{n=1}^{\infty} r_4(n)q^n = 1 + 8 \sum_{n=1}^{\infty} \sigma'(n)q^n. \tag{9.12}$$

For k squares, $k \geqslant 6$, the results are somewhat more complicated and the divisor sums depend not only on the divisors themselves, but also on their cofactors. Specifically,

$$\sum_{m=1}^{\infty} (-1)^m \frac{(2m-1)^2 q^{2m-1}}{1-q^{2m-1}} = \sum_{m=1}^{\infty} (-1)^m (2m-1)^2 \sum_{r=1}^{\infty} q^{r(2m-1)}$$

$$= \sum_{n=1}^{\infty} q^n \sum_{2m-1|n} (-1)^m (2m-1)^2$$

$$= \sum_{n=1}^{\infty} q^n \left\{ \sum_{\substack{d|n \\ d\equiv 3}} d^2 - \sum_{\substack{d|n \\ d\equiv 1}} d^2 \right\}$$

and

$$\sum_{m=1}^{\infty} \frac{m^2 q^m}{1+q^{2m}} = \sum_{m=1}^{\infty} m^2 \sum_{s=0}^{\infty} (-1)^s q^{(2s+1)m} = \sum_{n=1}^{\infty} q^n \sum_{(2s+1)m=n} (-1)^s m^2$$

$$= \sum_{n=1}^{\infty} q^n \left\{ \sum_{\substack{d|n \\ r\equiv 1}} d^2 - \sum_{\substack{d|n \\ r\equiv 3}} d^2 \right\},$$

where $r = 2s+1$ is the cofactor of d in n, i.e., $rd = n$. It follows that

$$\begin{aligned} \theta_3^6 &= 1 + \sum_{n=1}^{\infty} r_6(n)q^n \\ &= 1 + \sum_{n=1}^{\infty} q^n \left\{ 4\left(\sum_{\substack{d|n \\ d\equiv 3}} d^2 - \sum_{\substack{d|n \\ d\equiv 1}} d^2 \right) + 16\left(\sum_{\substack{d|n \\ r\equiv 1}} d^2 - \sum_{\substack{d|n \\ r\equiv 3}} d^2 \right) \right\}. \end{aligned} \tag{9.13}$$

For $k = 8$, we proceed as for $k = 4$, and the operations are actually simpler. Indeed,

$$(-1)^m \frac{m^3 q^{2m}}{1-q^{2m}} + \frac{m^3 q^m}{1-q^{2m}} = \frac{m^3 q^m(1+(-1)^m q^m)}{1-q^{2m}} = \frac{m^3 q^m}{1-(-1)^m q^m};$$

hence, for m even, we obtain $m^3 \sum_{r=1}^{\infty} q^{rm} = \sum_{n=1}^{\infty} q^n \sum_{m|n} m^3$. For m odd, we obtain

$$m^3 \sum_{r=1}^{\infty} q^{rm}(-1)^{r-1} = \sum_{n=1}^{\infty} q^n \sum_{\substack{rm=n \\ m \text{ odd}}} (-1)^{r-1} m^3$$

$$= \sum_{n=1}^{\infty} q^n \left\{ \sum_{\substack{m \text{ odd} \\ r \text{ odd} \\ mr=n}} m^3 - \sum_{\substack{m \text{ odd} \\ r \text{ even} \\ mr=n}} m^3 \right\}.$$

In all three sums, we observe that m^3 is added when m is of the same parity with $n = mr$, and it is subtracted when m is of a parity opposite to that of n. Hence, we can write the coefficient of q^n as $\sum_{d|n}(-1)^{n+d}d^3$ and obtain

$$\theta_3^8 = 1 + \sum_{n=1}^{\infty} r_8(n) q^n = 1 + 16 \sum_{n=1}^{\infty} q^n \sum_{d|n} (-1)^{n+d} d^3. \tag{9.14}$$

For $k \geqslant 10$ a new difficulty appears. Indeed, in (9.10) we observe functions that are not expanded into Lambert series. Let us set

$$\theta_2^4 \theta_3^2 \theta_4^4 = 16 \sum_{n=1}^{\infty} a_n q^n; \tag{9.15}$$

then, although we don't know the a_n, it follows from Theorem 8.3 that the a_n are all integers. We shall ignore these a_n for a moment and proceed to expand the Lambert series into divisor functions:

$$\sum_{m=1}^{\infty} (-1)^{m-1} \frac{(2m-1)^4 q^{2m-1}}{1-q^{2m-1}} = \sum_{m=1}^{\infty} (-1)^{m-1}(2m-1)^4 \sum_{r=1}^{\infty} q^{r(2m-1)}$$

$$= \sum_{n=1}^{\infty} q^n \sum_{2m-1|n} (-1)^{m-1}(2m-1)^4$$

$$= \sum_{n=1}^{\infty} q^n \left\{ \sum_{\substack{d|n \\ d\equiv 1}} d^4 - \sum_{\substack{d|n \\ d\equiv 3}} d^4 \right\},$$

$$\sum_{m=1}^{\infty} \frac{m^4 q^m}{1+q^{2m}} = \sum_{m=1}^{\infty} m^4 \sum_{s=0}^{\infty} (-1)^s q^{(2s+1)m} = \sum_{n=1}^{\infty} q^n \sum_{(2s+1)m=n} (-1)^s m^4$$

$$= \sum_{n=1}^{\infty} q^n \left\{ \sum_{r\equiv 1} d^4 - \sum_{r\equiv 3} d^4 \right\},$$

where $r = 2s + 1$ and $rd = n$. By substituting these values, as well as (9.15) in (9.10), we obtain

$$\begin{aligned}\theta_3^{10} &= 1 + \sum_{n=1}^{\infty} r_{10}(n) q^n \\ &= 1 + \sum_{n=1}^{\infty} q^n \left\{ \frac{4}{5} \left(\sum_{\substack{d \mid n \\ d \equiv 1}} d^4 - \sum_{\substack{d \mid n \\ d \equiv 3}} d^4 \right) \right. \qquad (9.16) \\ &\quad \left. + \frac{64}{5} \left(\sum_{\substack{d \mid n \\ r \equiv 1}} d^4 - \sum_{\substack{d \mid n \\ r \equiv 3}} d^4 \right) + \frac{32}{5} a_n \right\}.\end{aligned}$$

For $k = 12$, we proceed, essentially, as for $k = 10$. Let

$$\left(\frac{\theta_1'}{\pi} \right)^4 = 16 \sum_{n=1}^{\infty} b_n q^n; \tag{9.17}$$

then, by Theorem 8.3, the b_n are rational integers. To evaluate the Lambert series, observe that

$$\frac{m^5 q^m}{1 - q^{2m}} - \frac{(-1)^m m^5 q^{2m}}{1 - q^{2m}} = \frac{m^5 q^m (1 - (-1)^m q^m)}{1 - q^{2m}} = \frac{m^5 q^m}{1 + (-1)^m q^m}.$$

For m odd, this becomes

$$\frac{m^5 q^m}{1 - q^{2m}} = m^5 \sum_{r=1}^{\infty} q^{rm} = \sum_{n=mr}^{\infty} q^n \sum_{\substack{r \mid n \\ m \text{ odd}}} m^5;$$

for m even, we have

$$\begin{aligned}\frac{m^5 q^m}{1 + q^m} &= m^5 \sum_{r=1}^{\infty} (-1)^{r-1} q^{rm} = \sum_{n=rm}^{\infty} q^n \sum_{m \text{ even}} (-1)^{r-1} m^5 \\ &= \sum_{n=rm} q^n \left\{ \sum_{\substack{r \text{ odd} \\ m \text{ even}}} m^5 - \sum_{\substack{r \text{ even} \\ m \text{ even}}} m^5 \right\}.\end{aligned}$$

Here, a divisor d leads to a summand $+d^5$, except when d and r (hence also $n = dr$) are even, when it leads to a summand $-d^5$.

One verifies that one way to select the correct sign is, to set it equal to $(-1)^{n+d+r-1}$. Consequently, in the last identity of (9.10), the sums over m yield $\sum_{n=1}^{\infty} (-1)^{n-1} q^n \sum_{dr=n} (-1)^{d+r} d^5$. If we substitute this sum and the expansion (9.17) in the last identity of (9.10), we obtain

$$\theta_3^{12} = 1 + \sum_{n=1}^{\infty} r_{12}(n)q^n = 1 + \sum_{n=1}^{\infty} q^n \left\{(-1)^{n-1}8 \sum_{dr=n} (-1)^{d+r}d^5 + 16b_n\right\}. \tag{9.18}$$

§6. The Values of the $r_k(n)$ for Even $k \leqslant 12$

For ease of reference, we state here the results that we obtain if we identify the coefficients of q^n in the identities (9.11), (9.12), (9.13), (9.14), (9.16), (9.17), and (9.18). In all cases, the formal identity requires us to set $r_k(0) = 1$, and in what follows, we assume $n \geqslant 1$:

$$\begin{aligned}
r_2(n) &= 4(d_1(n) - d_3(n)),\\
r_4(n) &= 8\sigma'(n),\\
r_6(n) &= 16\left\{\sum_{\substack{rd=n\\ r\equiv 1}} d^2 - \sum_{\substack{rd=n\\ r\equiv 3}} d^2 - \frac{1}{4}\left(\sum_{\substack{d\mid n\\ d\equiv 1}} d^2 - \sum_{\substack{d\mid n\\ d\equiv 3}} d^2\right)\right\},\\
r_8(n) &= (-1)^n 16 \sum_{d\mid n} (-1)^d d^3,\\
r_{10}(n) &= \frac{64}{5}\left\{\sum_{\substack{d\mid n\\ r\equiv 1}} d^4 - \sum_{\substack{d\mid n\\ r\equiv 3}} d^4 + \frac{1}{16}\left(\sum_{\substack{d\mid n\\ d\equiv 1}} d^4 - \sum_{\substack{d\mid n\\ d\equiv 3}} d^4\right)\right\} + \frac{32}{5}a_n,\\
r_{12}(n) &= (-1)^{n-1}8 \sum_{dr=n} (-1)^{d+r}d^5 + 16b_n.
\end{aligned} \tag{9.19}$$

The value of $r_2(n)$ had already been obtained in Chapter 2. If $n = 2^a n_1 n_2'$, then we recall that $d_1(n) = d_3(n)$ and $r_2(n) = 0$, unless $n_2' = \prod_{q_j\mid n,\, q_j\equiv 3} q_j^{b_j}$ is a perfect square. The proof given in Chapter 2 that if $n_2' = n_2^2$ then $d_1(n) - d_3(n) = d(n_1)\,(n_1 = \prod_{p_j\mid n,\, p_j\equiv 1} p_j^{a_j})$ need not be repeated here. Also, the second formula, $r_4(n) = 8\sigma'(n)$, has already been obtained (see Chapter 3).

§7. The Size of $r_k(n)$ for Even $k \leqslant 8$

As we saw in Chapters 2 and 4, for $k = 2$ and $k = 3$, the functions $r_k(n)$ oscillate wildly; they vanish for arbitrarily high values of n and also increase indefinitely. In these cases, it made more sense to consider the average values $\bar{r}_k(x) = x^{-1}\sum_{n\leqslant x} r_k(n)$, and this was done in Chapters 2 and 4. For $k \geqslant 4$, $r_k(n) > 0$ for all $n > 0$. For $k = 4$, on the one hand, $r_4(n) = 8\sigma'(n) = 8\sigma(n)$ for $4 \nmid n$, and (see e.g., [98], Chapter XVIII]) for infinitely many odd n we have $\sigma(n) > (e^\gamma - \varepsilon)n \log\log n$, so that $r_4(n) > Cn \log\log n$ for an infinite sequence of integers n and any constant C less than $8e^{\gamma}$ ($\simeq 14.24\ldots$. On the other hand, $r_4(2^m) = 8(1+2) = 24$ also holds for the infinite set of integers $n = 2^m$. It follows that also the function $r_4(n)$ oscillates rather wildly and it is more informative to consider the average $\bar{r}_4(x) = x^{-1}\sum_{n \leqslant x} r_4(n) \simeq (\pi^2/2)x$, already obtained in Chapter 3.

The situation is completely different for $k \geqslant 5$. We shall see (in Chapter 12) that for $k \geqslant 5$, $r_k(n) = c_k n^{k/2-1}\mathscr{S}_k(n)(1+o(1))$, where c_k is a constant that depends only on k, while $\mathscr{S}_k(n)$ is a coefficient that depends, for fixed k, only on the arithmetic nature of the integer n, but has a positive lower bound and is bounded above. It follows that for each $k \geqslant 5$, there are two positive constants, say $K_1(k)$ and $K_2(k)$, such that, for any n,

$$K_1(k)n^{k/2-1} \leqslant r_k(n) \leqslant K_2(k)n^{k/2-1}. \tag{9.20}$$

Here we shall use the results of §6 to illustrate (9.20) for 6, 8, 10, and 12 squares.

By (9.19),

$$r_6(n) = 16\left\{\sum_{\substack{r\mid n\\ r\equiv 1}}\frac{n^2}{r^2} - \sum_{\substack{r\mid n\\ r\equiv 3}}\frac{n^2}{r^2} - \sum_{\substack{rd=n\\ d\equiv 1}}\frac{n^2}{(2r)^2} + \sum_{\substack{rd=n\\ d\equiv 3}}\frac{n^2}{(2r)^2}\right\}.$$

First, let n be odd. Then $r_6(n) = 16n^2(\sum_{r\mid n} a_r/r^2 - \frac{1}{4}\sum_{r\mid n} c_r/r^2)$, where $a_1 = 1$; $a_r = \pm 1$ according as $r \equiv \pm 1 \pmod 4$; $c_r = a_r$ for $n \equiv 1 \pmod 4$; and $c_r = -a_r$ for $n \equiv 3 \pmod 4$. Consequently, for $n \equiv 1 \pmod 4$, $r_6(n) = 16n^2\cdot\frac{3}{4}\sum_{r\mid n} a_r/r^2$, and for $n \equiv 3 \pmod 4$, $r_6(n) = 16n^2\cdot\frac{5}{4}\sum_{r\mid n} a_r/r^2$. Next, we observe that $\sum_{r\mid n} a_r/r^2 \leqslant 1 + \sum_{r\geqslant 2} 1/r^2 = \zeta(2) = \pi^2/6$ and $\sum_{r\mid n} a_r/r^2 \geqslant 1 - \sum_{r\geqslant 2} 1/r^2 = 2 - \zeta(2)$. Consequently,

$$2(12-\pi^2)n^2 = 12\left(2 - \frac{\pi^2}{6}\right)n^2 \leqslant r_6(n) \leqslant 20n^2\frac{\pi^2}{6} = \frac{10\pi^2}{3}n^2.$$

If $n = 2^a n_1$, n_1 odd, $a \geqslant 1$, then $c_r = 0$ unless $r = 2^a r_1$, $r_1 \mid n_1$; hence, $|\sum_{r\mid n} c_r/r^2| \leqslant (1/4^a)\sum_{r_1\mid n_1} 1/r_1^2 \leqslant \zeta(2)/4^a$. Consequently, $2 - \zeta(2) - \zeta(2)/4^{a+1} \leqslant \sum_{r\mid n} a_r/r^2 - \frac{1}{4}\sum_{r\mid n} c_r/r^2 \leqslant \zeta(2) + \zeta(2)/4^{a+1}$. As $a \geqslant 1$, we have $2 - \zeta(2)(1 + 4^{-a-1}) \geqslant 2 - \frac{17}{16}\zeta(2)$ and, for n even,

$$16(2 - \tfrac{17}{16}\zeta(2))n^2 \leqslant r_6(n) \leqslant 16\zeta(2)\cdot\tfrac{17}{16}n^2.$$

It follows that, regardless of the parity of n, (9.20) holds for $k = 6$, with $K_1(6) = (192 - 17\pi^2)/6$ ($\simeq 4.036$) and $K_2(6) = 10\pi^2/3$ ($\simeq 32.898$).

For $k = 8$ the method is the same and the computation is still simpler. By (9.19),

$$r_8(n) = (-1)^n 16\sum_{d\mid n}(-1)^d d^3 = 16\sum_{d\mid n}(-1)^{n+n/d}\frac{n^3}{d^3} = 16n^3\sum_{d\mid n}\frac{a_d}{d^3},$$

with $a_1 = 1$, $a_d = \pm 1$. It follows that $\sum_{d|n} a_d/d^3 \leqslant \sum_{d=1}^{\infty} 1/d^3 = \zeta(3)$ and $\sum_{d|n} a_d/d^3 \geqslant 1 - \sum_{d\geqslant 2} 1/d^3 = 2 - \zeta(3)$, so that

$$16(2 - \zeta(3))n^3 \leqslant r_8(n) \leqslant 16\zeta(3)n^3.$$

This proves (9.20) for $k = 8$, with $K_1(8) = 16(2 - \zeta(3))$ ($\simeq 12.76$) and $K_2(8) = 16\zeta(3)$ ($\simeq 19.23$).

§8. An Auxilliary Lemma

For the discussion of the cases $k = 10$ and $k = 12$ we have to know something about the coefficients a_n and b_n defined by the Taylor expansions (9.15) and (9.17). These coefficients are, in a very precise sense, small with respect to $r_{10}(n)$ and $r_{12}(n)$, respectively, as the following lemma shows.

Lemma 1. *For all integers* $n \geqslant 1$, $a_n = O(n^3)$ *and* $b_n = O(n^3 \log\log n)$.

Proof. By (9.15) and Corollary 8.3, we obtain $16\sum_{n=1}^{\infty} a_n q^n = (\theta_2\theta_3\theta_4)^2\theta_2^2\theta_4^2 = (\pi^{-1}\theta_1')^2\theta_2^2\theta_4^2$. Next, we take the derivative of $\theta_1(v\,|\,\tau)$ in (8.2), expand the squares in the exponents, and replace all occurring minus signs by plus. This change can only increase the coefficients; hence, the coefficients of q^n in $16\sum_{n=1}^{\infty} a_n q^n$ are no larger than those of the product

$$4q\left(\sum_{n=0}^{\infty}(2n+1)q^{n(n+1)}\right)^2\left(\sum_{n=-\infty}^{\infty} q^{n(n+1)}\right)^2\left(\sum_{n=-\infty}^{\infty} q^{n^2}\right)^2$$

$$= 4q \sum_{n_1,n_2,n_3,n_4} (2n_1+1)(2n_2+1)q^{n_1(n_1+1)+n_2(n_2+1)}q^{n_3(n_3+1)+n_4(n_4+1)}q^{n_5^2+n_6^2}$$

$$= 4q\sum_{n=0}^{\infty} c_n q^n, \qquad \text{say.}$$

Here $c_n = \sum(2n_1+1)(2n_2+1)$, with the sum extended over all nonnegative integers n_1, n_2 that satisfy

$$n_1(n_1+1) + n_2(n_2+1) + n_3(n_3+1) + n_4(n_4+1) + n_5^2 + n_6^2 = n \tag{9.21}$$

and where n_3, n_4, n_5, n_6 are rational integers. In equation (9.21), we may complete the square and obtain the equivalent equation

$$\sum_{i=1}^{4}(2n_i+1)^2 + (2n_5)^2 + (2n_6)^2 = 4n + 4. \tag{9.22}$$

By the inequality satisfied by the geometric and arithmetic means, we find that

$$(2n_1 + 1)(2n_2 + 1) \leqslant \tfrac{1}{2}\{(2n_1 + 1)^2 + (2n_2 + 1)^2\} \leqslant \tfrac{1}{2}(4n + 4) = 2n + 2, \tag{9.23}$$

where use has been made of (9.22). Also, (9.22) is a representation of the integer $4n + 4$ as a sum of six squares. We know already that the number $r_6(4n + 4)$ of such representations cannot exceed $K_2(6)(4n + 4)^2 = 16K_2(6)(n + 1)^2$. This is then an upper bound for the number of solutions of (9.22), and hence of (9.21), i.e., the number of terms in the sum that defines c_n. By (9.23), each of these terms is at most equal to $2(n + 1)$, so that $c_n \leqslant 32K_2(6)(n + 1)^3$. The coefficient a_{n+1} of q^{n+1} in $16\sum_{n=1}^{\infty} a_n q^n$ is no larger than that in $4q\sum_{n=0}^{\infty} c_n q^n$, i.e., it is at most $4c_n \leqslant 4\cdot 32K_2(6)(n + 1)^3$, and so $16|a_n| \leqslant 128K_2(6)n^3$, or $|a_n| \leqslant 8K_2(6)n^3$, as claimed.

The proof for b_n is similar. By (9.17), $(\pi^{-1}\theta_1')^4 = 16\sum_{n=1}^{\infty} b_n q^n$, and by (8.2) with $v = 0$, $\pi^{-1}\theta_1' = 2q^{1/4}\sum_{m=0}^{\infty}(-1)^m(2m + 1)q^{m^2+m}$, so that the coefficients b_n are less than, or at most equal to, the coefficients of the expansion $q\{\sum_{m=0}^{\infty} \times (2m + 1)q^{m(m+1)}\}^4$, where all minus signs have been replaced by plus signs. The expression in braces is the sum $\sum_{m_i}(2m_1 + 1)(2m_2 + 1)(2m_3 + 1)(2m_4 + 1)q^s$, $s = \sum_i m_i(m_i + 1)$. Here i ranges over 1, 2, 3, and 4, and in both summations the m_i range independently over all nonnegative integers. If we write the quadruple sum as $\sum_{n=0}^{\infty} d_n q^n$, then $d_n = \sum(2m_1 + 1)\cdots(2m_4 + 1)$, with the summation extended over all nonnegative integers m_1, m_2, m_3, m_4 for which the exponent of q equals n, i.e., such that $\sum_{i=1}^{4} m_i(m_i + 1) = n$. Just as in the computation of the a_n, by completing the square, the last equation becomes

$$\sum_{i=1}^{4} (2m_i + 1)^2 = 4n + 4. \tag{9.24}$$

Using the inequality satisfied by the geometric and arithmetic means and (9.24), we obtain $\prod_{i=1}^{4}(2m_i + 1) \leqslant (\frac{1}{4}\sum_{i=1}^{4}(2m_i + 1)^2)^2 = (n + 1)^2$. Next, the number of terms in the sum that defines d_n is the number of solutions of (9.24), i.e., at most $r_4(4n + 4)$. However, $r_4(4n + 4) = 8\sigma'(4n + 4) \leqslant 8\cdot 4 \times (n + 1)\log\log(4(n + 1))(e^\gamma + \varepsilon) \leqslant C(n + 1)\log\log(n + 1)$, with an easily computed constant $C = C(\varepsilon)$. Consequently, the coefficient of q^{n+1} in $(\theta_1'/\pi)^4$ is at most $C(n + 1)^3\log\log(n + 1)$, whence $16b_n \leqslant Cn^3\log\log n$ and $b_n \leqslant C_0 n^3\log\log n$, and the lemma is proved. □

§9. Estimate of $r_{10}(n)$ and $r_{12}(n)$

Now we are able to complete the verification of (9.20) for $k = 10$ and $k = 12$.

For $k = 10$, we recall that, by (9.19), we can write

$$r_{10}(n) = \frac{64}{5}\left\{\frac{1}{16}\left(\sum_{\substack{d\mid n\\ d\equiv 1}} d^4 - \sum_{\substack{d\mid n\\ d\equiv 3}} d^4\right) + \sum_{\substack{rd=n\\ r\equiv 1}} d^4 - \sum_{\substack{rd=n\\ r\equiv 3}} d^4\right\} + \frac{32}{5}a_n,$$

with the a_n defined by (9.15). We now consider the sum in braces. By proceeding as before, it may be written as $n^4 \sum c_d/d^4$, where the coefficients c_d may take only the values 0, ± 1, $\pm\frac{1}{16}$, $\pm\frac{17}{16}$, or $\pm\frac{15}{16}$. In particular, $c_1 = \frac{17}{16}$ if $n \equiv 1 \pmod 4$, $c_1 = \frac{15}{16}$ if $n \equiv 3 \pmod 4$. In any case, $|c_d| \le \frac{17}{16}$. It follows that the sum under consideration satisfies $\sum_{d\mid n} c_d/d^4 \leqslant \frac{17}{16}\sum_{d=1}^{\infty} 1/d^4 = \frac{17}{16}\zeta(4)$, and also,

$$\sum_{d\mid n} \frac{c_d}{d^4} \geqslant \frac{15}{16} - \frac{17}{16}\sum_{\substack{d\mid n\\ d\geqslant 2}} \frac{1}{d^4} \geqslant \frac{15}{16} - \frac{17}{16}\sum_{d=2}^{\infty}\frac{1}{d^4}$$

$$= \frac{15}{16} + \frac{17}{16} - \frac{17}{16}\sum_{d=1}^{\infty}\frac{1}{d^4} = 2 - \frac{17}{16}\zeta(4).$$

Moreover, it follows from Lemma 1 that $\frac{32}{5}|a_n| \le Cn^3$ for some constant C. Consequently, by recalling also that $\zeta(4) = \pi^4/90$, the following string of inequalities has been established:

$$C_4 n^4 \leqslant \frac{64}{5}\left(2 - \frac{17\pi^4}{1440}\right)\left(1 - \frac{C_3}{n}\right)n^4 = \frac{64}{5}\left(2 - \frac{17}{16}\frac{\pi^4}{90}\right)n^4 - Cn^3$$

$$\leqslant r_{10}(n) \leqslant \frac{64}{5}\frac{17}{16}\frac{\pi^4}{90}n^4 + Cn^3$$

$$= \frac{34\pi^4}{225}\left(1 + \frac{C_1}{n}\right)n^4 \leqslant C_2 n^4.$$

For sufficiently large, computable n_0 and $n \geqslant n_0$, $C_4 = \frac{64}{5}(2 - (17\pi^4/1440)(1 - \varepsilon)) \simeq 10.88 - \varepsilon'$ and $C_2 = (34\pi^4/225)(1 + \varepsilon) \simeq 14.72 + \varepsilon'$. We now define $C_5 = \min_{n\leqslant n_0}(r_{10}(n)/n^4)$, $C_6 = \max_{n\leqslant n_0}(r_{10}(n)/n^4)$; then $K_1(10) = \min(C_4, C_5)$ and $K_2(10) = \max(C_2, C_6)$ clearly satisfy (9.20) for every n.

Finally, for $k = 12$, by (9.19), $r_{12}(n) = 8\sum_{dr=n}(-1)^{n+d+r-1}d^5 + 16b_n$. The sum may be written as $n^5\sum_{dr=n}(-1)^{(d+1)(r+1)}1/r^5 = n^5\sum_{r\mid n} a_r/r^5$, with $a_1 = 1$, $a_r = \pm 1$ for $r \mid n$, $r > 1$. Hence, as seen before, $2 - \zeta(5) \leqslant \sum_{r\mid n} a_r/r^5 \leqslant \zeta(5)$, and, by use of Lemma 1,

$$8(2 - \zeta(5))(1 - \varepsilon)n^5 = 8(2 - \zeta(5))n^5 - Cn^3 \log\log n$$

$$\leqslant r_{12}(n) \leqslant 8\zeta(5)n^5 + Cn^3 \log\log n = 8\zeta(5)(1 + \varepsilon)n^5,$$

valid for arbitrarily small $\varepsilon > 0$ and $n \geqslant n_0(\varepsilon)$. Now $K_1(12)$ and $K_2(12)$ are determined as for $k = 10$, and this completes the verification of (9.20).

§10. An Alternative Approach

As already mentioned in Chapter 8, recently Andrews [6] gave proofs of many of our present results, as applications of bilateral basic hypergeometric functions. Much of this work is based on the properties of the function

$$K_{\lambda,k,i}(a_0, a_1, \ldots, a_\lambda; z; q) = \sum_{n=-1}^{\infty} (-1)^{n(\lambda+1)} z^{(k+1)n} q^v \Phi,$$

where $v = \frac{1}{2}\{(2k - \lambda + 1)n^2 - 2in + (\lambda + 1)n\}$ and

$$\Phi = \frac{1 - z^i q^{2ni}}{1 - z} \sum_{r=0}^{\lambda} a_r^{-n} \frac{(a_r)_n}{(zq/a_r)_n},$$

in which $(a)_n = (a; q)_n$ is an abbreviation for $(1 - a)(1 - aq) \cdots (1 - aq^{n-1})$.

The function $K_{\lambda,k,i}$ may appear hopelessly complicated, and it depends on a large number of parameters. Nevertheless, it satisfies some very simple identities, and by substituting for the parameters certain particular values, one obtains many theorems in a beautifully unified way. Also, the introduction of this function is perhaps less artificial than may appear at first sight. For example, $K_{3,1,1}(b, c, d, e; z; q)$ reduces to the well-known bilateral basic hypergeometric function ${}_6\psi_6$, for appropriate (admittedly somewhat complicated) values of its arguments. For a definition of these functions and their exact relations to $K_{\lambda,k,i}$ we have to refer the reader to [6]. We state, however, two identities satisfied by the $K_{\lambda,k,i}$, in order to convince the reader that these are indeed quite simple. For any values of the arguments $a_0, a_1, \ldots, a_\lambda$ for which these functions are defined, we have

$$K_{\lambda,k,0} = 0 \quad \text{and} \quad K_{\lambda,k,-i} = -z^{-i} K_{\lambda,k,i}.$$

The main theorem of Andrews (Theorem 3.3. of [6]) consists in showing that the above mentioned function $K_{3,1,1}(b, c, d, e; z; q)$ admits, besides its representation by ${}_6\psi_6$, which is a series, also another one as an infinite product. The reader will remember that a similar step was crucial also in the theory of theta functions.

It may be stated quite frankly that the proofs of the identities for $K_{\lambda,k,i}$, as well as of the fundamental Theorem 3.3 of [6], are not simple and, in particular, require some rather delicate passages to the limit as certain parameters increase to infinity.

To give an example of the use of the present method, we observe that if in the fundamental theorem one keeps $|q| < 1$ and $z \neq 0$ constant, takes the limit for $b, c, d, e \to \infty$, and then replaces q by q^2 and a by $-zq$, one obtains the result

$$\sum_{n=-\infty}^{\infty} q^{4n^2} z^{2n} (1 + zq^{4n+1}) = \prod_{m=1}^{\infty} (1 - q^{2m})(1 + zq^{2m-1})(1 + z^{-1} q^{2m-1}).$$

However, the first member equals $\sum_{n=-\infty}^{\infty} q^{(2n)^2} z^{2n} + \sum_{n=-\infty}^{\infty} q^{(2n+1)^2} z^{2n+1} = \sum_{n=-\infty}^{\infty} q^{n^2} z^n$, and if we combine these two results, we obtain Jacobi's triple product identity, i.e., our Lemma 8.1.

Andrews then proceeds to prove (9.19) for $k = 2, 4$, and 8, with hints how to prove (9.19) for $k = 6$. He also expresses the hope that also the general case of $r_{2k}(n)$ can be obtained by this method.

§11. Problems

1. Prove formally the last identity of (9.1).
2. Prove the analogues of (9.1) for $f_2(v \mid \tau)$ and $f_3(v \mid \tau)$.
3. Use cyclical permutations of subscripts to obtain $f_i''(u)$ and $f_i'''(u)$ for $i = 3$ and 4.
4. Complete the proofs of the six identities (9.8).
5. Prove the equivalence of (3.14) with the second identity of (9.10).
6.
   (a) Use the results of §6 to compute $r_k(n)$ for $k = 2, 4, 6$, and 8 and for $n = 5$, 10, 12, 100.
   (b) Find the number of essentially distinct representations of $n = 5, 10, 12$, and (if you are very patient and determined) 100 as sums of 2, 4, 6, and 8 squares. Use these results to compute the corresponding $r_k(n)$, and compare with the results of (a).
7. For $k = 10$ and $k = 12$, find $r_k(n)$ by direct decomposition into squares for $n = 5$ and 10 (and, if you have the stamina, also 12). Compare the values obtained with the leading terms of (9.16) and (9.18), and find in this way the values of a_n and b_n. Verify (i) that these are indeed integers, and (ii) that they are quite small compared to the corresponding $r_k(n)$.

Chapter 10

Various Results on Representations as Sums of Squares

§1. Some Special, Older Results

In [54] Dickson lists about 70 mathematicians who have made contributions to the problem of representations of integers by sums of five or more squares and to that of relations between the numbers of those representations. It is not possible to quote all these results, but in order to convey the flavor of some of them, we shall mention a few without proofs.

Eisenstein stated (see [64]; for proofs see [253] or [178]) the following: Set $s = \sum (r/m)r$, $\sigma = \sum (-1)^r (r/m)r$, with both summations over $r = 1, 2, \ldots, (m-1)/2$; then

$$r_5(m) = -80s,\ -80\sigma,\ -112s, \text{ or } 80\sigma,$$

according as $m \equiv 1, 3, 5$, or $7 \pmod 8$.

According to Liouville [166] (see also Humbert [117]), if m is odd, then $r_{12}(2m) = 264 \sum_{d \mid m} d^5$ and, more generally, $r_{12}(2^a m) = \frac{24}{31}(21 + 10 \cdot 32^a) \times \sum_{d \mid m} d^5$.

Perhaps the most interesting result of this type is the following theorem due to Stieltjes.

Theorem 1 (Stieltjes [259]). *Let* $d_0(n) = \sum_{d \mid n,\, d \text{ odd}} d$, *and define* a, n_1 *by* $n = 2^a n_1$ *with* n_1 *odd; then*

$$r_5(n) = 10 \frac{2^{3a+3} - 1}{2^3 - 1} \{d_0(n_1^2) + 2d_0(n_1^2 - 2^2) + 2d_0(n_1^2 - 4^2) + \cdots\}$$

$$= 10 \frac{2^{3a+3} - 1}{2^3 - 1} \sum d_0(m'm''),$$

with the summation extended over the positive odd integers m', m'' *with* $m' + m'' = 2n_1$.

On the basis of many numerical examples, Stieltjes then made the following

Conjecture (Stieltjes [260]). *For all odd primes p,*

$$r_5(p^2) = 10(p^3 - p + 1) \quad \text{and} \quad r_5(p^4) = 10\{p(p^2 - 1)(p^3 + 1) + 1\}.$$

While Stieltjes himself was not able to prove his conjecture, Hurwitz, by using Stieltjes's own Theorem 1, proved a more general result from which the conjecture follows.

Theorem 2 (Hurwitz [118]). *If* $n = 2^a p_1^{b_1} p_2^{b_2} \cdots p_t^{b_t}$, *then*

$$r_5(n^2) = 10 \frac{2^{3a+3} - 1}{2^3 - 1} \prod_{p \mid n} \frac{p^{3b+3} - p^{3b+1} + p - 1}{p^3 - 1}.$$

Hurwitz also proved

Theorem 3 (Hurwitz [119]). *Let* $n = 2^a m \prod_{j=1}^{s} q_j^{a_j}$, *with the primes* $q_j \equiv 3 \pmod 4$ *and* $p \equiv 1 \pmod 4$ *for all primes* $p \mid m$. *Then* $r_3(n^2) = 6m \prod_{q \mid n} (q_j^{a_j} + (q_j^{a_j} - 1)/(q_j - 1))$.

By using Theorem 3, one easily obtains a simple proof of Theorem 6.5.

Among the contributors to these problems, Dickson also quotes in [54] the following: Jacobi (see also Chapters 8 and 9), Lebesgue, Catalan, Glaisher, Torelli, Lemoine, Hermite, Pépin (who gave a proof of Stieltjes's conjecture), Minkowski, Cesaro, H. J. S. Smith, Gegenbauer, G. B. Mathews, Bachmann, Meissner, Jacobsthal, Sierpiński, G. Humbert, Dubouis [60], Uspenski, Boulyguine, Ramanujan, Hardy, Mordell, Goormaghtigh, E. T. Bell, Aida Ammei (ca. 1810), Ajima Chokuyen (1791), E. Lukas, Dickson, Landau, and many more.

§2. More Recent Contributions

The number of contributions made after 1920 on the problem of representations by sums of squares and related topics is so large that it appears hopeless to try to trace and mention all of them. Some will be listed here, with bibliographic references and a brief indication of their subject matter; many more will be found in the bibliographies of the cited papers.

The special problem of representation by an *even* number of squares, found in Ramanujan [223], with proofs provided (as already mentioned) by Mordell [182, 183], is taken up again by K. Ananda Rau [4].

In addition to his work mentioned later (in Chapter 14), G. Pall has two papers, one on rational automorphs [205] and one on the arithmetic of quaternions [206] (for automorphs, see Chapter 14).

C. D. Olds's work on representations of squares in [202] and [203] should be compared with Theorem 3 and with Stieltjes's conjecture. Further work on similar topics is due to H. F. Sandham, who gives [234, 235] formulae for $r_k(n^2)$ when $k = 7, 9, 11$, or 13.

Th. Skolem [251, 252] studies lattice points on spheres, by use of the arithmetic of quaternions; this leads him to results on representations as sums of two, three, and four squares.

Lattice points on spheres, with emphasis on their distribution within pre-assigned cones (with centers at the origin) is the subject of a sequence of papers by Yu. V. Linnik [161–163].

G. Benneton discusses special problems of representations of natural integers by sums of four and, more generally, of 2^m squares [23].

H. Gupta [90] and M. C. Chakrabarti [37] have partially overlapping results on integers that are *not* sums of fewer than four squares.

The general problem of representation of a natural integer by an arbitrary number $k \geqslant 2$ of squares has been studied, following Hardy, Ramanujan, and Littlewood, by a large number of mathematicians, both by theta functions and by the "circle method." The latter will be discussed in some detail in Chapters 12 and 13; nevertheless, as no details of the work itself need to be given here, it appears more appropriate to list the corresponding contributions at this point, rather than later.

It was soon observed that only a slight generalization was needed to yield, instead of the number of solutions of the single equation $\sum_{i=1}^{k} x_i^2 = n$, the number $t_k(n, m)$ of solutions of the system $\sum_{i=1}^{k} x_i^2 = n$, $\sum_{i=1}^{k} x_i = m$. A fairly large number of mathematicians, many of them Dutch or Russian, made contributions to these two related problems. Among them are H. D. Kloostermann [129]; van der Blij in five papers [27], P. Bronkhorst [30], B. van der Pol [216], L. Seshu [240], G. A. Lomadze, and A. Z. Walfisz.

Starting out with the determination of fully explicit formulae for $r_k(n)$ with k odd and $9 \leqslant k \leqslant 23$ in [169], Lomadze ended up by giving, in a series of papers [170–172], explicit, exact formulae for $r_k(n)$ for all k in the interval $9 \leqslant k \leqslant 32$. He also evaluated in finite terms many formulae for $r_k(n)$ and $r_k(n, m)$, previously known only as infinite series (see also [168]).

Walfisz, in a long sequence of partly expository papers (see [275] to [278]; [276] is an English translation) studies both $r_k(n)$ and $r_k(n, m)$.

E. Krätzel uses theta functions to obtain explicit formulae for $r_k(n)$ when $k \equiv 0 \pmod 4$ in [138], and when $k \equiv 2 \pmod 4$ in [139].

R. A. Rankin [225] uses elementary methods, as well as the theory of elliptic functions, to obtain identities that involve $r_k(n)$ for $k \equiv 0 \pmod 4$.

A somewhat different problem is considered by R. P. Bambah and S. Chowla in [16]. The authors show that, for some absolute constant c, every interval $x \leqslant n \leqslant x + c\sqrt{x}$ contains at least one integer n that is a sum of two squares.

R. Sprague shows [256] that, with 32 exceptions (all in the interval $2 \leqslant n \leqslant 128$), all integers are sums of *distinct* squares.

L. Carlitz [32, 34] uses elliptic and theta functions to obtain $r_k(n)$ for $k = 4$, 6, and 24 (see also [33]).

I. Ebel indicates [62] the number of solutions of the equation $\sum_{i=1}^{3} a_i x_i^2 = n$ subject to congruence conditions on the x_i.

A. G. Postnikov [218] considers the system $\sum_{i=1}^{k} x_i^2 = \sum_{i=1}^{k} x_i = n$ and obtains asymptotic expansions for $r_k(n, n)$ as k increases.

A. V. Malyshev [175] studies the distribution of lattice points on the 4-dimensional sphere.

In [226] to [228], R. A. Rankin uses theta functions and cusp forms to study the number of representations of a natural integer by a sum of a large number of squares, with or without additional conditions, such as bounds set on the largest square, etc. In particular, a simple formula is obtained for $r_{20}(n)$.

As corollary of a very general result, H. Petersson obtains [209] certain relations between $r_5(n)$, $r_7(n)$, and the class number of a certain quadratic field.

D. Pumplün uses the mentioned results of Petersson and obtains [220] arithmetic properties of the numbers $r_{2k+1}(n)$ for odd, squarefree n.

I. Niven and H. S. Zuckerman show in [199] that, for $k \geqslant 5$, every $n \geqslant n_0(k)$ is a sum of k nonvanishing squares (see Chapter 6) and also that for sufficiently large n one can solve the Diophantine equation $n = a^2 + b^2 + c^2 + d^k$ in integers, with $abcd \neq 0$, where $k = 2, 4, 6$ or is odd, while the result is false for even $k \geqslant 8$.

§3. The Multiplicativity Problem

A problem of a very different kind was raised and solved by P. T. Bateman [19]. He observed that if we set $f_k(n) = r_k(n)/(2k)$, then $f_k(n)$ is multiplicative if $k = 1, 2, 4$, and 8, and asked for what other values of k, if any, $f_k(n)$ is multiplicative in the sense that if $(n, m) = 1$, then $f_k(nm) = f_k(n) f_k(m)$. The perhaps unexpected answer is

Theorem 4. *The function $f_k(n)$ is multiplicative if and only if $k = 1, 2, 4$, or 8.*

Remark 1. As $r_0(n) = 0$ for $n \geqslant 1$, the function $f_0(n)$ is undefined for all $n \geqslant 1$. The proof of Theorem 4 suggests defining $f_0(n) = 1$ for all $n \geqslant 1$; if we do that, then the value $k = 0$ may be added to the list of values of k in Theorem 4.

Proof of Theorem 4. In order to show that, for a given k, $f_k(n)$ is not multiplicative, a single counterexample $(n, m) = 1$ with $f_k(n) f_k(m) - f_k(nm) \neq 0$ suffices. The simplest possibility—$n = 2$, $m = 3$—already works. To compute $r_k(2)$, we observe that, for $k \geqslant 2$, $\sum_{j=1}^{k} x_j^2 = 2$ has only the solutions $x_j = \pm 1$ for two values of j, $x_j = 0$ for all others, so that $r_k(2) = 2^2 \binom{k}{2} = 2k(k-1)$. This formula holds also for $k = 1$ when both sides vanish. In exactly the same way we obtain that for $k \geqslant 3$, $r_k(3) = 2^3 \binom{k}{3} = \frac{4}{3} k(k-1)(k-2)$, and, as before, we verify that this formula remains valid also in the omitted cases $k = 1$ and $k = 2$, when both sides of the equality vanish. It follows that $f_k(2) = k - 1$, $f_k(3) = \frac{2}{3}(k-1) \times (k-2)$, and $f_k(2) f_k(3) = \frac{2}{3}(k-1)^2(k-2)$. For $k \geqslant 6$, $\sum_{j=1}^{k} x_j^2 = 6$ has as solutions either $x_j = \pm 2$ for one single value of j and $x_j = \pm 1$ for two other values of j, with all other $x_j = 0$, which leads to $2^3 \cdot 3 \cdot \binom{k}{3}$ solutions; or $x_j = \pm 1$ for six values of j and all other $x_j = 0$, which leads to

$$2^6\binom{k}{6} = 64\frac{k(k-1)\cdots(k-5)}{6!} = \tfrac{4}{45}k(k-1)(k-2)(k-3)(k-4)(k-5)$$

solutions. This leads, for $k \geqslant 6$, to a total number of solutions

$$r_k(6) = \tfrac{4}{45}k(k-1)(k-2)(45 + (k-3)(k-4)(k-5)).$$

For $k = 1$ and $k = 2$, both sides of the equality vanish, and for $k = 3, 4$, or 5, only the first type of solutions exists, so that $r_k(6) = 4k(k-1)(k-2)$ and the formula is valid also in this case. In conclusion, we find that for all $k \geqslant 1$

$$f_k(6) = \tfrac{2}{45}(k-1)(k-2)(k^3 - 12k^2 + 47k - 15).$$

In all cases in which $f_k(n)$ is multiplicative, we must have, in particular, $F(k) \underset{\text{def}}{=} f_k(6) - f_k(2)f_k(3) = 0$. However,

$$\begin{aligned} F(k) &= \tfrac{2}{45}(k-1)(k-2)(k^3 - 12k^2 + 47k - 15) - \tfrac{2}{3}(k-1)^2(k-2) \\ &= \tfrac{2}{45}(k-1)(k-2)(k^3 - 12k^2 + 47k - 15 - 15k + 15) \\ &= \tfrac{2}{45}(k-1)(k-2)(k^3 - 12k^2 + 32k) \\ &= \tfrac{2}{45}k(k-1)(k-2)(k-4)(k-8) \end{aligned}$$

and can vanish only for $k = 0, 1, 2, 4$, or 8, as claimed (see Remark 1).

It is easy to verify that in all these cases, $f_k(n)$ (defined for $k = 0$ as $f_0(n) = 1$ for $n \geqslant 1$) is indeed multiplicative. In fact, $r_1(n) = 2\delta(n)$, with $\delta(n) = 1$ if $n = u^2$, $\delta(n) = 0$ otherwise. Then $f_1(n) = \delta(n)$ and $f_1(n)f_1(m) = \delta(n)\delta(m) = \delta(nm) = f_1(nm)$, where all terms vanish unless n and m are both squares, when all equal 1; indeed, for $(n, m) = 1$, nm is a square if and only if n and m are both squares.

Similarly, if $n = 2^a \prod_{i=1}^r p_i^{b_i} \prod_{j=1}^s q_j^{c_j}$, set $\delta(n) = 1$ if all c_j are even, $\delta(n) = 0$ otherwise. Then, by Theorem 2.3, $r_2(n) = 4\prod_{i=1}^r (1 + b_i)\delta(n)$ and $f_2(n) = \prod_{i=1}^r (1 + b_i)\delta(n)$. If $m = 2^{a'} \prod_{i=1}^{r'} (p_i')^{b_i'} \prod_{j=1}^{s'} (q_j')^{c_j'}$, $(m, n) = 1$, then

$$f_2(n)f_2(m) = \prod_{i=1}^{r} (1 + b_i) \prod_{i=1}^{r'} (1 + b_i')\delta(n)\delta(m) = f_2(nm),$$

where all terms vanish unless all c_j and all c_j' are even, in which case $\delta(n) = \delta(m) = \delta(nm) = 1$ and $f_2(nm) = \prod_{i=1}^r (1 + b_i) \prod_{i=1}^{r'} (1 + b_i')$.

Remark 2. The multiplicativity of $f_2(n)$ holds under the weaker condition that (n, m) is a power of 2.

The verifications that also $f_4(n)$ and $f_8(n)$ are multiplicative are similar and are left to the problems (see Problem 4).

An extremely readable and very beautiful presentation of many questions connected with representations of integers as sums of squares is due to 0. Taussky [263]. Anybody who considers embarking on more extensive investigations of this topic is well advised to start by consulting this paper.

Among the numerous other contributions, we list the following ones, in alphabetic order (of the first author in the case of joint authorship): N. C. Ankeny [7, 8]; D. P. Banerjee [17]; B. Derasimovič [52]; J. D. Dixon [57]; J. Drach [58]; T. Estermann [73]; T. Kano [126]; A. A. Kiselev and I. S. Slavutskii [127]; I. P. Kubilyus and Yu. V. Linnik [144]; J. H. van Lint [164]; Ju I. Manin [176]; M. Newman [196]; G. Pall and O. Taussky [208]; L. Reitan [229]; U. Richards [231]; A. Schinzel [239]; W. Sierpiński [250]; R. Spira [255]; M. V. Subba Rao [261]; J. V. Uspenskii [269]; C. Waid [274].

§4. Problems

1. Prove Stieltjes's conjecture by the use of Theorem 2.
2. Use Theorems 2 and 3 to compute $r_5(9)$, $r_5(10)$, and $r_5(12)$.
3. Prove Theorem 6.5 by the use of Theorem 3.
4. Verify the multiplicativity of $f_k(n)$ for $k = 4$ and $k = 8$.
5. Justify the statement of Remark 2.

Chapter 11

Preliminaries to the Circle Method and the Method of Modular Functions

§1. Introduction

We have presented in Chapters 2, 3, and 4 the contributions made by mathematicians, from antiquity until well into the 19th century, to the problem of representations of natural integers by sums of $k = 2$, 4, and 3 squares. Theorems were obtained that presented the number $r_k(n)$ of these representations as sums of divisor functions (for $k = 2$ and 4) or sums over Jacobi symbols (for $k = 3$). Next, in Chapters 8 and 9, we studied a method developed mainly by Jacobi, based on the use of theta functions, by which one obtains similar results, involving divisor functions, for an *even* number of squares. The method is entirely successful for $k \leqslant 8$, while for $k \geqslant 10$ the formulae contain, besides sums of divisor functions, also more complicated additional terms. Even for $k \leqslant 8$, the cases $k = 5$ and $k = 7$ were bypassed. Nevertheless, as mentioned earlier (see Chapter 10), Eisenstein presented formulae depending (as in the case $k = 3$) on Jacobi symbols for $r_5(n)$, and formulae for $r_5(n)$ of a different type were proposed and proved by Stieltjes (1856–1894) (see [259], [260]), and by Hurwitz [118]. Eisenstein also stated formulae for $r_7(n)$, expressed, like those for $r_5(n)$, with the help of the Legendre-Jacobi symbol. Proofs of Eisenstein's formulae were given by Smith [253, 254], who deduced them from his own, of a different appearance, and by Minkowski (1864–1909) (see [178]).

Although Smith's work had been published in 1867, the Académie des Sciences of Paris, apparently unaware of it, set up a prize (Grand Prix des Sciences Mathématiques—1882), for the proof of the formula for $r_5(n)$, stated by Eisenstein. Smith and Minkowski (the latter then just 18 years old) both submitted entries with the complete solution of the proposed problem—and the Académie decided not to split the prize, but to award the full prize to each of the submitted memoirs.

Liouville, by the use of elliptic functions, obtained formulae for $r_{12}(n)$ and also for the number of such representations as sums of 12 squares with certain side conditions (see Chapter 10). Much of this work (especially for $k = 5, 6, 7$, and 8 squares) was made easily accessible by the largely expository presentation of Bachmann [12]. Often the methods used to determine $r_k(n)$ worked only for integers n of a specific type (e.g., for $n = m^2$), or only for one or a few values of k. However, Jacobi and Eisenstein had already used elliptic functions, and Jacobi's treatment of $r_k(n)$ by the study of its generating function $\{\theta_3(q)\}^k = 1 + \sum_{n=1}^{\infty} r_k(n)q^n$ represents a unified approach and was quite successful, at least for even values of k.

There is, however, another way to study $\theta_3(q)$, $q = e^{\pi i\tau}$, namely by considering it as a function of τ and by observing that $\theta_3(q)$ transforms in a simple way when we replace τ by $(a\tau + b)/(c\tau + d)$, $a, b, c, d \in \mathbb{Z}$, $ad - bc = 1$.

The transformations $\tau \to M\tau = (a\tau + b)/(c\tau + d)$, with the indicated side conditions, form a group under composition, the *modular group*, mentioned already in Chapter 1 (see §4 for a precise definition). *Modular functions*, i.e., functions invariant under modular transformations, were used in the early twentieth century by Ramanujan in connection with the problem of representations as sums of squares [223]. However, many of Ramanujan's statements were presented without proofs. These were supplied, shortly afterwards, by Mordell [183], by the use of modular functions. Mordell proceeded to obtain in a unified way $r_k(n)$ for all even k (see [182]), always by the method of modular functions. As Hardy observed a few years later [96], Mordell's method is valid and works also for odd k, and in fact, in a subsequent paper [184], Mordell extended his work to this case. However, he was able to do so largely because in the meantime Hardy had solved the problem for all k in a unified way, by his then recently invented "circle method." This provided an *a posteriori* motivation for Mordell's approach and, in fact, permitted him to guess correctly the solution of the problem also for odd k, which he then would prove to be correct, by his own method of modular forms. In his paper [96], Hardy gives ample credit to Mordell for his contribution to the problem in [182], and much of this work is in fact incorporated in [96]. On the other hand, Mordell could then use Hardy's results of [96], which were not published in full untill 1920–1921, to complete in 1919 (see [184]) his earlier results of [182].

There is no doubt that the most satisfactory approach is that of Hardy; important contributions to that so-called "circle method" were made by Ramanujan, Littlewood (1885–1977), Rademacher (1892–1969), and, later on, Davenport (1907–1969) and others. Its drawback is its complexity. Many attempts were made to simplify it. Perhaps the most successful is that of Estermann in [72], which gives a complete, self-contained proof of the main theorem in some 32 pages. This approach, however, has the disadvantage that the operations performed appear to be unmotivated.

Here we shall start with a presentation of Hardy's "circle method," in order to motivate what is done. In a subsequent chapter, an account is given both of Mordell's approach using modular forms and functions following to a large extent the presentation in Knopp [133]) and of Estermann's streamlined version of the circle method.

In order not to interrupt the main trend of the exposition, we recall in the following sections of the present chapter some well-known facts about Farey series and Gaussian sums, and some of the basic theory on modular forms and functions. In the present chapter, as well as in those that follow, lengthy computations, easily available elsewhere, will not be reproduced.

§2. Farey Series

Let n be a positive integer. Then we have the following

Definition 1. The set of rational fractions h/k with $(h,k)=1$, $0 \leqslant h/k \leqslant 1$, $0 \leqslant h \leqslant k \leqslant n$, ordered by size, is called the *Farey series* of order n and is denoted by $\mathscr{F}_n$.

To give an example, the Farey series of order 4 consists of

$$\frac{0}{1} < \frac{1}{4} < \frac{1}{3} < \frac{1}{2} < \frac{2}{3} < \frac{3}{4} < \frac{1}{1}.$$

Let $h'/k' < h/k < h''/k''$ be three consecutive terms of $\mathscr{F}_n$; then we call the rational number $(h'+h)/(k'+k)$ the *mediant* between h'/k' and h/k, and similarly $(h+h'')/(k+k'')$ the mediant between h/k and h''/k''. The following theorem holds.

Theorem 1. *With the above notation,*

(i) $k'h - kh' = h''k - k''h = 1$,

(ii) $\dfrac{h}{k} = \dfrac{h'+h''}{k'+k''}$,

(iii) $n < k'+k < 2n$ *and* $n < k+k'' < 2n$,

(iv) $k' \neq k \neq k'' \neq k'$,

(v) $\dfrac{h'}{k'} < \dfrac{h'+h}{k'+k} < \dfrac{h}{k} < \dfrac{h+h''}{k+k''} < \dfrac{h''}{k''}$.

Examples. In $\mathscr{F}_4$ we verify, e.g., that for $\frac{1}{2} < \frac{2}{3} < \frac{3}{4}$, we have $2\cdot 2 - 1\cdot 3 = 3\cdot 3 - 2\cdot 4 = 1$, $\frac{2}{3} = (1+3)/(2+4)$, $4 < 2+3 < 8$, $4 < 3+4 < 8$. Also, $1 \neq 4 \neq 3 \neq 2 \neq 3 \neq 4 \neq 1$, and finally,

$$\frac{0}{1} < \frac{0+1}{1+4} < \frac{1}{4} < \frac{1+1}{4+3} < \frac{1}{3} < \frac{1+1}{3+2} < \frac{1}{2}$$

$$< \frac{1+2}{2+3} < \frac{2}{3} < \frac{2+3}{3+4} < \frac{3}{4} < \frac{3+1}{4+1} < \frac{1}{1}.$$

The proof of Theorem 1 is easy (see Problem 1, or [98]).

One often identifies the points of the real line, considered modulo one, with the points on the unit circle, by the correspondence $\alpha \leftrightarrow e^{2\pi i \alpha}$. In particular, one assigns to the rational h/k the point $e^{2\pi ih/k}$ on the unit circle. Starting with the Farey series of order n, we consider a dissection of the circle such that exactly one point corresponding to each h/k is on each arc. The most natural way to do this is to dissect the circle at the mediants of the fractions $h/k \in \mathscr{F}_n$. Then, if $h'/k' < h/k < h''/k''$ are three consecutive terms of $\mathscr{F}_n$, h/k belongs to the arc between the medians:

$$\frac{h'+h}{k'+k} < \frac{h}{k} < \frac{h+h''}{k+k''}.$$

Moreover,

$$\frac{h}{k} - \frac{h+h'}{k+k'} = \frac{1}{k(k+k')}, \qquad \frac{h+h''}{k+k''} - \frac{h}{k} = \frac{1}{k(k+k'')},$$

and we infer from Theorem 1 the following theorem.

Theorem 2. *The distance d between h/k and any endpoint of its subarc on the dissection of order n satisfies*

$$\frac{1}{2kn} < \frac{1}{k(2n-1)} \leqslant d \leqslant \frac{1}{k(n+1)} < \frac{1}{kn}.$$

§3. Gaussian Sums

Definition 2. Let $h, k \in \mathbb{Z}$, $k > 0$, $(h, k) = 1$; then the sum $\sum_{m \pmod{k}} e^{2\pi ihm^2/k}$ is called a *Gaussian sum* and is denoted by $G(h, k)$.

The Gaussian sums have a useful multiplicative property. Namely, for $(k_1, k_2) = 1$, one has $G(h, k_1k_2) = G(hk_1, k_2)G(hk_2, k_1)$. The proof follows from the remark that, if m_1 runs modulo k_1 and m_2 runs modulo k_2, then, for $(k_1, k_2) = 1$, $m = k_1m_2 + k_2m_1$ runs modulo k_1k_2 and

$$\exp\left(\frac{2\pi ihm^2}{k_1k_2}\right) = \exp\left(\frac{2\pi ih(k_1^2m_2^2 + 2k_1k_2m_1m_2 + k_2^2m_1^2)}{k_1k_2}\right)$$

$$= \exp\left(\frac{2\pi ihk_1m_2^2}{k_2}\right)\exp\left(\frac{2\pi ihk_2m_1^2}{k_1}\right).$$

By summing for $m_1 \pmod{k_1}$ and for $m_2 \pmod{k_2}$, we obtain the result.

Let $(h, k) = 1$; then the Gaussian sum $G(h, k)$ may be computed with the help of the following theorem.

Theorem 3. *For $(h, k) = 1$ the Gaussian sums satisfy the following relations:*

(i) $G(h, 1) = 1$, $G(h, 2) = 0$, *and*

$$G(h, 2^a) = \begin{cases} (1 + i^h)2^{a/2} & \text{for even } a > 0, \\ e^{\pi i h/4} 2^{(a+1)/2} & \text{for odd } a > 1. \end{cases}$$

(ii) *For k odd, $G(h, k) = (\frac{h}{k})G(1, k)$.*

(iii) *Regardless of the parity of k, $G(1, k) = \varepsilon_k \sqrt{k}$, with $\varepsilon_k = (1 + i) \times (1 + i^{-k})/2 = 1 + i$, 1, 0, or i, according as $k \equiv 0$, 1, 2, or 3 (mod 4); in particular, for $k \equiv 1 \pmod 2$, $|G(h, k)| = \sqrt{k}$ and $|G(h, k)| \leqslant \sqrt{2k}$ for all k.*

(iv) *For squarefree k, $G(h, k) = \sum_{m \,(\mathrm{mod}\, k)} (m/k) e^{2\pi i h m/k}$.*

(v) *The Gaussian sums are multiplicative, in the sense that for $(k_1, k_2) = 1$, $G(h, k_1 k_2) = G(hk_1, k_2)G(hk_2, k_1)$.*

Proof. Property (v) has already been proved, and we can use it to reduce the computation of any Gaussian sum to that of $G(h, p^a)$. Next, if for $a \geqslant 2$ and odd p we put $m = up^{a-1} + v$, $u = 0, 1, \ldots, p - 1$, and $v = 0, 1, \ldots, q$, $q = p^{a-1} - 1$, then

$$G(h, p^a) = \sum_{u=0}^{p-1} \sum_{v=0}^{q} \exp\left\{\frac{2\pi i h}{p^a}(u^2 p^{2a-2} + 2uvp^{a-1} + v^2)\right\}$$

$$= \sum_{v=0}^{p^{a-1}-1} \exp\left(\frac{2\pi i h v^2}{p^a}\right) \sum_{u=0}^{p-1} \exp\left(\frac{2\pi i h \cdot 2uv}{p}\right).$$

Here the inner sum vanishes unless $p \mid 2hv$; but $p \nmid 2h$, so that one needs $p \mid v$, in which case the inner sum equals p and $G(h, p^a) = pG(h, p^{a-2})$. If a is even, $G(h, p^a) = p^{a/2}G(h, 1) = p^{a/2}$. Here use has been made of $G(h, 1) = 1$; the proof of this relation, as well as of $G(h, 2) = 0$, is left to the reader (see Problem 3). If a is odd, then $G(h, p^a) = p^{(a-1)/2}G(h, p)$. It remains to compute $G(h, p)$.

We have $G(h, p) = \sum_{m \,(\mathrm{mod}\, p)} e^{2\pi i h m^2/p} = 1 + 2\sum e^{2\pi i h r/p}$, where r runs over the quadratic residues modulo p, so that

$$G(h, p) = \sum_{m \bmod p} \left(1 + \left(\frac{m}{p}\right)\right) e^{2\pi i h m/p} = \sum_{m=0}^{p-1} e^{2\pi i h m/p} + \sum_{m=0}^{p-1} \left(\frac{m}{p}\right) e^{2\pi i h m/p}.$$

The first sum vanishes, while the second is just (iv) for $k = p$, and may be written as

$$G(h, p) = \left(\frac{h}{p}\right) \sum_{m=0}^{p-1} \left(\frac{mh}{p}\right) e^{2\pi i h m/p} = \left(\frac{h}{p}\right) \sum_{n=0}^{p-1} \left(\frac{n}{p}\right) e^{2\pi i n/p} = \left(\frac{h}{p}\right) G(1, p).$$

To finish the proof of (ii), we now only have to invoke the multiplicative property (v) and use induction.

If we set $\Gamma(h,k)=\sum_{m \,(\mathrm{mod}\, k)}\left(\frac{m}{k}\right)e^{2\pi i h m/k}$, then we have proved that $G(h,p)=\Gamma(h,p)$. It should be observed that, generally, $G(h,k)=\Gamma(h,k)$ is *false* (counterexample: $G(h,p^2)=p$, $\Gamma(h,p^2)=\sum_{(n,p)=1,\,0<n<p^2}e^{2\pi i n/p^2}=0$). For squarefree k, however, observe that if $k=k_1k_2$, then just as for $G(h,k)$, also $\Gamma(h,k_1k_2)=\Gamma(hk_1,k_2)\Gamma(hk_2,k_1)$ holds. From this and $G(h,p)=\Gamma(h,p)$, (iv) follows.

It is easy to show that $|G(1,k)|=\sqrt{k}$ for odd k, but difficult to determine the proper argument. From $G(-1,k)=\left(\frac{-1}{k}\right)G(1,k)$ it follows that

$$\begin{aligned}
G(1,k)^2 &= \left(\frac{-1}{k}\right)G(1,k)G(-1,k)=\left(\frac{-1}{k}\right)\sum_{m\,(\mathrm{mod}\,k)}e^{2\pi i m^2/k}\sum_{n\,(\mathrm{mod}\,k)}e^{-2\pi i n^2/k}\\
&= \left(\frac{-1}{k}\right)\sum_{m\,(\mathrm{mod}\,k)}\sum_{n\,(\mathrm{mod}\,k)}e^{2\pi i(m+n)(m-n)/k}\\
&= \left(\frac{-1}{k}\right)\sum_{n\,(\mathrm{mod}\,k)}\sum_{d\,(\mathrm{mod}\,k)}e^{2\pi i d(d+2n)/k},
\end{aligned}$$

where $d=m-n$ also runs modulo k. We obtain $G(1,k)^2=\left(\frac{-1}{k}\right)\sum_{d\,(\mathrm{mod}\,k)}\times e^{2\pi i d^2/k}\sum_{n=0}^{k-1}e^{4\pi i d n/k}$. Here the inner sum vanishes except for $d=0$, when it reduces to k and the outside factor is 1. We obtain $G(1,k)^2=\left(\frac{-1}{k}\right)k$, and for $k\equiv 1 \pmod 4$, $G(1,k)=\pm\sqrt{k}$, whereas for $k\equiv 3 \pmod 4$, $G(1,k)=\pm i\sqrt{k}$. The fact that in both these cases the upper sign holds is quite nontrivial (see, e.g., Landau [146], where four essentially different proofs are given). Both formulae may be condensed into one by setting $G(1,k)=i^{((k-1)/2)^2}\sqrt{k}$.

The case $p=2$ has to be considered separately. The cases $k=2^a$ with $a=0$ or $a=1$ are settled. Otherwise, one can reduce the exponent as in the case of odd $k=p^a$ and verify separately for a even and a odd the claim of Theorem 3(i). By combining the results for k a prime power, either even or odd, we complete the proof of (iii). All claims of Theorem 3 are herewith proved (given the omitted steps). □

§4. The Modular Group and Its Subgroups

The variable τ used in Chapter 8 was introduced as the ratio $\tau=\omega_2/\omega_1$ of two periods of a doubly periodic function. These periods were selected for convenience in a definite, canonical way, so that, in particular $\operatorname{Im}\tau>0$, i.e., τ belongs to the upper half plane, which we shall denote by $\mathscr{H}$. These two periods are the generators of the lattice of periods $\Omega=\{m\omega_2+n\omega_1\}$, $m,n\in\mathbb{Z}$. However, if we set $\tau'=\omega_2'/\omega_1'$, with

$$\begin{pmatrix} \omega_2' \\ \omega_1' \end{pmatrix} = \begin{pmatrix} a & b \\ c & d \end{pmatrix} \begin{pmatrix} \omega_2 \\ \omega_1 \end{pmatrix},$$

where $a, b, c, d \in \mathbb{Z}$ and $ad - bc = \pm 1$, then it is immediate that

$$\begin{pmatrix} \omega_2 \\ \omega_1 \end{pmatrix} = \begin{pmatrix} d & -b \\ -c & a \end{pmatrix} \begin{pmatrix} \omega_2' \\ \omega_1' \end{pmatrix},$$

so that Ω is spanned equally well by any pair (ω_2', ω_1') of periods related to the original one by

$$\begin{pmatrix} \omega_2' \\ \omega_1' \end{pmatrix} = M \begin{pmatrix} \omega_2 \\ \omega_1 \end{pmatrix} \quad \text{with } \det M = \pm 1.$$

We observe that $\tau' = (a\tau + b)/(c\tau + d)$ has $\operatorname{Im} \tau' = (ad - bc)(\operatorname{Im} \tau)/|c\tau + d|^2$, and the requirement $\operatorname{Im} \tau' > 0$ forces us to restrict attention to the case $ad - bc = +1$. The set of matrices

$$M = \left\{ \begin{pmatrix} a & b \\ c & d \end{pmatrix} \middle| \, a, b, c, d \in \mathbb{Z}, ad - bc = 1 \right\}$$

forms a group $\bar{\Gamma}$ under matrix multiplication, with identity element $I = \left(\begin{smallmatrix} 1 & 0 \\ 0 & 1 \end{smallmatrix}\right)$ and center $C = \{I, -I\}$. On the other hand, if $\tau' = M\tau$ and $\tau'' = M'\tau'$, then $\tau'' = M'(M\tau) = (M'M)\tau = M''\tau$, say, so that composition of these transformations of τ, $\operatorname{Im} \tau > 0$, corresponds to the multiplication of the corresponding matrices. Nevertheless, if $\tau' = M\tau = (a\tau + b)/(c\tau + d)$, then also

$$(-I)M\tau = \frac{-a\tau - b}{-c\tau - d} = \frac{a\tau + b}{c\tau + d},$$

so that, while the set of transformations $\tau \to M\tau$ also forms a group under composition, this group, say Γ, is not isomorphic to $\bar{\Gamma}$, but we have instead $\Gamma \simeq \bar{\Gamma}/\{I, -I\} \simeq \bar{\Gamma}/C$.

Definition 3. The group $\Gamma \simeq \bar{\Gamma}/C$ of transformations $\tau \to \tau' = M\tau$, $M = \left(\begin{smallmatrix} a & b \\ c & d \end{smallmatrix}\right)$, a, b, c, $d \in \mathbb{Z}$, $ad - bc = 1$, is called the *modular group*.

Let us now consider an elliptic function on Ω, normalized to $(\omega_1, \omega_2) = (1, \tau)$, say $f(z; 1, \tau)$. This function may be considered as a function of τ and remains invariant under the transformations $\tau \to \tau' = M\tau$, $M \in \Gamma$ (for simplicity and by abuse of notation, we do not write $M \in \bar{\Gamma}$, because we are interested merely in the transformation and do not want to distinguish between M and $-M$). Concentrating upon the dependence on τ only, we shall consider functions $f(\tau)$ such that $f(M\tau) = f(\tau)$ for $M \in \Gamma$. Such functions are called *modular functions*, provided that they satisfy also some regularity conditions to be mentioned later (see Section 5 for precise definition). Many functions of interest do not stay invariant under *all* transformations of the modular group, but only under a certain subset $S \subset \Gamma$. Obviously, $f(I\tau) = f(\tau)$, and if $f(M_1\tau) = f(\tau)$ and $f(M_2\tau) = f(\tau)$, then $f(M_2(M_1\tau)) = f(M_2(\tau')) = f(\tau') = f(M_1\tau) = f(\tau)$, so that $f(\tau)$ stays invariant under the subgroup, say G, of Γ generated by S (with M also $M^{-1} \in S$).

Two complex numbers τ_1, τ_2 in the upper half plane $\mathscr{H} = \{\tau \mid \operatorname{Im} \tau > 0\}$ are called *equivalent* under a subgroup $G \subset \Gamma$ if there exists a matrix $M \in G$ $M \neq I$ such that $\tau_2 = M\tau_1$.

Definition 4. A set $\mathscr{F}$ of points, $\mathscr{F} \subset \mathscr{H}$, with the property that no two points of $\mathscr{F}$ are equivalent under a given group G of transformations, but such that every $\tau \in \mathscr{H}$ is equivalent to (exactly) one point of $\mathscr{F}$, is called a *fundamental region* (abbreviated FR) for G.

Such a fundamental region need not even be connected (hence, it is not necessarily a region), but for practical reasons one prefers to work with connected fundamental regions, whose description is as simple as possible. In particular, for Γ one has $S = \left(\begin{smallmatrix} 1 & 1 \\ 0 & 1 \end{smallmatrix}\right) \in \Gamma$, so that $\tau \to \tau + 1$ is a transformation of the modular group, and so is $T = \left(\begin{smallmatrix} 0 & -1 \\ 1 & 0 \end{smallmatrix}\right) \in \Gamma$, which corresponds to $\tau \to -1/\tau$. The first fact tells us that a fundamental region may be placed inside a vertical strip of width 1, say $|\operatorname{Re} \tau| \leqslant \frac{1}{2}$. The second tells us that it is sufficient to restrict a fundamental region either to $|\tau| \leqslant 1$ or to $|\tau| \geqslant 1$. This shows that a fundamental region may be placed inside $\mathscr{F}_1 = \{\tau \mid |\operatorname{Re} \tau| \leqslant \frac{1}{2}, |\tau| \geqslant 1\}$. In fact, one may show that no two *interior* points of $\mathscr{F}_1$ are equivalent under Γ, and it is standard to choose $\mathscr{F}_1$ as the fundamental region for Γ. Strictly speaking, $\mathscr{F}_1$ is not a fundamental region according to Definition 4, because the points on the boundary line $\operatorname{Re} \tau = -\frac{1}{2}$ are equivalent to the corresponding ones with $\operatorname{Re} \tau = \frac{1}{2}$. Hence, we delete from $\mathscr{F}_1$ the points of the boundary of $\mathscr{F}_1$ with $\operatorname{Re} \tau = -\frac{1}{2}$. Next, on the arc $|\tau| = 1$, the points with $\operatorname{Re} \tau > 0$ are equivalent, under $\tau \to -1/\tau$, to points with $|\tau| = 1$, $\operatorname{Re} \tau < 0$, and consequently we may delete them and set $\mathscr{F} = \{\tau \mid -\frac{1}{2} < \operatorname{Re} \tau \leqslant \frac{1}{2},\ |\tau| > 1 \text{ for } \operatorname{Re} \tau < 0,\ |\tau| \geqslant 1 \text{ for } \operatorname{Re} \tau > 0\}$. This is still not quite what we want, because of the point $\tau = i$. This point corresponds to itself under $\tau \to -1/\tau$; hence, it is a double boundary point. If we keep it in $\mathscr{F}$, then it is no longer the case that no point of $\mathscr{F}$ has also its image in $\mathscr{F}$; if we eliminate it from $\mathscr{F}$, then no point of the set $(ai + b)/(ci + d)$, $ad - bc = 1$, $a, b, c, d \in \mathbb{Z}$, has an equivalent point in $\mathscr{F}$. We decide to keep it in $\mathscr{F}$; fortunately, we shall not have to worry about the consequences of this blemish. We shall call $\mathscr{F}$ a *standard fundamental region* (SFR) for Γ. The reader interested in a more thorough treatment of fundamental regions may wish to consult the books of J. Lehner [157, 158] or Klein and Fricke [128].

If we write Γ as the union of the cosets of a subgroup $G \subset \Gamma$, it is easy to verify that a fundamental region for G is the union of j fundamental regions for

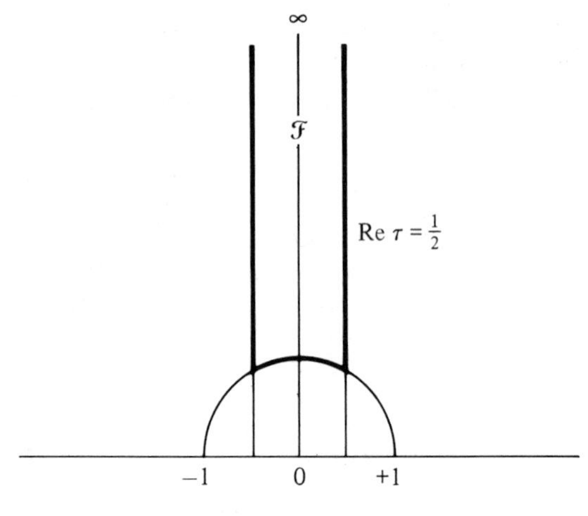

Figure 1

Γ, where j is the index of G in Γ. A function invariant under Γ is, in particular, periodic with period 1, so that if it is meromorphic in $\mathcal{H}$ and has only finitely many poles in $\operatorname{Im}\tau > 0$, it has a Fourier series

$$f(\tau) = \sum_{n=-\infty}^{\infty} a_n e^{2\pi i n\tau}. \tag{11.1}$$

Let $A = \left(\begin{smallmatrix} a & b \\ c & d \end{smallmatrix}\right) \in \Gamma$, and select, as we did for Γ, a FR for G, say $\mathcal{G}$, such that $\infty \in \mathcal{G}$. Then

$$A(\infty) = \lim_{\operatorname{Im}\tau\to\infty} \frac{a\tau + b}{c\tau + d} = \frac{b}{d}.$$

This is a real (indeed, a rational) point. If b/d happens to belong also to $\mathcal{G}$ (which was not the case for Γ, but is the case for many $G \subset \Gamma$), then it may be shown that it occurs as a point where two circles, both part of the boundary of $\mathcal{G}$, touch, while intersecting the real axis; such a point is a geometric cusp of $\mathcal{G}$. It is convenient to consider also $\tau = \infty$ (intersection of the straight boundary lines $\operatorname{Re}\tau = a$ and $\operatorname{Re}\tau = a + m$ at infinity) as a cusp.

Examples.

1. Figure 1 shows the SFR for Γ; it has the single cusp at $\tau = \infty$.
2. Consider the set of transformations generated by $S^2 = \left(\begin{smallmatrix} 1 & 2 \\ 0 & 1 \end{smallmatrix}\right)$ and $T = \left(\begin{smallmatrix} 0 & -1 \\ 1 & 0 \end{smallmatrix}\right)$. It generates a subgroup $\Gamma_\theta \subset \Gamma$, where the notation will soon become clear (see §7). It is easy to see that if we multiply the generators in any order and any number of times (the computations are made easier by observing that $T^2 = -I \in C$ and acts as identity on Γ), the resulting matrices have $a \equiv d \equiv b - 1 \equiv c - 1 \equiv \varepsilon \pmod 2$, where $\varepsilon = 0$ or 1. A

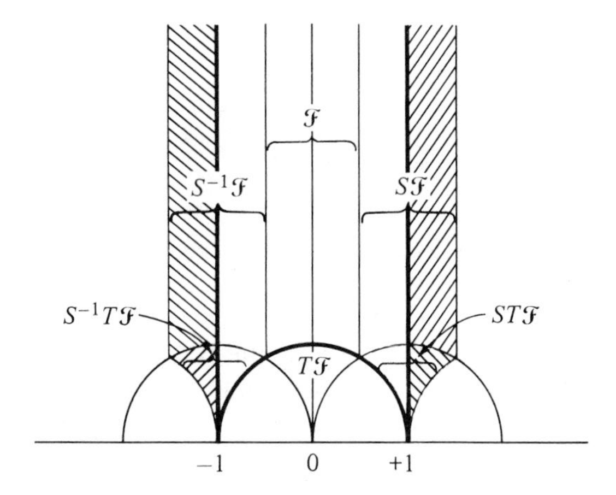

Figure 2. Parts of $S\mathcal{F}$ and $ST\mathcal{F}$ are replaced by equivalent parts of $S^{-1}\mathcal{F}$ and $S^{-1}T\mathcal{F}$ respectively, in order to obtain a symmetric fundamental region for Γ_θ. Eliminated portions are cross-hatched.

coset decomposition of Γ modulo Γ_θ is, i.e., $\Gamma_\theta \cup \Gamma_\theta S \cup \Gamma_\theta TS$, so that $[\Gamma : \Gamma_\theta] = 3$, and it follows that a possible fundamental region for Γ_θ is like that in Fig. 2. It has, besides the cusp at ∞, the two cusps at -1 and at $+1$. These, however, are equivalent under $\tau \to \tau + 2 \in \Gamma_\theta$, so that only one of them, say $\tau = -1$, is retained in the SFR for Γ_θ.

§5. Modular Forms

The requirement $f(M\tau) = f(\tau)$ is a very strong one and eliminates many important functions. A weaker requirement, satisfied by many functions of interest, is that for a fixed real number r, if $M \in G$, $M = \left(\begin{smallmatrix} a & b \\ c & d \end{smallmatrix}\right)$, then $f(M\tau) = (c\tau + d)^r f(\tau)$ for $\tau \in \mathcal{H}$ and all $M \in G$. In general, this requirement leads to no particular difficulties for r an even integer. Otherwise, however—especially for nonintegral r, and regardless of the initial determination of arguments for $(c\tau + d)^r$—the arguments of $f(M\tau)$ and of $(c\tau + d)^r f(\tau)$ may fail to stay equal, as M ranges over G. For that reason, it is better to consider a relation like

$$f(M\tau) = v(M)(c\tau + d)^r f(\tau), \tag{11.2}$$

where $|v(M)| = 1$, and $\arg v(M)$ yields precisely what is needed to keep (11.2) valid. It turns out that, in fact, $v(M)$ depends only on the lower row (c, d) of M (see [133] or [157] for proofs).

If we iterate the transformations, i.e., consider $\tau'' = M''\tau'$, $\tau' = M'\tau$, so that $\tau'' = M\tau$ with $M = M''M'$, two applications of (11.2) show that $v(M''M')$ cannot be taken arbitrarily, but has to satisfy certain compatibility conditions related to $v(M')$ and $v(M'')$, and also involving r.

Definition 5. A set of complex numbers $v(M)$ with $|v(M)| = 1$ and that satisfy the needed compatibility conditions, called *consistency relations*, on the matrices of some group G and with relation to some degree $-r$ (r as in (11.2)), is called a *system of multipliers* for the group G and the degree $-r$.

It should be observed that for $r \in \mathbb{Z}$, the consistency relations reduce to $v(M'M'') = v(M')v(M'')$, so that v is a character on the matrix group. In general, if $r \notin 2\mathbb{Z}$ then $v(-I) \neq v(I)$, so that, in general, v is a function on the matrix group and not on the group of transformations.

If G is a proper subgroup of Γ, then a function $f(\tau)$, meromorphic in $\mathscr{H}$, that satisfies (11.2) for a given system of multipliers (and has only finitely many poles in $\operatorname{Im} \tau > 0$) also has a "Fourier expansion" at $\tau = \infty$. This resembles (11.1), but is slightly more complicated, namely

$$f(\tau) = \sum_{n=-\infty}^{\infty} a_n e^{2\pi i (n+\kappa)\tau/m}; \tag{11.3}$$

here m is the smallest integer for which $f(\tau + m) = e^{2\pi i \kappa} f(\tau)$, or equivalently,

$$m = \min\left\{t \,\middle|\, S^t = \begin{pmatrix} 1 & t \\ 0 & 1 \end{pmatrix} \in G\right\}.$$

It is clear that one may fit a SFR $\mathscr{F}$ into a strip of width m, for which reason m is also called *the width of the cusp at* ∞. If q_j is a cusp, $q_j \neq \infty$, then there exists a real matrix A_j of determinant 1 such that $q_j = A_j(\infty)$. In that case, $f(\tau)$ has an expansion at q_j of the form

$$f(\tau) = (\tau - q_j)^{-r} \sum_{n=-\infty}^{\infty} a_n(j) e^{2\pi i (n+\kappa_j) A_j^{-1}\tau/\lambda_j}. \tag{11.4}$$

If only finitely many terms with $n < 0$ occur in (11.3), or in (11.4), we say that $f(\tau)$ is *meromorphic* at ∞, or at $\tau = q_j$, respectively.

Let $f(\tau)$ be meromorphic at q_j, and let $a_j(n)$ be the first nonvanishing coefficient. If this occurs at $n = -n_0 < 0$, we say that $f(\tau)$ has a *pole of order* $n_0 - \kappa_j$ at $\tau = q_j$; and if $n = n_0 \geqslant 0$, we say that $f(\tau)$ is regular, or holomorphic at q_j, with a *zero of order* $n_0 + \kappa_j$ at $\tau = q_j$. The corresponding statements for $q_j = \infty$ are similar. The reason for this terminology are obvious in (11.3), where for $\tau = x + iy$, $y \to \infty$, we have $|f(\tau)| \simeq a_{n_0} e^{2\pi i (n_0+\kappa) iy/m} = a_{n_0} e^{-(2\pi/m)(n_0+\kappa)y} = a_{n_0} z^{-(n_0+\kappa)}$, since $z = e^{-(2\pi/m)y} \to 0$ for $y \to \infty$. The same reasoning holds also in (11.4), by observing that for $\tau \to q_j$, $A_j^{-1} q_j \to i\infty$.

We are now prepared to give the necessary definitions.

Definition 6. For given $G \subset \Gamma$, let $v(M)$ be a multiplier system for G and the degree $-r$. A function $f(\tau)$, meromorphic in $\mathscr{H}$, is called a *modular form of degree* $-r$, with multiplier system v, with respect to the group G, if $f(\tau)$ satisfies

(11.2) and G possesses a FR $\mathscr{F}$ with at most finitely many poles of $f(\tau)$ in $\bar{\mathscr{F}}$, at each of which $f(\tau)$ has expansions of the type (11.3) or (11.4) with at most finitely many negative entries n.

Definition 7. If $f(\tau)$ is a modular form of degree $-r$ and furthermore $f(\tau)$ is regular in $\mathscr{H}$ as well as at all cusps of some SFR $\mathscr{F}$ of G, it is said to be an *entire* modular form.

Definition 8. If $f(\tau)$ is a modular form with respect to G, regular in $\mathscr{H}$ and with zeros of positive orders at all cusps of some SFR, then $f(\tau)$ is called a *cusp form.*

Definition 9. If $f(\tau)$ is a modular form with $r = 0$ and $v(M) = 1$ for all $M \in G$, then $f(\tau)$ is called a *modular function* with respect to G.

Remarks.

1. The change of sign of r in the terminology "degree $-r$" has historical origins and is not universally accepted. Terms like "dimension" or "weight" are also in use (the latter for $+r/2$ or r, rather than for $-r$).
2. Previous definitions refer to a specific $\mathscr{F}$; but if they hold with respect to any fundamental region, they hold with respect to all.
3. Very similar results hold in a much more general formulation and not only for subgroups of the modular group Γ. If $G \not\subset \Gamma$, it is customary to replace the term "modular" by "automorphic."

§6. Some Theorems

Here we shall quote, without proofs, some important theorems that will be needed in Chapter 12. Proofs may be found in [133], [157], [158], [128], [201], or [89].

Theorem 4. *Every entire modular function reduces to a constant.*

Theorem 5. *Let $f(\tau)$ be a cusp form of degree $-r$ with respect to G. Then the expansion of $f(\tau)$ at ∞ is of the form*

$$f(\tau) = \sum_{n+\kappa>0} a_n e^{2\pi i(n+\kappa)\tau/\lambda}$$

with some κ, $0 \leqslant \kappa < 1$, and $|a_n| = O(n^{r/2})$.

Comment. The first statement is obvious, because unless $n + \kappa > 0$, $f(\tau)$ could not have a zero of positive order at ∞ (i.e., for $\operatorname{Im}\tau \to \infty$). The statement $|a_n| = O(n^{r/2})$ is not obvious. It depends on the following lemma (due to Hecke [106]):

Lemma 1. *If $f(\tau)$ is a cusp form of degree $-r$ with respect to G, then $|f(x+iy)| \leqslant Cy^{-r/2}$ for all $\tau = x+iy \in \mathscr{H}$.*

We do not reproduce the lengthy proof of the lemma, but, using it, sketch the proof of the very important Theorem 5.

Proof of Theorem 5. By the use of Lemma 1, we obtain, integrating along the horizontal line from $z = x+iy$ to $z+\lambda$, where λ is the width of the cusp at ∞,

$$\int_z^{z+\lambda} f(\zeta)e^{-2\pi i(n+\kappa)\zeta/\lambda}\,d\zeta = \int_z^{z+\lambda}\left(\sum_{m+\kappa>0} a_m e^{2\pi i(m+\kappa)\zeta/\lambda}\right)e^{-2\pi i(n+\kappa)\zeta/\lambda}\,d\zeta$$

$$= \sum_{m+\kappa>0}\int_z^{z+\lambda} a_m e^{2\pi i(m-n)\zeta/\lambda}\,d\zeta = \lambda a_n.$$

Consequently, $|a_n| = \lambda^{-1}\,|\int_z^{z+\lambda} f(\zeta)e^{-2\pi i(n+\kappa)\zeta/\lambda}\,d\zeta| \leqslant \lambda^{-1}\cdot Cy^{-r/2}\cdot\lambda e^{2\pi(n+\kappa)y/\lambda}$. The maximum of $y^{-r/2}e^{2\pi(n+\kappa)y/\lambda}$ is easily found to be the constant $(4\pi(n+\kappa)e/\lambda r)^{r/2}$, taken for $y = \lambda r/4\pi(n+\kappa)$. Here λ depends only on the group G (and not on τ), and so

$$|a_n| \leqslant C\left(\frac{4\pi(1+(\kappa/n))e}{\lambda r}\right)^{r/2} n^{r/2} < C\left(\frac{8\pi e}{\lambda r}\right)^{r/2} n^{r/2}. \quad \square$$

It should be mentioned that the estimate $|a_n| = O(n^{r/2})$ has been improved recently (proof of Deligne of the Ramanujan–Petersson conjecture [51]) to $|a_n| = O(n^{(r-1+\varepsilon)/2})$ for any $\varepsilon > 0$. This improvement is highly significant, but we shall not need it here.

§7. The Theta Functions as Modular Forms

Let $\theta(\tau)$ stand for $\theta_3(0\,|\,\tau)$, so that $\theta(\tau) = \sum_{m=-\infty}^{\infty} e^{\pi i\tau m^2}$ and

$$\theta^s(\tau) = 1 + \sum_{n=1}^{\infty} r_s(n)e^{\pi i\tau n}. \tag{11.5}$$

It is clear that $\theta(\tau+2) = \sum_{m=-\infty}^{\infty} e^{\pi i(\tau+2)m^2} = \sum_{m=-\infty}^{\infty} e^{\pi i\tau m^2}e^{2\pi i m^2} = \theta(\tau)$. Also, (8.1), with $t = -i\tau = -i(x+iy) = -ix+y$, $y>0$, shows that $\theta(\tau) = \theta(it) = \sum_{m=-\infty}^{\infty} e^{-\pi t m^2} = \Psi(t,0) = t^{-1/2}\Psi(t^{-1},0) = (-i\tau)^{-1/2}\sum_{m=-\infty}^{\infty} e^{-\pi m^2/(-i\tau)} = (-i\tau)^{-1/2}\sum_{m=-\infty}^{\infty} e^{-i\pi m^2/\tau} = (-i\tau)^{-1/2}\theta(-1/\tau)$. Consequently,

$$\theta(M\tau) = v(M_{c,d})(c\tau+d)^{1/2}\theta(\tau) \tag{11.6}$$

holds for the two generators of Γ_θ, with $v(S^2) = 1$, $v(T) = (-i)^{-1/2}$. By using the consistency relations of $v(M)$ in Γ_θ, we may show that (11.6) holds for all $M \in \Gamma_\theta$ and a system of multipliers that we denote by $v_\theta(M)$. We now recall that Γ_θ has a SFR with two nonequivalent cusps, one at $\tau = \infty$ and one at $\tau = -1$. The expansion of $\theta(\tau)$ at $\tau = \infty$ is clearly (11.5), with $s = 1$ and $r_1(n) = 2$.

We can also expand $\theta(\tau)$ at $\tau = -1$. The computations are nontrivial, and we suppress them here (see [133] for a complete proof), but state the result as

Theorem 6. *At $\tau = -1$, $\theta(\tau)$ has the expansion*

$$\theta(\tau) = (\tau + 1)^{-1/2} \sum_{n=0}^{\infty} b_n \exp\left\{2\pi i(n + \tfrac{1}{8})\left(\frac{-1}{\tau + 1}\right)\right\}, \qquad b_n \neq 0.$$

From (11.5), (11.6), and Theorem 6 it follows that $\theta(\tau)$ is a modular form of degree $-\frac{1}{2}$ under Γ_θ. The reason for the term "theta group" and for the notation are now clear.

For later use we also quote (from [133]) the following theorem, which gives the explicit value of the multiplier system $v_\theta(M)$ for the function $\theta(\tau)$ and $M \in \Gamma_\theta$.

Theorem 7. *For $M = \left(\begin{smallmatrix} a & b \\ c & d \end{smallmatrix}\right) \in \Gamma_\theta$,*

$$v_\theta(M) = \begin{cases} \left(\dfrac{d}{|c|}\right) e^{-\pi i c/4} \\ \qquad \text{if } a \equiv b - 1 \equiv c - 1 \equiv d \equiv 0 \pmod 2 \text{ with } d \neq 0, \\ 1 \text{ if } d = 0, \\ \left(\dfrac{c}{|d|}\right)(-1)^{(\operatorname{sgn} c - 1)(\operatorname{sgn} d - 1)/4} e^{\pi i (d-1)/4} \\ \qquad \text{if } a - 1 \equiv b \equiv c \equiv d - 1 \pmod 2 \text{ if also } c \neq 0, \\ \operatorname{sgn} d \qquad \text{if } c = 0. \end{cases}$$

§8. Problems

1. Prove Theorem 1.
2. Fill in the details of the proof of the multiplicativity of Gaussian sums.
3. Prove part (i) of Theorem 3.
4. Justify the identity $\sum_{(n,p)=1,\, 0<n<p^2} e^{2\pi i n/p^2} = 0$.
5. Show that if $M \in \Gamma$ and $f(M\tau) = f(\tau)$, then also $f(M^{-1}\tau) = f(\tau)$.
6. Prove the multiplicativity (as defined in the text) of $\Gamma(h, k)$ for k squarefree.
7. Determine a simply connected fundamental region for

$$\Gamma(5) = \left\{M = \begin{pmatrix} a & b \\ c & d \end{pmatrix} \in \Gamma \,\middle|\, a \equiv d \equiv 1 \pmod 5,\ b \equiv c \equiv 0 \pmod 5\right\}.$$

8. Determine a simply connected fundamental region for

$$\Gamma_0(5) = \left\{ M = \begin{pmatrix} a & b \\ c & d \end{pmatrix} \in \Gamma \,\middle|\, c \equiv 0 \pmod{5} \right\}.$$

   In particular, find the width of the cusp at $\tau = \infty$.

9. Show (e.g., by following the suggestions of §5) that if $f(\tau)$ satisfies (11.2), then the multipliers $v(M)$ must satisfy the consistency relation

$$v(M_1 M_2)(c_3\tau + d_3)^r = v(M_1)v(M_2)(c_1 M_2\tau + d_1)^r(c_2\tau + d_2)^r,$$

   where (c_i, d_i) is the lower row of M_i and $M_3 = M_1 M_2$.

Chapter 12

The Circle Method

§1. The Principle of the Method

We mentioned in Chapter 1 that the number $r_s(n)$ of solutions of the Diophantine equation

$$\sum_{k=1}^{s} x_i^2 = n \tag{12.1}$$

is the coefficient of x^n in the Taylor expansion of the function $\{\theta(x)\}^s = 1 + \sum_{n=1}^{\infty} r_s(n)x^n$. Here, as in Chapter 8, we write $\theta(x)$ for $\theta_3(1; x)$ and we shall suppress the first entry, which will always be $z = 1$. From (12.1); it follows, by Cauchy's theorem, that

$$r_s(n) = \frac{1}{2\pi i}\int_{\mathscr{C}} x^{-n-1}\theta^s(x)\,dx, \tag{12.2}$$

where, we recall,

$$\theta(x) = \sum_{-\infty}^{\infty} x^{n^2} = 1 + 2\sum_{n=1}^{\infty} x^{n^2} = \sum_{k=0}^{\infty} a_k x^k, \qquad \text{say}, \tag{12.3}$$

and $\mathscr{C}$ is a sufficiently small circle around the origin.

The Taylor series for $\theta(x)$, and hence also for $\theta^s(x)$, converges for $|x| < 1$, as we know from Chapter 8. However, (12.3) shows that the gaps between successive powers of x with nonvanishing coefficients are rapidly increasing, while the coefficients remain constant; hence $\lim_{k\to\infty}(a_k/k^2) = \lim_{k\to\infty}(2/k^2) = 0$, and by Fabry's Theorem (see, e.g., [114]), $\theta(x)$ has the unit circle as its natural boundary. This shows that in (12.2) we have to take for $\mathscr{C}$ a closed curve that turns once around the origin and is contained entirely inside the unit circle. This can be achieved by setting $x = \exp(-2\pi\delta + 2\pi i\psi)$, where $\delta > 0$ is fixed and $0 \leqslant \psi < 1$. At the same time, one may observe (see Problem 1) that as $|x| \to 1^-$, $|\theta(x)|$ becomes large particularly quickly if x approaches a root of unity $e^{2\pi ih/k}$ with small denominator k. This suggests the following approach: the circle $|x| = e^{-2\pi\delta}$ is subdivided into a certain number of arcs, having such points $e^{2\pi ih/k}$ with $1 \leqslant k \leqslant N$ at or close to their centers; here N is a certain natural integer that we may select appropriately. Then a saddle point method, applied to each subarc, may permit an evaluation of the integral, and by summing over all arcs we may hope to obtain the integral in (12.2) with acceptable accuracy.

This reasoning motivates the following procedure: Let N be a fixed (large) positive integer, and consider the Farey series of order N. Then, if $h_1/k_k < h/k < h_2/k_2$ are three consecutive terms, let the mediants be $\psi_1 = \psi_1(h,k) = (h_1 + h)/(k_1 + k)$ and $\psi_2 = \psi_2(h,k) = (h + h_2)/(k + k_2)$, and consider the arc $\psi_1 \leqslant \psi = (h/k) + \phi \leqslant \psi_2$. As seen in §11.2, we then have

$$-\phi_1 = \frac{-1}{k(k+k_1)} = \frac{h_1+h}{k_1+k} - \frac{h}{k} \leqslant \phi \leqslant \frac{h+h_2}{k+k_2} - \frac{h}{k} = \frac{1}{k(k+k_2)} = \phi_2.$$

If we denote this interval (and, by abuse of notation, the corresponding arc) by $\phi_{h,k}$, equation (12.2) may be written as

$$r_s(n)\frac{1}{2\pi i} \sum_{\substack{h,k \\ 0 \leqslant h < k \leqslant N \\ (h,k)=1}} \int_{\phi_{h,k}} \frac{\theta^s(x)}{x^{n+1}}\,dx$$

$$= \sum \int_{\phi_{h,k}} \theta^s(e^{-2\pi\delta + 2\pi ih/k + 2\pi i\phi}) e^{-n(-2\pi\delta + 2\pi ih/k + 2\pi i\phi)}\,d\phi$$

$$= \sum e^{-2\pi inh/k} \int_{\phi_{h,k}} \theta^s(e^{-2\pi\delta + 2\pi ih/k + 2\pi i\phi}) e^{2\pi n\delta - 2\pi in\phi}\,d\phi.$$

Here and in what follows, we do not repeat the summation letters and conditions, which, unless changed explicitly, remain those of the first sum. Also, we shall omit the symbol $\phi_{h,k}$ whenever possible. Finally, it is reasonable to expect that, for different values of N, different values of δ will lead to the best results, so that we should think of δ really as depending on N, and we sometimes write δ_N.

Next, we recall that, by Theorem 11.2, $1/2kN \leqslant \phi_i = 1/k(k_i + k) \leqslant 1/kN$. The expression of $r_s(n)$ can be simplified if we introduce the complex variable $z = \delta - i\phi$, $\operatorname{Re} z = \delta > 0$; we then obtain

$$r_s(n) = \sum e^{-2\pi ihn/k} \int \theta^s(e^{2\pi ih/k - 2\pi z}) e^{2\pi nz}\,d\phi.$$

Also $\theta(e^{2\pi ih/k - 2\pi z}) = \sum_{m=-\infty}^{\infty} e^{m^2(2\pi ih/k - 2\pi z)}$; this suggests setting $m = qk + j$, $j = 0, 1, \ldots, k-1$, $-\infty < q < \infty$, and

$$\theta(e^{2\pi ih/k - 2\pi z}) = \sum_{j=0}^{k-1} \sum_{q=-\infty}^{\infty} \exp\left\{(q^2k^2 + 2qkj + j^2)\frac{2\pi ih}{k}\right\} \cdot e^{-2\pi(qk+j)^2 z}$$

$$= \sum_{j=0}^{k-1} e^{2\pi ihj^2/k} \sum_{q=-\infty}^{\infty} e^{-2\pi k^2(q+j/k)^2 z}.$$

The inner sum may be transformed using Poisson's formula. We recall that for $\operatorname{Re} t > 0$ and arbitrary $\alpha \in \mathbb{C}$, the application of Theorem 8.1 to $f(x) = e^{-\pi(x+\alpha)^2 t}$ leads to

$$\Psi(t, \alpha) = \sum_{n=-\infty}^{\infty} e^{-\pi(n+\alpha)^2 t} = t^{-1/2} \sum_{k=-\infty}^{\infty} e^{2\pi i k\alpha - \pi k^2/t}.$$

Hence, if we take $\alpha = j/k$ and $t = 2k^2 z$ ($\operatorname{Re} 2k^2 z = 2k^2 \delta > 0$), the inner sum equals $(2k^2 z)^{-1/2} \sum_{m=-\infty}^{\infty} e^{2\pi i m j/k - (\pi m^2/2k^2 z)}$ and we have

$$\begin{aligned}
\theta(e^{2\pi i h/k - 2\pi z}) &= (2z)^{-1/2} \frac{1}{k} \sum_{j=0}^{k-1} e^{2\pi i h j^2/k} \sum_{m=-\infty}^{\infty} e^{-(\pi m^2/2k^2 z) + 2\pi i m j/k} \\
&= \frac{1}{k\sqrt{2z}} \sum_{j=0}^{k-1} e^{2\pi i h j^2/k} \left\{ 1 + \sum_{m \neq 0} e^{-(\pi m^2/2k^2 z) + 2\pi i j m/k} \right\} \\
&= \frac{1}{k\sqrt{2z}} \left\{ G(h, k) + \sum_{m \neq 0} e^{-\pi m^2/2k^2 z} T_k(h, m) \right\} \\
&= \frac{1}{k\sqrt{2z}} \{ G(h, k) + H(h, k; z) \} \qquad \text{say,}
\end{aligned}$$

where $G(h, k) = \sum_{j=0}^{k-1} e^{2\pi i h j^2/k}$ is the Gaussian sum, $T_k(h, m) = \sum_{j=0}^{k-1} e^{2\pi i (h j^2 + m j)/k}$, and

$$H(h, k; z) = \sum_{m \neq 0} e^{-\pi m^2/2k^2 z} T_k(h, m). \tag{12.4}$$

Next, we proceed to show that

$$|T_k(h, m)| \leqslant \sqrt{2k}, \tag{12.5}$$

which will be sufficient for our purposes. Indeed,

$$T_k \bar{T}_k = |T_k|^2 = \sum_j e^{2\pi i (h j^2 + m j)/k} \sum_r e^{-2\pi i (h r^2 + m r)/k} = \sum_j \sum_r e^{(2\pi i/k)(j - r)(h(j + r) + m)},$$

where j and r both run modulo k. If we denote $j - r$ by d, then also d runs modulo k, and

$$|T_k|^2 = \sum_d \sum_r e^{(2\pi i/k) d (h(2r + d) + m)} = \sum_d e^{(2\pi i/k) d (d h + m)} \sum_r e^{(4\pi i/k) d h r}.$$

Since $(h, k) = 1$, the inner sum vanishes unless $k \mid 2d$, when it equals k. If k is odd and $0 \leqslant d < k$, only $d = 0$ gives a contribution and $|T_k|^2 = k$. If k is even, the inner sum equals k for both $d = 0$ and $d = k/2$. In the first case, $e^{(2\pi i/k) d (d h + m)} = 1$, and in the second it equals $e^{\pi i (k h/2 + m)} = \pm 1$, so that $|T_k|^2 = 0$ or $|T_k|^2 =$

$2k$. However, in all cases (12.5) holds. It now follows from (12.4) and (12.5) that

$$|H(h,k;z)| \leqslant 2\sqrt{2k}\sum_{m=1}^{\infty} e^{-\operatorname{Re}(\pi m^2/2k^2 z)}$$

$$= \sqrt{8k}e^{-(\pi\sigma/2k^2)}\sum_{m=1}^{\infty} e^{-\pi(m^2-1)\sigma/2k^2},$$

where

$$\sigma = \operatorname{Re}\frac{1}{z} = \operatorname{Re}\frac{\delta + i\phi}{\delta^2 + \phi^2} = \frac{\delta}{\delta^2 + \phi^2}.$$

However, $|\phi| \leqslant \max(\phi_1, \phi_2) < 1/kN$; hence, $|\phi k| \leqslant N^{-1}$ and

$$\frac{\sigma}{k^2} = \frac{\delta}{k^2\delta^2 + k^2\phi^2} \geqslant \frac{\delta}{N^2\delta^2 + N^{-2}} = \frac{1}{\delta N^2 + (\delta N^2)^{-1}}.$$

We now make the announced choice for $\delta = \delta_N$, namely $\delta = N^{-2}$, which maximizes σ/k^2. For this choice of δ, $\sigma/k^2 \geqslant \frac{1}{2}$ and $|H(h,k;z)| \leqslant \sqrt{8k}e^{-(\pi\sigma/2k^2)}\sum_{m=1}^{\infty} e^{-\pi(m^2-1)/4}$. The values of $m^2 - 1$ in the exponent are 0, 3, and for $m \geqslant 3$ we have $(m+1)(m-1) \geqslant 4(m-1) > 3(m-1)$; hence, the sum is majorized by $\sum_{r=0}^{\infty} e^{-3\pi r/4} = (1 - e^{-3\pi/4})^{-1}$, so that $|H(h,k;z)| < Ce^{-\pi\sigma/2k^2}\sqrt{k} < C_1\sqrt{k}$ $(C_1 < 4)$. In what follows, C, C_1, etc., will stand for constants, not always the same, either absolute or depending only on s, which is itself given.

So far we have obtained (suppressing for simplicity of writing the obvious parameters) that $\theta = (k\sqrt{2z})^{-1}(G + H)$, whence $\theta^s = (k\sqrt{2z})^{-s}\{G^s + \binom{s}{1}G^{s-1}H + \cdots + H^s\}$. We now recall from Theorem 11.3 that $|G| \leqslant \sqrt{2k}$; hence, for $1 \leqslant r \leqslant s$,

$$|G^{s-r}H^r| < C(2k)^{s/2}e^{-\pi\sigma/2k^2}. \tag{12.6}$$

We conclude that

$$\left|\theta^s - \frac{G^s}{k^s(2z)^{s/2}}\right| \leqslant C\frac{e^{-\pi\sigma/2k^2}}{k^{s/2}|z|^{s/2}} \tag{12.7}$$

and

$$\begin{aligned} r_s(n) = {} & \sum_{h,k} e^{-2\pi i h n/k}\int\left\{\frac{G(h,k)}{k(2z)^{1/2}}\right\}^s e^{2\pi n z}\,d\phi \\ & + \sum_{h,k} e^{-2\pi i h n/k}\int\left\{\theta^s - \left(\frac{G}{k\sqrt{2z}}\right)^s\right\}e^{2\pi n z}\,d\phi. \end{aligned} \tag{12.8}$$

§2. The Evaluation of the Error Terms and Formula for $r_s(n)$

From (12.7) and (12.8) it follows that

$$\left| r_s(n) - \sum e^{-2\pi i h n/k} \int \left\{ \frac{G(h,k)}{k\sqrt{2z}} \right\}^s e^{2\pi n z}\, d\phi \right|$$

$$\leqslant C \sum k^{-s/2} \int |z|^{-s/2} e^{2\pi n\delta - \pi\sigma/2k^2}\, d\phi$$

$$= Ce^{2\pi n/N^2} \sum \int k^{-s/2} \frac{e^{-\pi\sigma/2k^2}}{(\delta^2 + \phi^2)^{s/4}}\, d\phi$$

$$= Ce^{2\pi n/N^2} \sum \int \frac{e^{-(\pi/2k^2)\delta/(\delta^2+\phi^2)}}{\{k^2(\delta^2 + \phi^2)\}^{s/4}}\, d\phi$$

$$< Ce^{2\pi n/N^2} \sum \delta^{-s/4} \int \left\{ \frac{\pi\delta}{2k^2(\delta^2 + \phi^2)} \right\}^{s/4} \exp\left\{ -\frac{\pi\delta}{2k^2(\delta^2 + \phi^2)} \right\} d\phi$$

$$= E_1, \qquad \text{say.}$$

We shall consider E_1 as an error term to be evaluated. Let $y = \pi\delta/(2k^2 \times (\delta^2 + \phi^2))$; then the integrand equals $y^{s/4}e^{-y}$. It vanishes in the limit for $y \to 0$ and for $y \to \infty$, and has a single maximum $(s/4e)^{s/4}$, reached for $y = s/4$. It follows that the integral is majorized by $C \int_{-\phi_1}^{\phi_2} d\phi$ and $\int_{-\phi_1}^{\phi_2} d\phi \leqslant \int_0^{2\pi} d\phi = 2\pi$, so that, by using $\delta = N^{-2}$, we have $|E_1| < Ce^{2\pi n/N^2}\delta^{-s/4} = Ce^{2\pi n/N^2}N^{s/2}$. For any given n, we now select $N = [\sqrt{n}]$; then $|E_1| < Ce^{2\pi}n^{s/4} = O(n^{s/4})$.

The main difficulty is now the computation of the integral in the principal term. Here one proceeds as usual in the saddle point method: towards the ends of the interval—and especially beyond them—the integrand becomes so small that the contribution to the integral is negligible. Thus we shall replace $\int_{-\phi_1}^{\phi_2}$ by $\int_{-\infty}^{\infty}$ and estimate the new error E_2 introduced by this change. The factor $\{G(h,k)/k\sqrt{2}\}^s$ is independent of ϕ; hence, we can factor it out of the integral and have to estimate only

$$\left| \int_{\phi_2}^{\infty} z^{-s/2} e^{2\pi n z}\, d\phi \right| \leqslant \int_{\phi_2}^{\infty} \frac{e^{2\pi n\delta}}{(\delta^2 + \phi^2)^{s/4}}\, d\phi$$

$$\leqslant e^{2\pi n/N^2} \int_{1/2kN}^{\infty} \frac{d\phi}{\delta^{s/2}(1 + (\phi/\delta)^2)^{s/4}}$$

$$= e^{2\pi n/N^2} \delta^{1-(s/2)} \int_{1/2kN\delta}^{\infty} \frac{d\psi}{(1 + \psi^2)^{s/4}},$$

with $\psi = \phi/\delta$ and where we used $\phi_2 > 1/2kN$. The integral converges for $s > 2$, and there it is majorized by

$$\int_{N^2/2kN}^{\infty} \frac{d\psi}{\psi^{s/2}} = \frac{(N/2k)^{1-(s/2)}}{(s/2)-1} \leqslant Cn^{(2-s)/4}k^{(s/2)-1},$$

so that

$$\left| \int_{\phi_2}^{\infty} z^{-s/2} e^{2\pi n z}\, d\phi \right| < CN^{s-2}\, n^{(2-s)/4} k^{(s-2)/2} \leqslant Cn^{(s-2)/4} k^{(s-2)/2}.$$

If we multiply this by the factor $\{G(h,k)/\sqrt{2}k\}^s \leqslant (C\sqrt{k}/k)^s = Ck^{-s/2}$, we obtain that the absolute value of each term of the sum $\sum e^{-2\pi i h n/k} \int_{\phi_2}^{\infty} (\frac{G}{k\sqrt{2z}})^s e^{2\pi n z}\, d\phi$ is at most $Cn^{s/4-1/2}k^{-1}$. The sum over h has $\phi(k) \leqslant k$ terms, all majorized, as seen, by $Cn^{s/4-1/2}k^{-1}$, so that

$$\sum_{0<k\leqslant N} \sum_{\substack{0\leqslant h\leqslant k \\ (h,k)=1}} < \sum_{0<k\leqslant N} Cn^{s/4-1/2} < CNn^{s/4-1/2} \leqslant Cn^{1/2}n^{s/4-1/2} = Cn^{s/4}.$$

We conclude that the error E_2 introduced by extending the integral from ϕ_2 to ∞ is $O(n^{s/4})$, and the same holds, of course, for the extension from $-\phi_1$ to $-\infty$. This error is of the same order as E_1, and we record the result obtained so far as

$$r_s(n) = \sum_{h,k} \left\{ \frac{G(h,k)}{\sqrt{2}k} \right\}^s e^{-2\pi i h n/k} \int_{-\infty}^{\infty} \frac{e^{2\pi n(\delta - i\phi)}}{(\delta - i\phi)^{s/2}}\, d\phi + O(n^{s/4}). \tag{12.9}$$

Let us denote the integral by I. In order to compute it, we replace the argument $\delta - i\phi$ by z, so that

$$I = i \int_{\delta+i\infty}^{\delta-i\infty} z^{-s/2} e^{2\pi n z}\, dz = -i \int_{\delta-i\infty}^{\delta+i\infty} z^{-s/2} e^{2\pi n z}\, dz$$

$$= -i(2\pi n)^{s/2-1} \int_{2\pi n\delta - i\infty}^{2\pi n\delta + i\infty} w^{-s/2} e^{w}\, dw,$$

where $w = 2\pi n z$. For $\operatorname{Re} w > 0$, the integrand has no singularities; hence, in the plane cut along the negative real axis, the line of integration may be deformed into the contour $\mathscr{C}$ (see Fig. 1), namely, from $-\infty$ along the negative real axis in the lower half plane to $-\varepsilon$; then along $|w| = \varepsilon$, around the origin to $-\varepsilon$ on the upper rim of the cut, and back to $-\infty$. Then, however, $(1/2\pi i)\int_{\mathscr{C}} w^{-s/2} e^{w}\, dw = \{\Gamma(s/2)\}^{-1}$, by the classical Hankel formula (see, e.g., [221] or [114]) and $I = -i(2\pi n)^{s/2-1} 2\pi i/\Gamma(s/2) = (2\pi)^{s/2} n^{(s/2)-1}/\Gamma(s/2)$. With this, (12.9) becomes

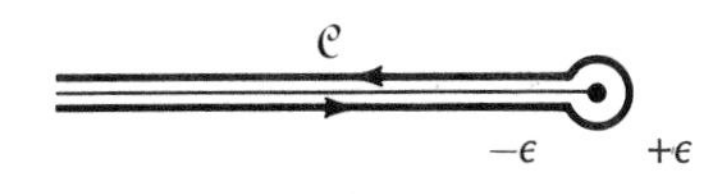

Figure 1

$$r_s(n) = \sum_{0<k\leqslant N} \sum_{\substack{0\leqslant h\leqslant k\\(h,k)=1}} \left\{\frac{G(h,k)}{\sqrt{2}k}\right\}^s e^{-2\pi i h n/k} \frac{(2\pi)^{s/2}}{\Gamma(s/2)} n^{(s/2)-1} + O(n^{s/4}).$$

There is one more simplification we can effect. If we extend the summation over k to infinity, we introduce a new error E_3; this is majorized by

$$\begin{aligned}\sum_{k>N} \sum_{\substack{0\leqslant h<k\\(h,k)=1}} \left|\frac{G(h,k)}{\sqrt{2}k}\right|^s \frac{(2\pi)^{s/2}}{\Gamma(s/2)} n^{(s/2)-1} &\leqslant C \sum_{k>\sqrt{n}} \frac{k^{s/2}}{k^s} n^{(s/2)-1} \sum_{0\leqslant h<k} 1\\ &= Cn^{(s/2)-1} \sum_{k>\sqrt{n}} k^{1-(s/2)}\\ &\leqslant Cn^{(s/2)-1} \int_{\sqrt{n}}^{\infty} u^{1-(s/2)}\,du\\ &= Cn^{(s/2)-1} \left.\frac{u^{2-(s/2)}}{(s/2)-2}\right|_{u=\sqrt{n}},\end{aligned}$$

for $s/2 > 2$, i.e., for $s > 4$. It follows that E_3 is $O(n^{(s/2-1} \cdot n^{1-(s/4)}) = O(n^{s/4})$, like the previous error terms, and we formulate the result obtained as

Theorem 1. *For $s \geqslant 5$, $r_s(n) = \rho_s(n) + O(n^{s/4})$, where*

$$\rho_s(n) = \frac{\pi^{s/2}}{\Gamma(s/2)} n^{(s/2)-1} \mathscr{S}(s,n), \qquad \mathscr{S}(s,n) = \sum_{k=1}^{\infty} A_k,$$

with

$$A_k = A_k(n,s) = k^{-s} \sum_{\substack{h \pmod k\\(h,k)=1}} G(h,k)^s e^{-2\pi i h n/k}.$$

The infinite series in the principal term of $r_s(n)$ is called, following Hardy, the *singular series*. In view of the fact that n and s do not change, we write simply A_k for the summands in $\mathscr{S}$ $(= \mathscr{S}(s,n))$.

For $s = 4$, the error term is of the same order, $O(n)$, as the principal term, so that, even if the theorem were valid, it would not have any meaning beyond the fairly trivial $r_4(n) = O(n)$.

§3. Evaluation of the Singular Series

One of the most interesting, although rather complicated, facets of this theory is that it is possible to sum the singular series. We start by using Theorem 11.3, in order to prove

Lemma 1. *A_k is a multiplicative arithmetical function.*

Proof. We have to show that, for $(k_1, k_2) = 1$, $A_{k_1 k_2} = A_{k_1} A_{k_2}$. Using the definition of A_k, we have

$$A_{k_1} A_{k_2} = (k_1 k_2)^{-s} \sum_{\substack{h_1 \pmod{k_1} \\ (h_1, k_1)=1}} \sum_{\substack{h_2 \pmod{k_2} \\ (h_2, k_2)=1}} \{G(h_1, k_1) G(h_2, k_2)\}^s$$

$$\times \exp\left\{-2\pi i n\left(\frac{h_1}{k_1} + \frac{h_2}{k_2}\right)\right\}.$$

Since $(k_1, k_2) = 1$, we see that $h_1' = h_1 k_2$ runs through a complete set of residues modulo k_1 precisely when h_1 does. Hence, if we set $h = h_1 k_2 + h_2 k_1$, it follows by Theorem 11.3 that the expression inside the braces equals $G(h, k_1 k_2)$. Also,

$$\frac{h_1}{k_1} + \frac{h_2}{k_2} = \frac{h_1 k_2 + h_2 k_1}{k_1 k_2} = \frac{h}{k_1 k_2},$$

and if h_1 runs through a complete residue system modulo k_1 and h_2 runs through a complete residue system modulo k_2, then $h = h_1 k_2 + h_2 k_1$ runs through a complete residue system modulo $k_1 k_2$. Moreover, $(h, k_1 k_2) = 1$ if and only if $(h_1, k_1) = 1$ and $(h_2, k_2) = 1$, so that $A_{k_1} A_{k_2} = (k_1 k_2)^{-s} \sum_{\substack{h \pmod{k_1 k_2} \\ (h, k_1 k_2)=1}} G(h, k_1 k_2) \exp(-2\pi i h / k_1 k_2) = A_{k_1 k_2}$. □

Lemma 2. $\mathscr{S} = \sum_{k=1}^{\infty} A_k = \prod_p \{1 + A_p + A_{p^2} + \cdots\}$, *or, if we set* $1 + \sum_{a=1}^{\infty} A_{p^a} = \mathscr{S}_p$, *then* $\mathscr{S} = \prod_p \mathscr{S}_p$.

Proof. The product over the primes equals $\sum A_{p_1^{a_1}} A_{p_2^{a_2}} \cdots A_{p_g^{a_g}}$, the sum being taken over all finite products, with one factor from each term $\mathscr{S}_p$. By Lemma 1, this product equals A_k, where $k = p_1^{a_1} p_2^{a_2} \cdots p_g^{a_g}$, and each product occurs exactly once, by the elementary theorem of uniqueness of factorization of rational integers into primes. □

It follows that it is sufficient to compute A_k for $k = p^a$, a prime power. Each of the series $\mathscr{S}_p = 1 + \sum_{a=1}^{\infty} A_{p^a}$ is in fact finite, because, as we shall presently show, $A_{p^a} = 0$ for sufficiently large a (and fixed n, s).

We shall have to make several case distinctions. The computations will depend on the parities of p, a, and s, and if p is odd, also of n.

Lemma 3. *If $p = 2$, then A_{2^a} has the following values:*

(i) $A_2 = 0$;
(ii) *for a even,*

$$A_{2^a} = \begin{cases} \dfrac{\cos\{(\pi/4)(2n_1 - s)\}}{2^{((s/2)-1)(a-1)}} & \text{if } n = 2^{a-2}n_1 \ (n_1 \text{ not necessarily odd}), \\ 0 & \text{if } 2^{a-2} \nmid n; \end{cases}$$

(iii) *for a odd, $a \geqslant 3$,*

$$A_{2^a} = \begin{cases} (-1)^{(n_2-s)/4} 2^{(1-s/2)(a-1)} & \text{if } n = 2^{a-3}n_2 \text{ and } n \equiv s \pmod 4, \\ 0 & \text{otherwise.} \end{cases}$$

Examples. For $a = 2$,

$$A_4 = \frac{\cos\{(\pi/2)(n - s/2)\}}{2^{(s/2)-1}}$$

always, because $2^{a-2} = 1$, $n = n_1$. Similarly, for $a = 3$, $A_8 = (-1)^{(n-s)/4} 2^{2-s}$ if $n \equiv s \pmod 4$, $A_8 = 0$ otherwise, because $2^{a-3} = 1$ and $n = n_2$.

Proof of Lemma 3. For $a = 1$, $G(h, 2) = 0$ by Theorem 11.3, and $A_2 = 0$ for all n and s. Otherwise, for a even, by Theorem 11.3, $G(h, 2^a) = (1 + i^h)2^{a/2}$, so that

$$A_{2^a} = \frac{2^{sa/2}}{2^{as}} \sum_{\substack{h \text{ odd} \\ h \pmod{2^a}}} (1 + i^h)^s e^{-2\pi i h n/2^a}$$

$$= 2^{-as/2} \left\{ (1 + i)^s \sum_{\substack{h \equiv 1 \pmod 4 \\ 0 < h < 2^a}} e^{-2\pi i h n/2^a} \right.$$

$$\left. + (1 - i)^s \sum_{\substack{h \equiv 3 \pmod 4 \\ 0 < h < 2^a}} e^{-2\pi i h n/2^a} \right\},$$

or, writing $h = 4j \pm 1$, $j = 1, 2, \ldots, 2^{a-2}$, in the two sums, we have

$$A_{2^a} = 2^{-sa/2} \left\{ (1 + i)^s e^{-2\pi i n/2^a} \sum_{j=1}^{2^{a-2}} e^{-2\pi i j n/2^{a-2}} \right.$$
$$\left. + (1 - i)^s e^{2\pi i n/2^a} \sum_{j=1}^{2^{a-2}} e^{-2\pi i j n/2^{a-2}} \right\}. \tag{12.10}$$

Here we replace $1 + i$ by $\sqrt{2}e^{\pi i/4}$, $1 - i$ by $\sqrt{2}e^{-\pi i/4}$, and observe that for n not divisible by 2^{a-2}, either sum vanishes unless $a = 2$, when it reduces to 1, so that

$$A_4 = 2^{-s}\{(\sqrt{2}e^{\pi i/4})^s e^{-\pi i n/2} + (\sqrt{2}e^{-\pi i/4})^s e^{\pi i n/2}\}$$

$$= 2^{-s/2}\{e^{(\pi i/2)(s/2-n)} + e^{-(\pi i/2)(s/2-n)}\}$$

$$= 2^{1-(s/2)}\cos\frac{\pi}{2}\left(\frac{s}{2} - n\right) = 2^{1-(s/2)}\cos\frac{\pi}{2}\left(n - \frac{s}{2}\right),$$

as claimed.

If $n = 2^{a-2}n_1$, then the sums in (12.10) are both equal to 2^{a-2}, so that

$$A_{2^a} = 2^{-sa/2+a-2}\{(\sqrt{2}e^{\pi i/4})^s e^{-2\pi i n/2^a} + (\sqrt{2}e^{-\pi i/4})^s e^{2\pi i n/2^a}\}$$

$$= 2^{(s/2)(1-a)+a-1}\cos\left(\frac{\pi s}{4} - \frac{2\pi n_1}{4}\right).$$

This finishes the proof for even exponent a. The proof for odd a is similar. □

Comment. In all cases, $|A_{2^a}| \leqslant 2^{(1-(s/2))(a-1)}$, so that the sum for $\mathcal{S}_2$ already converges for $s \geqslant 3$, even without taking into account the vanishing of most summands A_{2^a}.

Lemma 4. *If p is an odd prime and sa is even, then*

$$A_{p^a} = \begin{cases} (p-1)p^{-((s/2)-1)a-1}i^{((p^a-1)/2)^2 s} & \text{if } p^a \mid n, \\ -p^{-((s/2)-1)a-1}i^{((p^a-1)/2)^2 s} & \text{if } p^{a-1} \parallel n, \\ 0 & \text{if } p^{a-1} \nmid n; \end{cases}$$

if sa is odd, then

$$A_{p^a} = \begin{cases} 0 & \text{if either } p^a \mid n \text{ or } p^{a-1} \nmid n, \\ p^{-(as/2)+a-1/2}\left(\dfrac{n_1}{p}\right)i^{((p^a-1)/2)^2 s}\dfrac{(1-i)(1+i^p)}{2} & \\ & \text{if } n = p^{a-1}n_1,\ p \nmid n_1. \end{cases}$$

Proof. Let as be odd. In $A_{p^a} = p^{-as}\sum_{h \bmod p^a,\, p \nmid h} G^s(h, p^a)e^{-2\pi i h n/p^a}$, by Theorem 11.3, $G(h, p^a) = (\frac{h}{p^a})G(1, p^a) = (\frac{h}{p})i^{((p^a-1)/2)^2}p^{a/2}$ and $A_{p^a} = i^{((p^a-1)/2)^2 s}p^{-as/2} \times \sum_{h \pmod{p^a},\, p \nmid h}(\frac{h}{p})e^{-2\pi i h n/p^a}$. The condition $p \nmid h$ may be suppressed, because for $p \mid h$, $(\frac{h}{p}) = 0$. To compute the sum, set $h = vp + j$, $v = 0, 1, \ldots, p^{a-1} - 1$, and $j = 0, 1, \ldots, p - 1$; then the sum becomes $\sum_{v=0}^{p^{a-1}-1} e^{-2\pi i v n/p^{a-1}} \times$

$\sum_{j=0}^{p-1}(j/p)e^{-2\pi ijn/p^a}$. The first sum vanishes if $p^{a-1}\nmid n$. If $n = p^{a-1}n_1$, the first sum equals p^{a-1} and the second sum becomes $\sum_{j=0}^{p-1}(\frac{j}{p})e^{-2\pi ijn_1/p}$. If $p \mid n_1$, this sum equals $\sum_{j=0}^{p-1}(\frac{j}{p}) = 0$. Hence, $A_{p^a} \neq 0$ only if $n = p^{a-1}n_1$, $p \nmid n_1$. In this case,

$$\sum_{j=0}^{p-1} e^{-2\pi ijn_1/p} = \overline{G(n_1, p)} = \left(\frac{n_1}{p}\right)\overline{G(1, p)} = \left(\frac{n_1}{p}\right)\frac{(1-i)(1+i^p)}{2}\sqrt{p},$$

where use has been made of Theorem 11.3, in particular of the representation (iv) of $G(h, k)$ for squarefree k. Consequently,

$$A_{p^a} = p^{-as/2} i^{((p^a-1)/2)^2 s} p^{a-1}\left(\frac{n_1}{p}\right)\frac{(1-i)(1+i^p)}{2}\sqrt{p},$$

and the statement of Lemma 4 for as odd is proved. The case of as even in handled similarly, and the details may be suppressed. □

We observe that in all cases,

$$\left|1 + \sum_{a=1}^{\infty} A_{p^a}\right| \leqslant 1 + \sum_{a=1}^{\infty}\frac{p-1}{p^{(as/2)-a+1}}$$

$$= 1 + \frac{p-1}{p^{s/2}}\sum_{a=1}^{\infty}\frac{1}{p^{((s/2)-1)(a-1)}}$$

$$= 1 + \frac{p-1}{p^{s/2}}\frac{1}{1-p^{1-s/2}} \quad \text{for } s \geqslant 3.$$

In particular,

$$|\mathscr{S}_p - 1| = \left|\sum_{a=1}^{\infty} A_{p^a}\right| \leqslant \frac{p-1}{p^{s/2}-p} < \frac{1}{p^{(s/2)-1}-1}.$$

Similarly, recalling that $A_2 = 0$, we have

$$|\mathscr{S}_2 - 1| = \left|\sum_{a=2}^{\infty} A_{2^a}\right| \leqslant \sum_{a=2}^{\infty} 2^{(1-(s/2))(a-1)}$$

$$= 2^{1-s/2}(1 - 2^{-1+(s/2)})^{-1} = \frac{1}{2^{s/2}-1}.$$

It follows that for all p,

$$1 - \frac{1}{p^{(s/2)-1}-1} \leqslant \mathscr{S}_p \leqslant 1 + \frac{1}{p^{(s/2)-1}-1}.$$

In case $s \geqslant 5$, we have $p^{(s/2)-1} \geqslant p^{3/2}$, so that the products

$$\prod_p \left(1 \pm \frac{1}{p^{(s/2)-1} - 1}\right)$$

both converge absolutely to finite, nonvanishing values, say $c_1 > c_2 > 0$, with c_1, c_2 independent of the value of n. We conclude that for $s \geqslant 5$,

$$c_2(s) = \prod_p \left(1 - \frac{1}{p^{(s/2)-1} - 1}\right) \leqslant \mathscr{S} \leqslant \prod_p \left(1 + \frac{1}{p^{(s/2)-1} - 1}\right) = c_1(s).$$

In fact, the product converges also for $s = 4$. Indeed, for given n, there exist only finitely many prime divisors p of n, and for each of them only finitely many powers p^a, such that $p^{a-1} \mid n$. For all other primes $p \nmid n$, and only for $a = 1$ is it the case that $p^{a-1} \parallel n$. In this case $A_p(n, s) \simeq p^{-(s-1)/2}$ for s odd, $A_p(n, s) \simeq p^{-s/2}$ for s even. If $s = 4$, for $p \nmid n$, $A_p(n, 4) \simeq p^{-2}$ and $|\mathscr{S}| \leqslant C \sum_p p^{-2}$ converges for every n. However, $c_2(4)$ is no longer positive, and $\mathscr{S} = \mathscr{S}(4, n)$ is not bounded away from zero, but may become arbitrarily small. As for the error term, it can be reduced by more careful estimates to $O(n^{(s/4)-1/8+\varepsilon})$ for every $\varepsilon > 0$, so that in fact Theorem 1 holds, with the inferred error term, for all $s \geqslant 4$. See [18] for details.

§4. Explicit Evaluation of $\mathscr{S}$

In this section, we shall need some results from number theory. We recall that, for $\operatorname{Re} z > 1$, $\zeta(z) = \sum_{n=1}^{\infty} n^{-z}$ converges and $\zeta(z) = \prod_p (1 - p^{-z})^{-1}$. Furthermore (see, e.g., [266] or [86]), $\zeta(2m) = (-1)^{m-1}(2\pi)^{2m} B_{2m}/2(2m)!$, where the *Bernoulli numbers* B_{2m} are rational numbers. Specifically, they are the coefficients in the expansion

$$\frac{z}{2} \frac{e^z + 1}{e^z - 1} = \sum_{m=0}^{\infty} B_{2m} \frac{z^{2m}}{(2m)!} \qquad (B_0 = 1).$$

We also recall that (see, e.g., [97]) for arbitrary odd k, the *Ramanujan sums*, defined by $c_k(n) = \sum_{h \,(\mathrm{mod}\, k),\, (h,k)=1} e^{2\pi i h n/k}$, may be represented by

$$c_k(n) = \sum_{d \mid (n,k)} d\mu(k/d), \tag{12.11}$$

where $\mu(m)$ stands for the Möbius function.* Indeed,

$$c_k(n) = \sum_{h \,(\mathrm{mod}\, k)} e^{2\pi i h n/k} \sum_{d \mid (h,k)} \mu(d) = \sum_{d \mid k} \mu(d) \sum_{\substack{h = dh_1 \\ k = dk_1 \\ h_1 \,(\mathrm{mod}\, k_1)}} e^{2\pi i h n/k}.$$

---

* We recall the definition of μ. First, $\mu(1) = 1$. If $m \neq 1$, and $m = p_1 \cdots p_k$ with the p_i primes, then $\mu(m) = (-1)^k$ if the p_i are distinct, and $\mu(m) = 0$ otherwise.

Here the inner sum equals $\sum_{h_1=1}^{k_1} e^{2\pi i n h_1/k_1}$ and vanishes unless $k_1 \mid n$, in which case it equals k_1, so that $c_k(n) = \sum_{d \mid k,\, (k/d) \mid n} \mu(d)(k/d) = \sum_{\delta \mid (k,n)} \delta\mu(k/\delta)$.

(I) Case s Even

Computation of $\mathscr{S}_2$. By Lemma 3, the only odd power a with $A_{2^a} \neq 0$ is $a = 3$, when $A_8 = (-1)^{(n-s)/4} 2^{2-s}$ if $n \equiv s \pmod 4$, $A_8 = 0$ otherwise. Similarly, the only even a with $A_{2^a} \neq 0$ is $a = 2$, with $A_4 = \{\cos(\pi(n - s/2)/2)\}/2^{(s/2)-1}$, so that

$$\mathscr{S}_2 = 1 + 0 + \frac{\cos(\pi(n - s/2)/2)}{2^{(s/2)-1}} + \frac{(-1)^{(n-s)/4}}{4^{(s/2)-1}}\eta,$$

with $\eta = 1$ if $s \equiv n \pmod 4$, $\eta = 0$ otherwise.

In the same way we verify that, for $n = 2^b n_1$, n_1 odd, $a \geqslant 2$ even,

$$A_{2^a} = \frac{\cos\{(\pi/2)(n_1 \cdot 2^{b-a+2} - s/2)\}}{2^{(a-1)((s/2)-1)}}, \tag{12.12}$$

and for $a \geqslant 3$ odd,

$$A_{2^a} = \frac{(-1)^{(n_1 \cdot 2^{b-a+3} - s)/4}}{2^{(a-1)((s/2)-1)}}, \tag{12.12'}$$

if $n_1 \cdot 2^{b-a+3} \equiv s \pmod 4$, $A_{2^a} = 0$ otherwise. In case $s \equiv 0 \pmod 4$, this leads to $\mathscr{S}_2 = 1 + (-1)^{s/4}\{2^{1-s/2} + 2^{2(1-s/2)} + \cdots + 2^{(b-1)(1-s/2)} - 2^{b(1-s/2)}\}$. In particular, we verify that, for $b = 0$, the expression inside the braces vanishes and $\mathscr{S}_2 = 1$, in agreement with previous results for n odd, $s \equiv 0 \pmod 4$. In a similar way, we obtain for $s \equiv 2 \pmod 4$ that

$$\mathscr{S}_2 = 1 + (-1)^{(2n_1 - s)/4} 2^{(1-s/2)(b+1)}.$$

Computation of $\mathscr{S}_p$ *for p Odd*. In the proof of Lemma 4, we have obtained that

$$A_{p^a} = i^{((p^a-1)/2)^2 s} p^{-sa/2} \sum_{\substack{h \pmod{p^a} \\ p \nmid h}} \left(\frac{h}{p}\right)^{as} e^{-2\pi i h n/p^a}.$$

As s is even, the sum is the Ramanujan sum $\sum_{h \pmod{p^a},\, p \nmid h} e^{-2\pi i h n/p^a}$. By (12.11), with $k = p^a$, it equals

$$\sum_{d \mid (n, p^a)} d\mu(p^a/d) = \begin{cases} p^a - p^{a-1} & \text{if } (n, p^a) = p^a, \\ -p^{a-1} & \text{if } n = p^{a-1} n_1,\ p \nmid n_1, \\ 0 & \text{if } p^{a-1} \nmid n. \end{cases}$$

If $p \nmid n$, then $(n, p) = 1$, $d = 1$, and the sum vanishes except for $a = 0$ and $a = 1$, when it equals 1 and -1, respectively, so that $A_1 = 1$, $A_p = i^{s((p-1)/2)^2} p^{-s/2}$, and $\mathscr{S}_p = 1 - i^{s((p-1)/2)^2} p^{-s/2}$. If $n = p^b n_1$, $p \nmid n_1$, we obtain similarly

$$A_{p^a} = \begin{cases} i^{s((p^a-1)/2)^2} p^{-as/2}(p^a - p^{a-1}) & \text{if } a \leqslant b, \\ i^{s((p^a-1)/2)^2} p^{-as/2}(-p^{a-1}) & \text{if } a = b + 1, \\ 0 & \text{if } a > b + 1. \end{cases}$$

In particular, for $s \equiv 0 \pmod 4$,

$$\begin{aligned} \mathscr{S}_p &= 1 + \frac{p-1}{p^{s/2}} + \frac{p^2 - p}{p^{2(s/2)}} + \cdots + \frac{p^b - p^{b-1}}{p^{b(s/2)}} - \frac{p^b}{p^{(b+1)(s/2)}} \\ &= (1 - p^{-s/2})(1 + p^{1-s/2} + \cdots + p^{b(1-s/2)}). \end{aligned} \tag{12.13}$$

If $b = 0$, then $\mathscr{S}_p = 1 - p^{-s/2}$, in agreement with previous result for $p \nmid n$. If $s \equiv 2 \pmod 4$, then we obtain again (12.13) for $p \equiv 1 \pmod 4$, while, for $s \equiv 2 \pmod 4$, $p \equiv 3 \pmod 4$, we obtain $\mathscr{S}_p = (1 + p^{-s/2})(1 - p^{1-s/2} + p^{2(1-s/2)} - \cdots + (-1)^b p^{b(1-s/2)})$. Again, we verify that the result holds for $b = 0$, thus avoiding yet another case distinction.

Let us elaborate further the case $s \equiv 0 \pmod 4$. We have obtained that

$$\begin{aligned} \mathscr{S} &= \prod_p \mathscr{S}_p = \mathscr{S}_2 \prod_{p>2} \mathscr{S}_p = \mathscr{S}_2 \prod_{p>2} (1 - p^{-s/2}) \prod_{p \mid n} (1 + p^{1-s/2} + \cdots + p^{b(1-s/2)}) \\ &= \mathscr{S}_2 \mathscr{P}_1 \mathscr{P}_2, \qquad \text{say,} \end{aligned}$$

with $\mathscr{P}_1 = \prod_{p>2} (1 - p^{-s/2})$ and $\mathscr{P}_2 = \prod_{d \mid n,\, d \text{ odd}} d^{1-s/2}$. For $u > 1$, $\zeta(u) = \prod_p (1 - p^{-u})^{-1}$, so that for $s > 2$,

$$\prod_{p>2} (1 - p^{-s/2}) = \{\zeta(s/2)(1 - 2^{-s/2})\}^{-1} = \frac{2^{s/2}}{(2^{s/2} - 1)\zeta(s/2)}.$$

We also recall that $\zeta(2k) = (-1)^{k-1}(B_{2k}/2)(2\pi)^{2k}/(2k)!$, so that, for $s \equiv 0 \pmod 4$,

$$\begin{aligned} \mathscr{S} &= \mathscr{S}_2 \frac{2^{s/2}}{2^{s/2} - 1} \frac{2(-1)^{-1+s/4}}{B_{s/2}} \frac{(s/2)!}{(2\pi)^{s/2}} \sum_{\substack{d \mid n \\ d \text{ odd}}} d^{1-s/2} \\ &= \frac{2}{|B_{s/2}|} \frac{(s/2)!}{\pi^{s/2}(2^{s/2} - 1)} \mathscr{S}_2 \sum_{\substack{d \mid n \\ d \text{ odd}}} d^{1-s/2}. \end{aligned}$$

It now follows that $\rho_s(n)$, defined in Theorem 1, may be represented by

$$\rho_s(n) = \frac{\pi^{s/2}}{\Gamma(s/2)} \frac{n^{(s/2)-1}(s/2)!2}{|B_{s/2}|(2^{s/2}-1)\pi^{s/2}} \mathscr{S}_2 \sum_{\substack{d\mid n \\ d \text{ odd}}} d^{1-s/2}$$

$$= \frac{sn^{(s/2)-1}}{|B_{s/2}|(2^{s/2}-1)} \mathscr{S}_2 \sum_{\substack{d\mid n \\ d \text{ odd}}} d^{-(s/2)+1}.$$

Next, we recall that for $b = 0$—i.e., for $n = 2^b n_1 = n_1$ odd and $s \equiv 0 \pmod 4$—we have $\mathscr{S}_2 = 1$, while for $b > 0$, $\mathscr{S}_2 = 1 + (-1)^{s/4}\{2^{1-s/2} + \cdots + 2^{(b-1)(1-s/2)} - 2^{b(1-s/2)}\}$, and we observe that the product $n^{(s/2)-1}\mathscr{S}_2$ leads to summands $(n/d)^{(s/2)-1}$. Here $d = 2^c$ $(0 \leqslant c \leqslant b)$ are divisors of n, so that we obtain rational integers in the product, in fact, $\delta^{(s/2)-1}$, where δ is the complementary divisor of $d = 2^c$, $d \mid n$. If, furthermore, $s \equiv 0 \pmod 8$, so that $s/4$ is even, then $\mathscr{S}_2 = 1 + 2^{1-s/2} + \cdots + 2^{(b-1)(1-s/2)} - 2^{b(1-s/2)}$. Let $\mathscr{S}'_2 = \mathscr{S}_2 + 2\cdot 2^{b(1-s/2)}$ (i.e., change the last $-$ into a $+$); then $\mathscr{S}'_2 \sum_{d\mid n,\, d \text{ odd}} d^{1-s/2} = \sum_{t\mid n} t^{1-s/2}$, where the sum is taken over *all* divisors of n. In order to obtain instead $\mathscr{S}_2 \sum_{d\mid n,\, d\text{ odd}} d^{1-s/2}$, we have to subtract, rather than add, $t^{(s/2)-1}$ for those divisors $t = 2^b d$ (d odd) of n that contain the heighest power of 2. This is accomplished, for even n, by writing $\sum_{t\mid n}(-1)^{n/t}t^{1-s/2}$. For odd n, this leads to the correct numerical value, but with the wrong sign; however, $\mathscr{S}_2 \sum_{d\mid n} d^{1-s/2} = \sum_{t\mid n}(-1)^{n+(n/t)}t^{1-s/2}$ is correct in all cases where $s \equiv 0 \pmod 8$, and we obtain

$$\rho_s(n) = \frac{s}{(2^{s/2}-1)|B_{s/2}|} \sum_{\delta\mid n} (-1)^{\delta+n} \delta^{(s/2)-1},$$

where $\delta = n/d$ is the complementary divisor of d in n. For $s \equiv 4 \pmod 8$,

$$\mathscr{S}_2 = 1 - 2^{1-s/2} - \cdots - 2^{(b-1)(1-s/2)} + 2^{b(1-s/2)},$$

and, proceeding as before, we now obtain

$$\mathscr{S}_2 \sum_{\substack{d\mid n \\ d \text{ odd}}} d^{1-s/2} = \sum_{t\mid n} (-1)^{t+(n/t)+n+1} t^{1-s/2}.$$

In order to obtain a single formula in both cases, observe that the exponent may be written as $n + (n/t) + (s/4)(t+1)$, or, with $\delta = n/t$, $t = n/\delta$, as $n + \delta + (s/4)((n/\delta)+1)$. This leads, for $s \equiv 0 \pmod 8$, to the same result as does $n + \delta$; hence, we obtain in general, for all $s \equiv 0 \pmod 4$,

$$\rho_s(n) = \frac{s}{(2^{s/2}-1)|B_{s/2}|} \sum_{\delta\mid n} (-1)^{n+\delta+(s/4)((n/\delta)+1)} \delta^{(s/2)-1}.$$

EXAMPLES.

1. Let $s = 8$; then $B_{s/2} = B_4 = -\frac{1}{30}$, so that

$$\rho_8(n) = \frac{8}{(2^4 - 1)(\frac{1}{30})} \sum_{d \mid n} (-1)^{n+d} d^3$$

$$= \frac{240}{15} \sum_{d \mid n} (-1)^{n+d} d^3 = 16 \sum_{d \mid n} (-1)^{n+d} d^3,$$

in agreement with (9.19) for $r_8(n)$.

2. Let $s = 4$; then $B_{s/2} = B_2 = \frac{1}{6}$ and

$$\rho_4(n) = \frac{4}{(2^2 - 1)(\frac{1}{6})} \sum_{d \mid n} (-1)^{n+d+(n/d)+1} d$$

$$= 8 \sum_{d \mid n} (-1)^{n+d+(n/d)+1} d.$$

It is not difficult to show that this result is the value of $r_4(n)$, as given in (9.19) (see Problem 10).

(II) Case s Odd

As before, it is possibe to treat this case for every n. For simplicity, however, we restrict ourselves to n odd and squarefree.

Computation of $\mathcal{S}_2$. By Lemma 3, $A_2 = 0$ and for a even, $A_{2^a} = \cos\{(\pi/4)(2n_1 - s)\} \cdot 2^{(a-1)(1-s/2)}$ if $n = 2^{a-2} n_1$, $A_{2^a} = 0$ for $2^{a-2} \nmid n$. As n is now odd, $A_{2^a} \neq 0$ is possible only for $a = 2$, and $A_4 = \cos\{(\pi/4)(2n - s)\} \cdot 2^{1-s/2}$. Similarly, for a odd, only A_{2^3} can be different from zero, namely $A_8 = (-1)^{(n-s)/4} 2^{2(1-s/2)}$ if $n \equiv s \pmod 4$; otherwise $A_8 = 0$. We observe that, with n and s both odd,

$$(-1)^{(n-s)/4} \frac{(-1)^{(n-s)/2} + 1}{2} = (-1)^{(n-s)/2} \frac{(-1)^{(n-s)/4} + (-1)^{-(n-s)/4}}{2}$$

$$= (-1)^{(n-s)/2} \frac{e^{\pi i(n-s)/4} + e^{-\pi i(n-s)/4}}{2}$$

$$= (-1)^{(n-s)/2} \cos\frac{\pi}{4}(n - s),$$

and this equals $(-1)^{(n-s)/4}$ if $n \equiv s \pmod 4$ and vanishes otherwise; hence, $A_8 = 2^{2-s} \cos(\pi(n - s)/4)$ for all odd n and s. We conclude that $\mathcal{S}_2 = 1 + \cos(\pi(n - s)/4) \cdot 2^{1-s/2} + \cos(\pi(n - s)/4) \cdot 2^{2-s}$.

Computation for $\mathcal{S}_p$, *p Odd.* If $p \nmid n$, then, by Lemma 4 (see Problem 15), $A_p = (\frac{-1}{p})^{(s-1)/2} (\frac{n}{p}) p^{(1-s)/2} = (\frac{\varepsilon_s n}{p}) p^{(1-s)/2}$, with $\varepsilon_s = (-1)^{(s-1)/2}$, while $A_{p^a} = 0$ for $a > 1$. If $n = pn_1$, $p \nmid n_1$, then $A_{p^2} = -p^{1-s}$, while $A_{p^a} = 0$ for $a = 1$ and

$a \geqslant 3$ (we recall that n is squarefree, so that $p^2 \nmid n$). Consequently, for $p \nmid n$, $\mathscr{S}_p = 1 + (\frac{\varepsilon_s n}{p})p^{(1-s)/2}$, while for $p \mid n$, $\mathscr{S}_p = 1 - p^{1-s} = (1 - p^{(1-s)/2})(1 + p^{(1-s)/2})$. It follows that

$$\begin{aligned}
\mathscr{S} &= \mathscr{S}_2 \prod_{p \mid n} (1 - p^{1-s}) \prod_{p \nmid n} \left(1 + \left(\frac{\varepsilon_s n}{p}\right) p^{(1-s)/2}\right) \\
&= \mathscr{S}_2 \prod_{p \nmid n} \left(1 + \left(\frac{\varepsilon_s n}{p}\right) p^{(1-s)/2}\right) \prod_{p \geqslant 3} (1 - p^{1-s}) \\
&\quad \times \prod_{\substack{p \nmid n \\ p \geqslant 3}} \{(1 - p^{(1-s)/2})(1 + p^{(1-s)/2})\}^{-1} \\
&= \mathscr{S}_2 \prod_{p \geqslant 3} (1 - p^{1-s}) \prod_{\substack{p \nmid n \\ p \geqslant 3}} \left(1 - \left(\frac{\varepsilon_s n}{p}\right) p^{(1-s)/2}\right)^{-1} \\
&= \mathscr{S}_2 \mathscr{P}_1 \mathscr{P}_2, \qquad \text{say.}
\end{aligned}$$

Here

$$\begin{aligned}
\mathscr{P}_1 &= (1 - 2^{1-s})^{-1} \prod_p (1 - p^{1-s}) = \frac{2^{s-1}}{(2^{s-1} - 1)\zeta(s-1)} \\
&= \frac{2^{s-1}}{(2^{s-1} - 1)} \frac{2(s-1)!}{(2\pi)^{s-1} |B_{s-1}|} \\
&= \frac{2^s (s-1)!}{(2^{s-1} - 1) 2^{s-1} \pi^{s-1} |B_{s-1}|}.
\end{aligned}$$

Next, in $\mathscr{P}_2$ we may suppress the condition $p \nmid n$, because, for $p \mid n$ the Jacobi symbol $(\frac{\varepsilon_s n}{p})$ vanishes. Hence

$$\begin{aligned}
\mathscr{P}_2 &= \prod_{p \geqslant 3} \left(1 - \left(\frac{\varepsilon_s n}{p}\right) p^{(1-s)/2}\right)^{-1} \\
&= \prod_{p \geqslant 3} \left(1 + \sum_{v=1}^{\infty} \left\{\left(\frac{\varepsilon_s n}{p}\right) p^{(1-s)/2}\right\}^v\right) = \sum_{\substack{m \text{ odd} \\ m > 0}} \left(\frac{\varepsilon_s n}{m}\right) m^{(1-s)/2}.
\end{aligned}$$

It follows that

$$\begin{aligned}
\frac{\pi^{s/2}}{\Gamma(s/2)} \mathscr{S} &= \mathscr{S}_2 \frac{\pi^{s/2} 2^s (s-1)!}{2^{s-1}(2^{s-1} - 1)\pi^{s-1} |B_{s-1}|} \\
&\quad \times \frac{1}{1 \cdot 3 \cdots (s-2)\sqrt{\pi}/2^{(s-1)/2}} \sum_{\substack{m > 0 \\ m \text{ odd}}} \left(\frac{\varepsilon_s n}{m}\right) m^{(1-s)/2}.
\end{aligned}$$

The odd factors of $(s-1)!$ in the numerator are canceled by those in the denominator. The remaining even factors equal $2\cdot 4\cdots(s-1) = 2^{(s-1)/2} \times ((s-1)/2)!$, so that

$$\rho_s(n) = \frac{\pi^{s/2}}{\Gamma(s/2)} n^{(s/2)-1}\mathscr{S}$$

$$= \mathscr{S}_2 \frac{\pi^{s/2}2^{s+(s-1)/2}((s-1)/2)!2^{(s-1)/2}}{2^{s-1}(2^{s-1}-1)\pi^{s-1/2}|B_{s-1}|} n^{(s/2)-1} \sum_{\substack{m>0\\ m\ \mathrm{odd}}} \left(\frac{\varepsilon_s n}{m}\right) m^{(1-s)/2}$$

$$= \mathscr{S}_2 \frac{\pi^{(1-s)/2}((s-1)/2)!2^s n^{(s/2)-1}}{(2^{s-1}-1)|B_{s-1}|} D(n),$$

where $D(n)\,(= D_s(n))$ stands for the Dirichlet series $\sum_{m=1,\, m\ \mathrm{odd}}^{\infty} \left(\frac{\varepsilon_s n}{m}\right) m^{(1-s)/2}$. This series can be summed in finite terms, by following e.g., the methods developed by Dirichlet (see, e.g., [146]).

Example. Let $s = 5$; then $\varepsilon_5 = (-1)^{(5-1)/2} = 1$ and $\left(\frac{\varepsilon_5 n}{m}\right) = \left(\frac{n}{m}\right)$. Also, $\mathscr{S}_2 = 1 + 2^{-3/2}\cos\{(\pi/4)(2n-5)\} + \frac{1}{8}\cos\{(\pi/4)(n-5)\}$. In particular, for $n \equiv 1 \pmod 8$, $\mathscr{S}_2 = \frac{5}{8}$; for $n \equiv 5 \pmod 8$, $\mathscr{S}_2 = \frac{7}{8}$; and for $n \equiv 3 \pmod 4$, $\mathscr{S}_2 = \frac{5}{4}$. It follows that

$$\rho_5(n) = \mathscr{S}_2 \frac{2^5\pi^{-2}2!}{15|B_4|} n^{3/2} D(n)$$

$$= \mathscr{S}_2 \frac{2^6 n^{3/2}}{\pi^2\cdot 15(\frac{1}{30})} \sum_{\substack{m=1\\ m\ \mathrm{odd}}}^{\infty} (-1)^{(m-1)(n-1)/4}\left(\frac{m}{n}\right) m^{-2}$$

$$= \mathscr{S}_2 \frac{128}{\pi^2} L(2,\chi) n^{3/2},$$

where $D(n) = L(2,\chi) = \sum_{m=1}^{\infty} \chi(m) m^{-2}$, with $\chi(m)$ the character $(-1)^{(m-1)(n-1)/4} \times \left(\frac{m}{n}\right) = \left(\frac{n}{m}\right)$ for odd m, $\chi(m) = 0$ for even m. We obtain in particular that

$$\rho_5(n) = C(n)\pi^{-2}L(2,\chi)n^{3/2}, \tag{12.14}$$

where $C(n) = 128\mathscr{S}_2$, so that

$$C(n) = \begin{cases} \frac{5}{8}\cdot 128 = 80 & \text{for } n \equiv 1 \pmod 8, \\ \frac{7}{8}\cdot 128 = 112 & \text{for } n \equiv 5 \pmod 8, \\ \frac{5}{4}\cdot 128 = 160 & \text{for } n \equiv 3 \pmod 4, \end{cases}$$

and $L(2,\chi) = \sum_{m=1,\, m\ \mathrm{odd}}^{\infty} \left(\frac{n}{m}\right) m^{-2}$. It is easy to see that $|L(2,\chi)| \leqslant \sum_{m\ \mathrm{odd}} m^{-2} =$

$\frac{3}{4}\zeta(2) = \pi^2/8$ and also $|L(2, \chi)| \geqslant 1 - \sum_{m \text{ odd}, m \geqslant 2} m^{-2} = 2 - \pi^2/6 > \frac{1}{3}$, so that indeed $c_2 n^{3/2} < \rho_5(n) < c_1 n^{3/2}$, with $\frac{80}{3} < c_2 < c_1 < 160\pi^2/8 = 20\pi^2 < 200$. This proves for $k = 5$ the claim advanced for $k \geqslant 5$ in Section 9.7.

In fact, for every n, $L(2, \chi)$ can be summed in finite terms by several methods, but none is really simple. We shall sketch two of them. In either method, we have to observe that the character $(-1)^{(m-1)(n-1)/4}(\frac{m}{n})$ is periodic, with period n or $4n$ according as $n \equiv 1 \pmod 4$ or $n \equiv 3 \pmod 4$. Also, $m \equiv 1 \pmod 2$; hence, $\chi(m)$ is periodic, with period $2n$ (if $n \equiv 1 \pmod 4$), or $4n$ (if $n \equiv 3 \pmod 4$).

The first method of summation is due to B. Berndt [24]; it applies directly only in case χ is a primitive character mod n_1, where, as just seen, $n_1 = 2n$ or $n_1 = 4n$ according as $n \equiv 1$ or $n \equiv 3 \pmod 4$. This means that there exists no integer k satisfying $k \mid n_1$, $0 < k < n_1$, and such that for some character $\chi_1(n) \pmod k$, $\chi(m) = \chi_1(m)$ whenever $(n_1, m) = 1$. This, however, does not present a serious difficulty. Indeed, if χ is a character mod n_1 and χ_1 the (unique) primitive character mod k (where $k \mid n_1$) that coincides with χ whenever both are different from zero, then (see [146]) $L(s, \chi) = \prod_{p \mid n_1, p \nmid k} (1 - \chi_1(p)p^{-s}) \times L(s, \chi_1)$, and $L(s, \chi)$, in particular $L(2, \chi)$, may be found by the present method.

We recall from Chapter 11 that the sums $\Gamma(h, k) = \sum_{m \pmod k} (\frac{m}{k}) e^{2\pi i h m/k}$ coincide with the Gaussian sums $G(h, k)$ whenever k is squarefree. These sums may be generalized in two ways: one may replace (m/k) by an arbitrary character $\chi(m) \pmod k$, and one may also replace the integral variable h by a complex variable s. This leads us to consider sums of the form $\sum_{m \pmod k} \chi(m) e^{2\pi i m s/k}$. We call them also Gaussian sums, but recall that, for k not squarefree, these sums are *not* identical with the (correspondingly generalized) sums $\sum_{m \pmod k} e^{2\pi i s m^2/k}$ of Chapter 11. To distinguish them from the earlier ones, we denote them by $G(s, \chi)$.

Lemma 5.

(i) *For primitive characters $\chi \pmod k$ and $s = n \in \mathbb{Z}$, we have $G(s, \bar{\chi}) = \chi(s) G(\bar{\chi})$, where $G(\chi)$ stands for $G(1, \chi)$.*

(ii) *For all characters, $|G(\chi)|^2 = k$.*

(iii) *If $\chi(k)$ is even (i.e., if $\chi(-1) = \chi(1) = 1$), then $G(\bar{\chi}) = \overline{G(\chi)}$ and hence we have $G(\chi) G(\bar{\chi}) = k$.*

The easy proofs are left to the reader (see Problem 16).

We now consider the integral

$$I_N = \frac{1}{2\pi i} \int_{\mathscr{C}} \frac{\pi e^{-\pi i s}}{s^2 \sin \pi s} \frac{G(s, \bar{\chi})}{G(\bar{\chi})} ds$$

taken around the square $\mathscr{C}$ centered at the origin, with sides parallel to the coordinate axes, at the abscissae $x = \pm(N + \frac{1}{2})$ and ordinates $y = \pm(N + \frac{1}{2})$ (N a large integer). I_N equals the sum of the residues at the poles of the integrand inside $\mathscr{C}$, and these poles are the integers n, $-N \leqslant n \leqslant N$. We verify that on $\mathscr{C}$ we have $|e^{-i\pi s}G(s,\bar{\chi})/\sin \pi s| \leqslant M$, with M a constant independent of N, so that the integrand is less in absolute value than $\{M/(N+\frac{1}{2})^2\}/\pi/|G(\bar{\chi})|$. It follows that

$$|I_N| < \frac{M}{(N+\frac{1}{2})^2}\frac{\pi}{|G(\bar{\chi})|}\cdot 4(2N+1) \quad \text{and} \quad \lim_{N\to\infty} I_N = 0.$$

Consequently, the sum of all the residues vanishes and we have

$$\lim_{N\to\infty} \sum_{\substack{m=-N\\ m\neq 0}}^{N} \operatorname{Res}\left\{\frac{\pi e^{-i\pi s}}{s^2 \sin \pi s}\frac{G(s,\bar{\chi})}{G(\bar{\chi})}\right\}_{s=m}$$

$$+ \operatorname{Res}\left\{\frac{\pi e^{-i\pi s}}{s^2 \sin \pi s}\frac{G(s,\bar{\chi})}{G(\bar{\chi})}\right\}_{s=0} = 0. \tag{12.15}$$

By use of Lemma 5, for $s = m$ we have $G(m,\bar{\chi})/G(\bar{\chi}) = \chi(m)$, so that the residue of the simple pole at $s = m \neq 0$ equals $\chi(m)/m^2$, and the first term is $\lim_{N\to\infty}\sum_{n=-N}^{N}\chi(n)n^{-2}$. If $\chi(n)$ is an odd character, then the sum vanishes and this case is of no further interest. If the character is even, the sum is equal to $2\sum_{n=1}^{\infty}\chi(n)n^{-2} = 2L(2,\chi)$; henceforth, we shall assume that $\chi(m)$ is an even character. As for the last term, also $s = 0$ is a simple pole. Indeed,

$$\begin{aligned} G(s,\bar{\chi}) &= \sum_{m \,(\mathrm{mod}\, k)} \bar{\chi}(m)e^{2\pi i m s/k} \\ &= \sum_{m=1}^{k-1}\bar{\chi}(m)\left\{1 + \frac{2\pi i m s}{k} - \frac{4\pi^2 m^2 s^2}{2k^2} + O(s^3)\right\} \\ &= \sum_{m=1}^{k-1}\bar{\chi}(m) + \frac{2\pi i s}{k}\sum_{m=1}^{k-1} m\bar{\chi}(m) - \frac{2\pi^2 s^2}{k^2}\sum_{m=1}^{k-1} m^2\bar{\chi}(m) + O(s^3). \end{aligned}$$

The first sum vanishes; the second sum is the *first moment* of $\bar{\chi}$, $M_1(\bar{\chi}) = \sum_{m=1}^{k-1} m\bar{\chi}(m)$, and vanishes for $\bar{\chi}(m)$ even. Indeed,

$$\begin{aligned} M_1(\bar{\chi}) &= \sum_{m=1}^{k-1}(k-m)\bar{\chi}(k-m) = \sum_{m=1}^{k-1}(k-m)\bar{\chi}(m) \\ &= k\sum_{m=1}^{k-1}\bar{\chi}(m) - \sum_{m=1}^{k-1} m\bar{\chi}(m) = -M_1(\bar{\chi}). \end{aligned}$$

It follows that $G(s,\bar{\chi}) = -2\pi^2 s^2 k^{-2} M_2(\bar{\chi}) + O(s^3)$, where $M_2(\bar{\chi}) = \sum_{m=1}^{k-1} m^2\bar{\chi}(m)$ is the *second moment* of $\bar{\chi}(m)$. We conclude that

$$\frac{\pi e^{-i\pi s}}{s^2 \sin \pi s}\frac{G(s,\chi)}{G(\bar{\chi})} = -\frac{2\pi^2 s^2}{k^2}M_2(\bar{\chi})\frac{\pi s}{\sin \pi s}\frac{e^{-i\pi s}}{s^3}\frac{1}{G(\bar{\chi})}(1 + O(s)),$$

and for $s \to 0$ this becomes

$$-\frac{2\pi^2}{G(\bar{\chi})k^2}M_2(\bar{\chi})\frac{1}{s} + O(1).$$

If we also replace $G(\bar{\chi})$ by $k/G(\chi)$, we see that the residue at $s = 0$ equals $-(2\pi^2/k^3)G(\chi)M_2(\bar{\chi})$.

Equation (12.15) now reads

$$2L(2,\chi) - \frac{2\pi^2}{k^3}G(\chi)M_2(\bar{\chi}) = 0,$$

whence

$$L(2,\chi) = \frac{\pi^2 G(\chi)M_2(\bar{\chi})}{k^3}. \tag{12.16}$$

As an example, let $n = 5$, so that since $n \equiv 1 \pmod 4$, $n_1 = 2n = 10$. The character $\chi(m) = (\frac{m}{n})$ for m odd, $\chi(m) = 0$ for m even, so that $\chi(1) = \chi(9) = 1$, $\chi(3) = \chi(7) = -1$, $\chi(m) = 0$ for $m = 2, 4, 5, 6, 8, 10$. This character *is not* primitive. Indeed, it coincides with $\chi_1(m) = (m/5)$ for all m prime to 10. Hence, if we denote by $\chi_1(m)$ the primitive character mod 5 given by $\chi_1(m) = (\frac{m}{5})$, then $L(s,\chi) = (1 - \chi_1(2)\cdot 2^{-s})L(s,\chi_1) = (1 + 2^{-s})L(s,\chi_1)$. In particular, $L(2,\chi) = \frac{5}{4}L(2,\chi_1)$. As for $L(2,\chi_1)$, it is given by (12.16) with $k = 5$, $G(\chi) = \sum_{m=1}^{4}(\frac{m}{5})e^{2\pi i m/5} = \sqrt{5}$, and $M_2(\bar{\chi}) = M_2(\chi) = \sum_{m=1}^{4} m^2(\frac{m}{5}) = 1 - 4 - 9 + 16 = 4$, so that $L(2,\chi_1) = \pi^2\sqrt{5}(\frac{4}{125})$ and $L(2,\chi) = \pi^2/5\sqrt{5}$.

The second method is classical (it is due, essentially, to Dirichlet; see [146]) and easier to present; however, the effective computations are quite tedious. Consider the second Bernoulli polynomial $b_2(x) = x^2 - x + \frac{1}{6}$ and the corresponding Bernoulli function $B_2(x) = b_2(x - [x])$. It is periodic, of period 1, and $B_2(-x) = B_2(1 - x)$; also, because $(1 - x)^2 - (1 - x) + \frac{1}{6} = x^2 - x + \frac{1}{6}$, it follows that $B_2(-x) = B_2(x)$ is even. Therefore, its Fourier series is a cosine series and we easily find that, for $0 \leqslant x < 1$, one has $x^2 - x + \frac{1}{6} = (1/\pi^2)\sum_{k=1}^{\infty}(\cos 2\pi kx)/k^2$. Let us assume that $n \equiv 1 \pmod 4$; then $\chi(m) = (\frac{n}{m}) = (\frac{m}{n}) = (\frac{-m}{n})$ if $(n,m) = 1$, $\chi(m) = 0$ otherwise, so that also $\chi(m)$ is an even function of m, of the same even period $n_1 = 2n$ as $\cos(2\pi mj/n_1)$. We now determine constants a_j $(j = 1,2,\ldots,n = n_1/2)$, such that

$$\sum_{j=1}^{n_1/2} a_j \cos\left(\frac{2\pi mj}{n_1}\right) = \chi(m) \quad \text{for } m = 1, 2, \ldots, \frac{n_1}{2}. \tag{12.17}$$

It can be verified that the determinant of the matrix of cosines does not vanish, so that the values of the a_j are well determined. Also, if (12.17) holds for $m = 1, 2, \ldots, n_1/2$, then it also holds, by periodicity and evenness, for all integers m. It now follows that

$$\sum_{j=1}^{n_1/2} a_j B_2\left(\frac{j}{n_1}\right) = \sum_{j=1}^{n_1/2} a_j \pi^{-2} \sum_{m=1}^{\infty} m^{-2} \cos\left(\frac{2\pi mj}{n_1}\right)$$

$$= \pi^{-2} \sum_{m=1}^{\infty} m^{-2} \sum_{j=1}^{n_1/2} a_j \cos(2\pi mj/n_1) = \pi^{-2} \sum_{m=1}^{\infty} \chi(m) m^{-2},$$

so that the Dirichlet series $D(n) = L(2, \chi) = \sum_{m=1}^{\infty} \chi(m) m^{-2}$ is summed as the finite sum $\pi^2 \sum_{j=1}^{n_1/2} a_j B_2(j/n_1)$. A similar approach works for $n \equiv 3 \pmod 4$, except that now we take $n_1 = 4n$.

In order to complete the discussion in the case of odd s, the restrictions to n odd and squarefree have to be removed. For n even, but still squarefree, $n = 2n_1$, n_1 odd, and it turns out that (12.14) still holds, with $C(n) = 160$, for $n \equiv 2$ or $6 \pmod 8$. If, however, n is not squarefree, then $\rho_s(n)$ is related not to $r_s(n)$, but to $R_s(n)$. From $R_s(n)$ one then obtains $r_s(n)$ by Theorem 1.1, for which we obtain here

$$r_s(n) = \sum_{d^2 \mid n} \rho_s\left(\frac{n}{d^2}\right) + O(n^{s/4}).$$

See [96] for complete proofs.

§5. Discussion of the Density of Representations

It is possible to give a representation for $A_{p^a} = A_{p^a}(n, s)$, different from that in Theorem 1. Indeed, for $(h, k) = 1$, $G(h, k) = \sum_{m \,(\mathrm{mod}\, k)} e^{2\pi i h m^2/k}$, so that

$$G(h, k)^s = \prod_{j=1}^{s} \sum_{m_j \,(\mathrm{mod}\, k)} e^{2\pi i h m_j^2/k} = \sum_{\substack{m_j \,(\mathrm{mod}\, k) \\ 1 \leqslant j \leqslant s}} \exp\left\{2\pi i h \sum_{j=1}^{s} \frac{m_j^2}{k}\right\};$$

consequently, by taking $k = p^a$, we obtain

$$A_{p^a} = p^{-as} \sum_{\substack{h \,(\mathrm{mod}\, p^a) \\ (h, p)=1}} G(h, p^a)^s e^{-2\pi i h n/p^a}$$

$$= p^{-as} \sum_{\substack{h \,(\mathrm{mod}\, p^a) \\ (h, p)=1}} \sum_{\substack{m_j \,(\mathrm{mod}\, p^a) \\ 1 \leqslant j \leqslant s}} \exp\left\{\frac{2\pi i h}{p^a}\left(\sum_{j=1}^{s} m_j^2 - n\right)\right\}.$$

In order to avoid the awkward condition $(h, p) = 1$, we take the full sum for $h \pmod{p^a}$ and subtract from it the sum with $p \mid h$. We obtain, by exchanging also the order of the finite sums and by writing in the second sum hp for h,

$$A_{p^a} = p^{-as} \sum_{\substack{m_j \,(\mathrm{mod}\, p^a) \\ 1 \leqslant j \leqslant s}} \sum_{h=1}^{p^a} \exp\left\{\frac{2\pi i h}{p^a}\left(\sum_{j=1}^{s} m_j^2 - n\right)\right\}$$

$$- p^{-as} \sum_{\substack{m_j \,(\mathrm{mod}\, p^a) \\ 1 \leqslant j \leqslant s}} \sum_{h=1}^{p^{a-1}} \exp\left\{\frac{2\pi i h}{p^{a-1}}\left(\sum_{j=1}^{s} m_j^2 - n\right)\right\}.$$

In general, the inner sums vanish, the exceptions being the cases

$$\sum_{j=1}^{s} m_j^2 \equiv n \pmod{p^a} \tag{12.18}$$

for the first sum and

$$\sum_{j=1}^{s} m_j^2 \equiv n \pmod{p^{a-1}} \tag{12.18$'$}$$

for the second sum, when their values equal the number of their terms, i.e., p^a and p^{a-1}, respectively. Let us denote by $M_{p^a}(n, s)$ the number of solutions of (12.18) with $1 \leqslant m_j \leqslant p^a$, $1 \leqslant j \leqslant s$, and define $M_{p^{a-1}}(n, s)$ similarly. We observe that the range of the m_j in the second sum is $1 \leqslant m_j \leqslant p^a$ rather than $1 \leqslant m_j \leqslant p^{a-1}$, and this yields an extra factor p for each value of j. We may write the result obtained in the form

$$A_{p^a}(n, s) = p^{a(1-s)} M_{p^a}(n, s) - p^{(a-1)(1-s)} M_{p^{a-1}}(n, s).$$

It follows that $\mathscr{S}_{p,N} = 1 + \sum_{a=1}^{N} A_{p^a}$ becomes a telescoping series:

$$1 + (p^{(1-s)} M_p - 1) + (p^{2(1-s)} M_{p^2} - p^{(1-s)} M_p) + \cdots$$

$$+ (p^{a(1-s)} M_{p^a} - p^{(a-1)(1-s)} M_{p^{a-1}}) + \cdots$$

$$+ (p^{N(1-s)} M_{p^N} - p^{(N-1)(1-s)} M_{p^{N-1}}) = p^{N(1-s)} M_{p^N},$$

and $\mathscr{S}_p = \lim_{N\to\infty} \mathscr{S}_{p,N} = \lim_{N\to\infty} p^{N(1-s)} M_{p^N}$, a limit reached, of course, after a finite number of steps, because $\mathscr{S}_p$ is a finite sum.

On the other hand, we observe that in the sum $\sum_{j=1}^{s} m_j^2$, each m_j is allowed to take p^a values, so that, as each m_j runs independently modulo p^a, the sum takes p^{as} (not all distinct) values. In fact, each value occurs about equally often, and among them a fraction $1/p^a$ are congruent to n modulo p^a. It follows that, on the average, the number of solutions of (12.18) is $p^{as}/p^{a} = p^{a(s-1)}$. The true number of solutions is by definition M_{p^a}; the value $\mathscr{S}_p = \lim_{a\to\infty}\{M_{p^a}/p^{a(s-1)}\}$ can be considered as the local density at the prime p (i.e., in the sense of congruences) of the number of representations of n as a sum $\sum_{j=1}^{s} m_j^2$ of s squares. From $\mathscr{S}(s,n) = \prod_p \mathscr{S}_p$ it follows that $\mathscr{S}$ may be considered as the global arithmetic density of solutions. We may obtain it more directly as follows: Let $N_1, N_2, \ldots, N_g, \ldots$ be a sequence of integers, increasing unboundedly and containing an ever increasing number of primes. Such a sequence may be, e.g., that of factorials, $N_g = g!$, or the product of the first g primes, $N_g = 2\cdot 3\cdot 5\cdots p_g$, etc. We then consider the number $M_{N_g}(n,s)$ of solutions of the congruence $\sum_{j=1}^{s} m_j^2 \equiv n \pmod{N_g}$, with each m_j running through a complete residue system modulo N_g. Then $\mathscr{S}(=\mathscr{S}(s,n)) = \lim_{g\to\infty}\{M_{N_g}(n,s)/N_g^{s-1}\}$. The remaining factors of $\rho_s(n)$, namely $\{\pi^{s/2}/\Gamma(s/2)\}n^{(s/2)-1}$, have a similar interpretation. Let us ask for the number of solutions of the *inequality* $\sum_{j=1}^{s} m_j^2 \leqslant n$ in integers m_j. This will be approximately equal to the volume of the hypersphere of radius $R = \sqrt{n}$, which, in s-dimensional Euclidean space, equals $V(R) = \pi^{s/2}R^s/\Gamma(1+(s/2))$. In order to obtain the approximate number of solutions of the *equation* $\sum_{j=1}^{s} m_j^2 = n$, we subtract from $V(\sqrt{n})$ the corresponding volume $V(\sqrt{n-1})$, related to the number of solutions of the inequality $\sum_{j=1}^{s} m_j^2 \leqslant n-1$. We obtain

$$\frac{\pi^{s/2}}{\Gamma(1+(s/2))}(n^{s/2}-(n-1)^{s/2}) = \frac{\pi^{s/2}}{\Gamma(1+(s/2))}n^{s/2}(1-(1-n^{-1})^{s/2})$$

$$= \frac{\pi^{s/2}n^{s/2}}{(s/2)\Gamma(s/2)}\left\{1-\left(1-\frac{s}{2n}+O(n^{-2})\right)\right\}$$

$$= \frac{2\pi^{s/2}}{s\Gamma(s/2)}n^{s/2}\frac{s}{2n}(1+O(n^{-1}))$$

$$= \frac{\pi^{s/2}}{\Gamma(s/2)}n^{(s/2)-1}(1+O(n^{-1}) \simeq \frac{\pi^{s/2}}{\Gamma(s/2)}n^{(s/2)-1}.$$

This is consequently the average number of solutions to be expected, ignoring the arithmetic nature of n. It is the average number of solutions from the point of view of real variables. It follows that, up to an error term, $\rho_s(n)$ may be interpreted as the product of local densities, at all finite primes and—to use a terminology current in algebraic number theory—also at the infinite prime, i.e., in the real field.

This problem of the density of solutions has been considered in the more general framework of quadratic forms $Q(x_1, x_2, \ldots, x_s)$ by C. L. Siegel in a rightly celebrated sequence of papers [243] (see also Chapter 14).

§6. Other Approaches

While shorter than the circle method discussed in the present chapter, the proofs of the same results with modular functions by Knopp [133], Mordell [184], and others, and with the "simplified" circle method by Estermann [72], are of a comparable degree of complexity. It does not seem warranted to present them in great detail. It should suffice to give them in abbreviated form, in order to suggest at least the flavor of the respective approaches. This will be done in the next chapter.

A final remark: In Examples 1 and 2 of Section 4, we saw that $r_s(n) = \rho_s(n)$ for $s = 4$ and $s = 8$. This holds also for $s = 5$, although we could not verify it, because we could not compute $r_5(n)$ by elementary methods. While, generally speaking, a second and third proof of Theorem 1 are not expected to offer much new information, the added insight that we shall obtain from the study of $r_s(n)$ by the method of modular functions will lead to the proof of the following theorem.

Theorem 2. *For* $5 \leqslant s \leqslant 8$, $r_s(n) = \rho_s(n)$, *while for* $s > 8$, *the equality does not hold for all* n.

In fact, the equality of Theorem 2 holds also for $s = 4$ (and even for $s = 3$), but this will not follow from the proofs of Chapter 13. For a justification, see [18].

§7. Problems

1. Show that, for $x \to e^{2\pi i h/k}$ from the inside of the unit circle, $|\theta(x)|$ increases faster when k is small than when k is large.
2. Prove Lemma 3 for odd exponent $a \geqslant 3$.
3. Prove Lemma 4 for as even.
4. Show that $\liminf_{n\to\infty} r_4(n)/n = 0$ and infer from this that, as a function of n, $\mathscr{S}(4, n)$ is not bounded away from zero.
5. Prove equations (12.12), (12.12′).
6. Let $s \equiv 2 \pmod 4$ and $n = 2^b n_1$. Show that

$$\mathscr{S}_2 = 1 + (-1)^{(2n_1 - s)/4} 2^{(b+1)(1-s/2)}.$$

7. Compute A_{p^a} for $n = p^b n_1, p \nmid n_1$. In particular, show that (12.13) holds if $s \equiv 0 \pmod 4$.

8.

(a) Show that (12.13) holds for $s \equiv 2 \pmod 4$ and $p \equiv 1 \pmod 4$.
(b) Compute the corresponding formula for $s \equiv 2 \pmod 4$ and $p \equiv 3 \pmod 4$.

9. Verify the formula

$$\mathscr{S}_2 \sum_{\substack{d \mid n \\ d \text{ odd}}} d^{1-(s/2)} = \sum_{t \mid n} (-1)^{n+(n/t)+(s/4)(t+1)} t^{1-(s/2)}$$

for all $s \equiv 0 \pmod 4$.

10. Prove that $\sum_{d \mid n} (-1)^{n+d+(n/d)+1} d = \sum_{d \mid n,\, d \not\equiv 0 \pmod 4} d$. Conclude, as an illustration of Theorem 2, that the value of $r_4(n)$ coincides with $\rho_4(n)$.

*11. Sum the series $L(2, \chi)$ in finite terms, with $\chi(m) = (\frac{n}{m})$ for m odd, $\chi(m) = 0$ for m even.

12. Show that (12.14) holds for all squarefree n.

13. Use the method of the text and the results of Problems 6 and 8 to obtain $\rho_6(n)$; compare the result with the value of $r_6(n)$ obtained in Chapter 9.

14. Use the method of the text to obtain $\rho_7(n)$ for squarefree n. Use this and Theorem 2 to obtain $r_7(n)$. (You may wish to compare your results with those of [253], [178], and [12].)

15. Show that if p is odd and n is squarefree, then Lemma 4 implies that, if s is odd, for $p \nmid n$, $A_p = (\frac{\varepsilon_s n}{p}) p^{(1-s)/2}$, with $\varepsilon_s = (-1)^{(s-1)/2}$ and $A_{p^a} = 0$ for $a > 1$, while for $p \mid n$, $A_p = 0$, $A_{p^2} = -p^{1-s}$, and $A_{p^a} = 0$ for $a \geqslant 3$.

16. Prove Lemma 5.

17. Verify that on $\mathscr{C}$, $|e^{-i\pi s} G(s, \bar{\chi}) / \sin \pi s|$ is bounded.
(Hint: Consider the square of the absolute value and verify separately the extrema on the horizontal and on the vertical sides of $\mathscr{C}$.)

18. Let $n = 3$, so that $n_1 = 4n = 12$ and $\chi(m) = (-1)^{(m-1)(n-1)/4} (\frac{m}{n}) = (-1)^{(m-1)/2} (\frac{m}{3})$ for m odd. Compute $L(2, \chi)$.
(Hint: Verify whether $\chi(m)$ is or is not primitive.)
(Ans.: $\pi^2/6\sqrt{3}$.)

19. Expand $b_2(x) = x^2 - x + \frac{1}{6}$ into a Fourier cosine series valid in $0 \leqslant x \leqslant 1$.

20. Try to compute $L(2, \chi)$ by the second method, for $n = 3$ and/or $n = 5$. (Remark: do not be disappointed if you get bogged down in the computation of the a_j's—it is indeed difficult.)

Chapter 13

Alternative Methods for Evaluating $r_s(n)$

§1. Estermann's Proof

Theorem 12.1 is formulated in [72] as follows:

Theorem 1. *For integers h, k with* $(h, k) = 1$, *let* $\lambda(h/k) = (2k)^{-1} \sum_{q=1}^{2k} e^{2\pi i h q^2/2k}$, *and set* $A_k = \sum_{1 \leqslant h \leqslant 2k, (h,k)=1} \lambda^s e^{-2\pi i h n/2k}$. *Then, if* $S(n) = \sum_{k=1}^{\infty} A_k$ *and* $s = 5, 6, 7$, *or* 8, *one has for some constant* $c = c(s)$, *independent of n, that*

$$r_s(n) = c(s) n^{(s/2)-1} S(n).$$

It is clear that $\lambda(h/k) = k^{-1} G(h, k)$ (the factor $\frac{1}{2}$ compensates for the doubling of the summation interval), $S(n)$ is essentially our singular series $\mathscr{S} = \mathscr{S}(s, n)$, and $c(s)$ can only be equal to $\pi^{s/2}/\Gamma(s/2)$ if Theorems 12.1 and 12.2 are to hold.

The idea of the proof is roughly as follows: The functions $\theta_q^s(\tau)$ ($q = 2, 3, 4$; Estermann denotes $\theta_4(\tau)$ by $\theta_0(\tau)$, but we prefer to keep our previous notation) are defined, and their transformations under $\tau \to \tau + 1$ and $\tau \to -1/\tau$ are proved directly (i.e., without appeal to Poisson's summation formula). Next, the author introduces functions $\phi_q(\tau)$, by setting $\phi_3(\tau) = 1 + \sum_r \lambda^s(r)(ir - i\tau)^{-s/2}$ ($\sum_r$ means summation over all rationals; only absolutely convergent series occur) and shows (by a lengthy proof) that $\lambda(r) = \lim_{\operatorname{Im}\tau \to \infty} \times \{(-i\tau)^{-1/2} \theta_3(r - 1/\tau)\}$ for all rationals $r = h/k$; the functions $\phi_2(\tau)$ and $\phi_4(\tau)$ are then defined by $\phi_4(\tau) = \phi_3(\tau + 1)$ and $\phi_2(\tau) = (-i\tau)^{-s/2} \phi_4(-1/\tau)$.

The concept of a c.f. $-\alpha$ (for "comparison function of dimension $-\alpha$") of a given function $f(\tau)$ is introduced as follows. Given $f(\tau)$ holomorphic in $\operatorname{Im}\tau > 0$, with $\lim_{\operatorname{Im}\tau \to \infty} f(\tau) = L$, assume that for every rational r and some constant $\alpha > 0$, $\lim_{\operatorname{Im}\tau \to \infty} \{(-i\tau)^{-\alpha} f(r - 1/\tau)\} = l(r)$ exists. Then, if $\sum_r l(r)(ir - i\tau)^{-\alpha}$ converges absolutely to $h(\tau)$, the function $\phi(\tau) = L + h(\tau)$ is said to be *the* c.f. $-\alpha$ of $f(\tau)$ (if it exists at all for a given function $f(\tau)$ and α, a c.f. $-\alpha$ is obviously unique). Several properties of c.f. $-\alpha$ are proved. For $s \geqslant 5$, the author shows that the $\phi_q(\tau)$ are the c.f. $-s/2$ of the $\theta_q^s(\tau)$. Next, the author defines functions $g_q(\tau) = \phi_q(\tau)\{\theta_q(\tau)\}^{-s}$ ($q = 2, 3, 4$) and sets $F_1(\tau)$, $F_2(\tau)$, $F_3(\tau)$ equal to the fundamental symmetric functions of the $g_q(\tau)$ (so e.g., $F_1(\tau) = \sum_{q=2}^{4} g_q(\tau)$, etc.). The functions $F_q(\tau)$ are shown to satisfy $F_q(\tau) = F_q(\tau + 1) = F_q(-1/\tau)$ and also $F_q(-\bar{\tau}) = \overline{F_q(\tau)}$, and to be holomorphic in $\operatorname{Im}\tau > 0$. To facilitate the desired conclusion, the change of variables $e^{2\pi i\tau} = z$ is made, and we set $F_q((\log z)/2\pi i) = G_q(z)$.

A study of $G_q(z)$ shows that it has no singularities in $0 < |z| < e^{-\pi}$. Next, the new assumption $s \leqslant 8$ is introduced, so that, from here on, $s = 5, 6, 7$, or 8 only. Under these conditions, it is shown that $\lim_{z\to 0}\{zG_q(z)\} = 0$, so that $G_q(z)$ is holomorphic also at the origin. Next, consider the compact subset $\mathscr{A}$ of $\mathbb{C}$ defined by $\mathscr{A} = \{0\} \cup \{|z| < 1, |\log z| \geqslant 2\pi\}$. If $\mathscr{D}_1 = \{|z| < 1\}$, $\mathscr{D}_2 = \{|z| < e^{-2\pi}\}$, then $\mathscr{D}_2 \subset \mathscr{A} \subset \mathscr{D}_1$. A lemma is proved, to the effect that for any domains $\mathscr{D}_1, \mathscr{D}_2$ and compact set $\mathscr{A}$ with $\mathscr{D}_2 \subset \mathscr{A} \subset \mathscr{D}_1$ and $f(z)$ holomorphic in $\mathscr{D}_1$ and real on the boundary of $\mathscr{A}$, the function $f(z)$ reduces to a constant. These conditions are verified for the $G_q(z)$ with the previous definition of $\mathscr{A}$. It follows that the $G_q(z)$, and hence the $F_q(\tau)$, are really constants. The cubic $u^3 - F_1(\tau)u^2 + F_2(\tau)u - F_3(\tau) = 0$ therefore has constant coefficients, and its solutions, which are the $g_q(\tau)$, are also constants. In particular, $g_3(\tau)$ is constant. It is possible to compute that $\lim_{y\to\infty} g_3(x + iy) = 1$; hence, $g_3(\tau) = 1$ and $\phi_3(\tau)\theta_3(\tau)^{-s} = g_3(\tau)$ now prove that $\theta_3(\tau)^s = \phi_3(\tau)$. By the definition $\phi_3(\tau) = 1 + \sum_r \lambda^s(r)(ir - i\tau)^{-s/2}$ and the obvious identity $\sum_r f(r) = \sum_{0<r\leqslant 2}\sum_{q=-\infty}^{\infty} \times f(r + 2q)$, we now see, denoting $\sum_{q=-\infty}^{\infty}(2iq - i\tau)^{-s/2}$ by $F(\tau)$, that $\phi_3(\tau) = 1 + \sum_{0<r\leqslant 2}\lambda^s(r)F(\tau - r)$. However, $F(\tau)$ has period 2 and vanishes for $\operatorname{Im}\tau \to \infty$, so that it may be expanded into a Fourier series $F(\tau) = \sum_{n=1}^{\infty} b_n e^{2\pi i n\tau}$, with $b_n = \frac{1}{2}\int_{\tau_0}^{\tau_0+2} F(\tau)e^{-2\pi i n\tau}\,d\tau$ (τ_0 arbitrary in the upper half plane $\mathscr{H}$). The integral that determines b_n may be estimated; it reduces to $b_n = cn^{(s/2)-1}$, with $c = c(s) = \frac{1}{2}\int_{i-\infty}^{i+\infty}(-iw)^{-s/2}e^{-\pi i w}\,dw$, and depends only on s. Combining these results, $\theta_3^s(\tau) = 1 + \sum_{n=1}^{\infty} r_s(n)e^{2\pi i n\tau} = \phi_3(\tau) = 1 + \sum_{0<r\leq 2}\lambda^s(r)c(s) \times \sum_{n=1}^{\infty} n^{(s/2)-1}e^{2\pi i n(\tau - r)}$, or, by identification of coefficients,

$$r_s(n) = c(s)n^{(s/2)-1}\sum_{0<r\leqslant 2}\lambda^s(r)e^{-\pi i n r}.$$

Here the sum equals

$$\sum_{\substack{k=1\\(h,k)=1}}^{\infty}\ \sum_{0<h\leqslant 2k}\lambda^s\left(\frac{h}{k}\right)e^{-\pi i n h/k} = \sum_{k=1}^{\infty} A_k = S(n), \qquad \text{say,}$$

and the proof of Theorem 1 is complete.

Theorem 2. *For* $s = 8$ *we have* $r_8(n) = 16(\sigma_3(n) - 2\sigma_3(n/2) + 16\sigma_3(n/4))$, *where* $\sigma_3(n) = \sum_{d\,|\,n} d^3$.

Theorem 3. *For* $s = 5$, *the number* $R_5(n)$ *of primitive representations of* n *as a sum of five squares is given by* $R_5(n) = C(n)\pi^{-2}n^{3/2}D(n)$, *where* $D(n) = \sum_{m=1}^{\infty}\chi(m)m^{-2}$, $\chi(m) = \left(\frac{n}{m}\right)$ *for* m *odd*, $\chi(m) = 0$ *otherwise, and*

$$C(n) = \begin{cases} 80 & \text{for } n \equiv 0, 1, \text{ or } 4 \pmod 8,\\ 160 & \text{for } n \equiv 2, 3, 6, \text{ or } 7 \pmod 8,\\ 112 & \text{for } n \equiv 5 \pmod 8.\end{cases}$$

For every n, $r_5(n) = \sum_{d^2\,|\,n} R_5(n/d^2)$.

The proof of Theorem 2 starts with the trivial proofs that $\lambda(0) = 1$, $\lambda(1) = 0$, and $\lambda(r+2) = \lambda(r)$ and the nontrivial proof (using the transformation formula for $\theta_3(\tau)$ under $\tau \to -1/\tau$; see (8.1)) that for $r \neq 0$, $\lambda^2(-1/r) = (ir)^{-1}\lambda^2(r)$. Next, the following lemma is stated and proved.

Lemma 1. *Let* $\mathbf{A} \subset \mathbb{Q}$ *be a subset of the rationals such that* (i) $0 \in \mathbf{A}$ *and* $1 \in \mathbf{A}$; (ii) *if* $r \in \mathbf{A}$, *then also* $r \pm 2 \in \mathbf{A}$; (iii) *if* $0 \neq r \in \mathbf{A}$, *then also* $-r^{-1} \in \mathbf{A}$. *In that case* $\mathbf{A} = \mathbb{Q}$

This lemma is used to prove the following

Proposition. *Let* $f_m(r)$ $(m = 1, 2)$ *be defined over* $\mathbb{Q}$, *with* $f_1(0) = f_2(0)$, $f_1(1) = f_2(1)$, $f_m(r) = f_m(r+2)$, *and, for* $r \neq 0$, $f_m(-r^{-1}) = -r^{-2}f_m(r)$; *then* $f_1(r) = f_2(r)$ *for all* $r \in \mathbb{Q}$.

This proposition, in turn, permits us to prove that, for $r = h/k$ $(h, k \in \mathbb{Z})$,

$$\lambda^4(r) = \begin{cases} 0 & \text{if } 2 \nmid hk, \\ (-1)^{k-1}k^{-2} & \text{if } 2 \mid hk. \end{cases} \tag{13.1}$$

Indeed, it is sufficient to take as $f_1(r)$ and $f_2(r)$ the two sides in (13.1).

If $s = 8$, consider the Ramanujan sum $c_k(x) = \sum_{h \pmod k, (h,k)=1} e^{2\pi i h x/k}$, and observe that

$$\sum_{u \mid v} c_u(x) = \begin{cases} v & \text{if } v \mid x, \\ 0 & \text{if } v \nmid x. \end{cases}$$

Also, if $2 \nmid u$, then $c_{2u}(x) = (-1)^x c_u(x)$. From (13.1) it follows that $\lambda^8(h/k) = \mathrm{k}^{-4}$ if $2 \mid hk$, otherwise $\lambda^8 = 0$. This implies that

$$A_k = \sum_{\substack{h \leqslant 2k \\ (h,k)=1}} \lambda^8\left(\frac{h}{k}\right) e^{-2\pi i n h/2k} = \begin{cases} k^{-4}c_k(n) & \text{if } 2 \nmid k \\ k^{-4}c_{2k}(n) & \text{if } 2 \mid k. \end{cases}$$

Now set

$$S_1 = \sum_{u=1}^{\infty} u^{-4}c_u(n), \qquad S_2 = \sum_{u \text{ odd}} u^{-4}c_u(n), \qquad S_3 = \sum_{4 \mid u} u^{-4}c_u(n).$$

Then one verifies successively that

$$S(n) = S_2 + 16S_3,$$

$$S_1 - S_2 - S_3 = \sum_{v \equiv 2 \pmod 4} v^{-4}c_v(n) = \sum_{u \text{ odd}} (2u)^{-4}c_{2u}(n) = \frac{(-1)^n}{16} S_2,$$

so that $S(n) = 16S_1 - 15S_2 - (-1)^n S_2$. To evaluate S_1 and S_2, we compute $\zeta(4) = \sum_{n=1}^{\infty} n^{-4}$, $\sum_{v \text{ odd}} v^{-4} = \zeta(4) - \frac{1}{16}\zeta(4) = \frac{15}{16}\zeta(4)$,

$$\sum_{n=1}^{\infty} \frac{1}{n^4} S_1 = \sum_{n=1}^{\infty} \frac{1}{n^4} \sum_{u=1}^{\infty} \frac{1}{u^4} c_u(n) = \sum_{q=1}^{\infty} q^{-4} \sum_{u \mid q} c_u(n)$$

$$= \sum_{q \mid n} q^{-3} = n^{-3}\sigma_3(n),$$

and similarly,

$$\tfrac{15}{16}\zeta(4) S_2 = \sum_{\substack{q \text{ odd} \\ q \mid n}} q^{-3} = \sum_{q \mid n} q^{-3} - \sum_{\substack{q \mid n \\ q \text{ even}}} q^{-3} = n^{-3}(\sigma_3(n) - \sigma_3(n/2)).$$

For n odd, $\sigma_3(n/2) = 0$, so that

$$n^3 \cdot \tfrac{15}{16}\zeta(4) S(n) = \begin{cases} \sigma_3(n) & \text{if } 2 \nmid n \\ -\sigma_3(n) + 16\sigma_3(n/2) & \text{if } 2 \mid n. \end{cases}$$

If n is even and u_j $(j = 1, 2, \ldots, q)$ are the positive divisors of $n/2$ that do not divide $n/4$, then $2u_j$ are the positive divisors of n that do not divide $n/2$, so that $\sigma_3(n) - \sigma_3(n/2) = \sum_{m=1}^{q} (2u_m)^3 = 8\sum_{m=1}^{q} u_m^3 = 8(\sigma_3(n/2) - \sigma_3(n/4))$ for $2 \mid n$. This shows that $\frac{15}{16}\zeta(4) n^3 S(n) = \sigma_3(n) - 2\sigma_3(n/2) + 16\sigma_3(n/4)$, whence $r_8(n) = C\{\sigma_3(n) - 2\sigma_3(n/2) + 16\sigma_3(n/4)\}$. Setting $n = 1$, it follows that $C = 16$ and the proof of Theorem 2 is complete.

We shall omit here the discussion of the long (9 pages) and very computational proof of Theorem 3, which uses the same general ideas as those of Theorem 2. The interested reader is directed to [72] for the details of the proof.

§2. Sketch of the Proof by Modular Functions

Once it is known that, at least for $s \geqslant 5$, $r_s(n) = \rho_s(n) + O(n^{s/4})$, it is possible to prove this fact by the use of modular functions. The main idea of the proof is as follows.

We already know (see (11.5)) that

$$\theta^s(\tau) = 1 + \sum_{n=1}^{\infty} r_s(n) e^{\pi i \tau n}. \tag{13.2}$$

We now define the function

$$\psi_s(\tau) = 1 + \sum_{n=1}^{\infty} \rho_s'(n) e^{\pi i \tau n}. \tag{13.3}$$

Here $\rho_s'(n)$ is provisionally defined by (13.11) below, but subsequently it will be shown that $\rho_s'(n) = \rho_s(n)$, the function already known from Theorem 12.1.

To show that $\theta^s(\tau)$ and $\psi_s(\tau)$ are both modular forms of degree $-s/2$ under Γ_θ, we prove first that both functions satisfy

$$f(M\tau) = v_s(M)(c\tau + d)^{s/2} f(\tau) \tag{13.4}$$

for every $M = \left(\begin{smallmatrix} * & * \\ c & d \end{smallmatrix}\right) \in \Gamma_\theta$ and a system of multipliers v_s, the same for both functions. Then (13.2) and (13.3) may be considered as the expansions of $\theta^s(\tau)$ and of $\psi_s(\tau)$, respectively, at $\tau = \infty$, and it only remains to show that these functions also have the right kind of expansions at the other cusp of the usual fundamental domain of Γ_θ, i.e., at $\tau = -1$. Indeed, we find that

$$\theta^s(\tau) = (\tau + 1)^{-s/2} \sum_{m \geqslant 0} c_s(m) \exp\left\{2\pi i\left(m + \frac{s}{8}\right)\left(\frac{-1}{\tau + 1}\right)\right\} \tag{13.5}$$

and

$$\psi_s(\tau) = (\tau + 1)^{-s/2} \sum_{n + \{s/8\} > 0} a_s(n) \exp\left\{2\pi i\left(n + \left\{\frac{s}{8}\right\}\right)\left(\frac{-1}{\tau + 1}\right)\right\}, \tag{13.6}$$

which are the typical forms of expansions at the cusp $\tau = -1$. Here $\{x\} = x - [x]$ is the fractional part of x. In (13.6) the expansion starts with a term $\exp\{2\pi i\{s/8\}(-1)/(\tau + 1)\}$ if $s \not\equiv 0 \pmod 8$, and with $\exp\{2\pi i(-1)/(\tau + 1)\}$ if $s \equiv 0 \pmod 8$.

Two consequences of what we shall establish:

A. The difference $F(\tau) = \theta^s(\tau) - \psi_s(\tau) = \sum_{n=1}^{\infty} (r_s(n) - \rho_s'(n)) e^{\pi i n \tau}$ is a cusp form, so that if we define a_s by $F(\tau) = \sum_{n=1}^{\infty} a_s(n) e^{\pi i n \tau}$ then $r_s(n) - \rho_s'(n) = a_s(n) = O(n^{s/4})$, by Theorem 11.5.

B. The ratio $\Phi_s(\tau) = \psi_s(\tau)/\theta^s(\tau)$ is a modular *function*; indeed, both, numerator and denominator transform according to (13.4), so that their ratio transforms simply as $\Phi_s(M\tau) = \Phi_s(\tau)$ for all $M \in \Gamma_\theta$. From (13.2), (13.3) on the one hand and (13.5), (13.6) on the other, it follows that the expansions of $\Phi_s(\tau)$ at the cusps are of the forms

$$\Phi_s(\tau) = 1 + \sum_{m=1}^{\infty} g_m e^{\pi i m \tau} \quad \text{at } \tau = \infty; \tag{13.7}$$

and

$$\Phi_s(\tau) = \sum_{m + \{s/8\} > 0} d_m \exp\left\{2\pi i\left(m + \left\{\frac{s}{8}\right\} - \frac{s}{8}\right)\left(\frac{-1}{\tau + 1}\right)\right\} \quad \text{at } \tau = -1. \tag{13.8}$$

For $s = 5$, 6, or 7, we have $\{s/8\} = s/8$ and the first term of the expansion at

$\tau = -1$ is d_0, a nonvanishing constant. If $s = 8$, then $\{s/8\} = 0$ and the summation in (13.8) starts with $m = 1$, so that the first term is $d_1 \exp\{2\pi i(1-1)(-1)/(\tau+1)\} = d_1 \neq 0$. In both cases, no negative powers of $\exp\{2\pi i(-1/(\tau+1)\}$ occur. As $\Phi_s(\tau)$ is a modular *function*, holomorphic in $\mathscr{H}$ (because $\theta(\tau)$ vanishes only for real—actually, only rational—τ, as we know from (8.6)) and also at the cusps (which, incidentally, means that $\psi_s(\tau)$ has there a zero of order at least equal to that of $\theta^s(\tau)$), $\Phi_s(\tau)$ reduces to a constant, by Theorem 11.4. The value of that constant follows from (13.7), by observing that $\lim_{\mathrm{Im}\,\tau\to\infty} \Phi_s(\tau) = 1$. We conclude that, for $s = 5, 6, 7$, and 8, $\psi_s(\tau) = \theta^s(\tau)$, and in particular that $r_s(n) = \rho'_s(n)$. We show next that $\rho'_s(n)$, as defined by (13.11), is in fact identical with the $\rho_s(n)$ of Theorem 12.1.

On the other hand, for $s > 8$ one has $s = 8t + s_1$, say, with $t \geqslant 1$, $0 \leqslant s_1 \leqslant 7$, and $t \geqslant 2$ if $s_1 = 0$. Hence, in (13.8), $m + \{s/8\} - s/8 = m - t$, and the first term corresponds to $m = 0$, except for $s_1 = 0$, when $t \geqslant 2$ and the sum starts with the term corresponding to $m = 1$. In the first case, the first term is $d_0 \exp\{2\pi i(-t)(-1)/(\tau+1)\}$; in the second case it is $d_1 \exp\{2\pi i(1-t)(-1)/(\tau+1)\}$ and $1 - t \leqslant -1$. We have already remarked that $d_0 d_1 \neq 0$; hence, $\Phi_s(\tau) \to \infty$ as $\tau \to -1$ within $\mathscr{H}$, and $\Phi_s(\tau)$ is not constant. This means, in particular, that $r_s(n)$ cannot be equal to $\rho_s(n)$ for all n, as claimed. This will finish the proof of the following theorem.

Theorem 4. *With $\rho_s(n)$ as defined in Theorem 12.1 and for $s = 5, 6, 7$, and 8, $r_s(n) = \rho_s(n)$. For $s \geqslant 9$, $r_s(n) = \rho_s(n) + O(n^{s/4})$, where the error term is not identically zero.*

§3. The Function $\psi_s(\tau)$

To implement this program, one starts by defining the function $\psi_s(\tau)$ by

$$\psi_s(\tau) = \frac{1}{2}\sum_c \sum_d{}^{*}\, \bar{v}_s(M_{c,d})(c\tau + d)^{-s/2}; \tag{13.9}$$

here the asterisk means that the summation proceeds over all sets of integers c, d, with $c + d \equiv 1 \pmod 2$, $(c, d) = 1$. Furthermore, $M_{c,d}$ is any one of the matrices $M = \begin{pmatrix} a & b \\ c & d \end{pmatrix} \in \Gamma_\theta$, and $v_s(M_{c,d}) = \{v_\theta(M_{c,d})\}^s$, with $v_\theta(M)$ the multiplier of degree $-\frac{1}{2}$ for Γ_θ, defined in Section 11.7. We already know that $\theta(\tau)$ is a modular form of degree $-\frac{1}{2}$ for Γ_θ, with system of multipliers $v_\theta(M)$. It then immediately follows that $\theta^s(\tau)$ is a modular form for Γ_θ, of degree $-s/2$ and set of multipliers $v_\theta^s(M)$; we denote it simply by $v_s(M)$ for $M \in \Gamma_\theta$.

Next, one proves that $\psi_s(\tau)$ is holomorphic in $\mathscr{H}$; this is easy, by the absolute convergence, for $s > 4$, of the series in (13.9). Less obvious is the fact that, for all $M = \begin{pmatrix} a & b \\ c & d \end{pmatrix} \in \Gamma_\theta$.

$$\psi_s(M\tau) = v_s(M_{c,d})(c_\tau + d)^{s/2}\psi_s(\tau), \tag{13.10}$$

so that $\psi_s(\tau)$ satisfies (13.4).

The main tool in the proof of (13.10), which we do not reproduce here, is the use of the consistency conditions, as defined in §5 of Chapter 11 and satisfied by the $v_s(M)$ for $M \in \Gamma_\theta$.

Next, we recall that (see Chapter 11, §5) Γ_θ has a fundamental region with two inequivalent cusps, ∞ and -1. Our aim is to prove

Theorem 5. *The expansion of* $\psi_s(\tau)$ *at* ∞ *is*

$$\psi_s(\tau) = 1 + \sum_{n=1}^{\infty} \rho_s'(n) e^{\pi i n \tau},$$

where

$$\rho_s'(n) = \frac{e^{-\pi i s/4} \pi^{s/2}}{\Gamma(s/2)} n^{(s/2)-1} \sum_{c=1}^{\infty} \mathrm{B}_c(n) \tag{13.11}$$

and

$$B_c(n) = c^{-s/2} \sum_{0 \leqslant h < 2c}^{*} \bar{v}_s(M_{c,h}) e^{\pi i n h/c}. \tag{13.11'}$$

Proof. From $\theta(\tau + 2) = \theta(\tau)$, it follows that $v_s(S^2) = 1$ (recall that $S = \left(\begin{smallmatrix} 1 & 1 \\ 0 & 1 \end{smallmatrix}\right)$), and we verify by $\theta(M\tau) = \theta((-M)\tau)$ that $v_s(-M)(-c\tau - d)^{s/2} = v_s(M)(c\tau + d)^{s/2}$. This permits us to rewrite (13.9) as

$$\psi_s(\tau) = 1 + \sum_{c>0} \sum_{d}^{*} \bar{v}_s(M_{c,d})(c\tau + d)^{-s/2}.$$

Given c, we may set $d = h + 2mc$, $0 \leqslant h < 2c$, where h and m are uniquely determined by c and d. Since $h = d - 2mc$, we have $M_{c,h} = M_{c,d}(S^2)^{-m}$ and by the consistency conditions $v_s(M_{c,d}) = v_s(M_{c,h})$, so that

$$\psi_s(\tau) = 1 + \sum_{c>0} \sum_{0 \leqslant h < 2c}^{*} \bar{v}_s(M_{c,h}) \sum_{m=-\infty}^{\infty} (c\tau + h + 2mc)^{-s/2}.$$

By Theorem 8.2, the inner sum equals

$$\{\Gamma(s/2)\}^{-1} e^{-\pi i s/4} \pi^{s/2} \textstyle\sum_{n=1}^{\infty} n^{(s/2)-1} e^{\pi i n(\tau + h/c)} c^{s/2}$$

and

$$\psi_s(\tau) = 1 + \{\Gamma(s/2)\}^{-1} e^{-\pi i s/4} \pi^{s/2}$$

$$\times \sum_{c>0} c^{-s/2} \sum_{0 \leqslant h < 2h}^{*} \bar{v}_s(M_{c,h}) \sum_{n=1}^{\infty} n^{(s/2)-1} e^{\pi i n(\tau + h/c)}.$$

Observing that for $s \geqslant 5$ we have absolute convergence, this leads to

$$\psi_s(\tau) = 1 + \sum_{n=1}^{\infty} \rho_s'(n) e^{\pi i n \tau}$$

with

$$\rho_s'(n) = \frac{e^{-\pi i s/4}\pi^{s/2}}{\Gamma(s/2)} n^{(s/2)-1} \sum_{c=1}^{\infty} B_c(n),$$

where

$$B_c(n) = c^{-s/2} \sum_{0 \leqslant h < 2c}^{*} \bar{v}_s(M_{c,h}) e^{\pi i n h/c},$$

as claimed. This finishes the proof of Theorem 5. □

§4. The Expansion of $\psi_s(\tau)$ at the Cusp $\tau = -1$

We now consider the expansion of $\psi_s(\tau)$ at $\tau = -1$, the other cusp of the fundamental region of Γ_θ considered. We shall sketch a proof of the following theorem.

Theorem 6. *At $\tau = -1$, $\psi_s(\tau)$ has the expansion*

$$\psi_s(\tau) = (\tau + 1)^{-s/2} \sum_{n=1}^{\infty} (-1)^{s/2} \exp\left\{\frac{2\pi i n}{8}\left(\frac{-1}{\tau+1}\right)\right\} \cdot \phi_s(n), \tag{13.12}$$

where

$$\phi_s(n) = \frac{(2\pi)^{s/2}}{\Gamma(s/2)} \left(\frac{n}{8}\right)^{(s/2)-1} \sum_{\substack{d>0 \\ d \text{ odd}}} D_d(n)$$

and

$$D_d(n) = \tfrac{1}{8} d^{-s/2} e^{-\pi i d s/4} \sum_{\substack{0 \leqslant m < 8d \\ (m,d)=1}} \exp\left\{\frac{2\pi i n}{8d}(sd^2 - n)\right\}.$$

Sketch of Proof of Theorem 6. By (11.4), we know (see also [133], Chapter 2]) that, on account of (13.10), there exist constants κ, λ, and $a_s(n)$ such that

$$\psi_s(\tau) = (\tau + 1)^{-s/2} \sum_{-\infty}^{\infty} a_s(n) \exp\left\{\frac{2\pi i(n+\kappa)}{\lambda}\left(\frac{-1}{\tau+1}\right)\right\}.$$

The general theory of modular forms permits the computation of these constants, and one finds $\lambda = 1$, $\kappa = s/8 - [s/8]$, i.e., the fractional part of $s/8$, which we denote by $\{s/8\}$. It follows that

$$\psi_s(\tau) = (\tau + 1)^{-s/2} \sum_{-\infty}^{\infty} a_s(n) \exp\left\{2\pi i\left(n + \left\{\frac{s}{8}\right\}\right)\left(-\frac{1}{\tau + 1}\right)\right\}. \quad (13.13)$$

On the other hand, if we substitute $-1 - 1/\tau$ for τ in (13.9) and replace $v_s(M_{c,d})$ by its value from Theorem 11.7, we obtain, after routine manipulations and an application of Lipschitz's formula (Theorem 8.2), that

$$\psi_s\left(-1 - \frac{1}{\tau}\right) = \tau^{s/2} \sum_{n=1}^{\infty} \phi_s(n) e^{2\pi i n\tau/8}, \quad (13.14)$$

with $\phi_s(n)$ as defined in Theorem 6. We now replace in (13.14) τ by $-(\tau + 1)^{-1}$, observe that $\{-(\tau + 1)\}^{s/2} = (-1)^{s/2}(\tau + 1)^{s/2}$, and obtain (13.12). Theorem 6 is proved. □

If we compare (13.12) with (13.13) and recall the uniqueness of Laurent expansions, it follows that $a_s(n) = 0$ for $n + \{s/8\} \leqslant 0$ and also that $\phi_s(n) = 0$, unless n belongs to just one definite residue class modulo 8. Specifically, if $s \equiv \mu \pmod 8$, $0 \leqslant \mu < 8$, then $\kappa = \{s/8\} = \mu/8$, and $n/8$ in (13.12) has to lead to the same value of the exponential as $n + \{s/8\} = n + \mu/8$ in (13.13), whence $n \equiv \mu \equiv s \pmod 8$. Consequently,

$$\phi_s(m) = \begin{cases} (-1)^{s/2} a_s(n) & \text{if } m = 8n + s, \\ 0 & \text{otherwise,} \end{cases}$$

and (13.12) and (13.13) combine to yield

$$\begin{aligned} \psi_s(\tau) &= (-1)^{s/2}(\tau + 1)^{-s/2} \sum_{\substack{m=1 \\ n \equiv s \,(\mathrm{mod}\ 8)}}^{\infty} \phi_s(n) \exp\left\{\frac{2\pi i n}{8}\left(-\frac{1}{\tau + 1}\right)\right\} \\ &= (\tau + 1)^{-s/2} \sum_{n + \{s/8\} > 0} a_s(n) \exp\left\{2\pi i\left(n + \left\{\frac{s}{8}\right\}\right)\left(-\frac{1}{\tau + 1}\right)\right\}, \end{aligned} \quad (13.15)$$

so (13.6) is proved. From (13.10), Theorem 5, and (13.6), it follows that $\psi_s(\tau)$ is a modular form of degree $-s/2$, according to Definition 11.6. In fact, by Theorem 5 and (13.15), it follows that $\psi_s(\tau)$ stays bounded when τ approaches a cusp (from inside the fundamental region). Hence, by Definition 11.7, we have completed the proof of the following theorem.

Theorem 7. *The function $\psi_s(\tau)$ is an entire modular form of degree $-s/2$, with multiplier system $v_s(M)$, on the group Γ_θ.*

§5. The Function $\theta^s(\tau)$

From Theorem 11.6 it is easy to obtain (13.5), and from (11.6) one infers that $f(\tau) = \theta^s(\tau)$ satisfies (13.4). These results, together with (13.2), which is just (11.5), complete the proof of the following result.

Theorem 8. *The function $\theta^s(\tau)$ is an entire modular form of degree $-s/2$ on the group Γ_θ.*

§6. Proof of Theorem 4

Once we have justified (modulo the omitted computations) Theorem 7 and Theorem 8, the consequences (A) and (B) of §2 and the conclusions drawn from them now immediately follow, and in order to complete the proof of Theorem 4, it only remains to show that $\rho_s'(n) = \rho_s(n)$. By (13.11), this requires us to show that

$$\sum_{i=1}^{\infty} e^{-\pi i s/4} B_c(n) = \mathscr{S}(s, n), \tag{13.16}$$

where, by Theorem 12.1,

$$\mathscr{S}(s, n) = \sum_{c=1}^{\infty} A_c \quad \text{and} \quad A_c = c^{-s} \sum_{\substack{h \ (\mathrm{mod}\ c) \\ (h,c)=1}} \left(\sum_{\substack{m \ (\mathrm{mod}\ c) \\ (m,c)=1}} e^{2\pi i h m^2/c} \right)^s e^{-2\pi i h n/c}.$$

In fact, more is true and (13.16) holds termwise, because the following identity holds:

$$A_c(n) = e^{-\pi i s/4} B_c(n). \tag{13.17}$$

In order to prove (13.17), one has to use the explicit values of $\bar{v}_s(M_{c,h})$ from Theorem 11.7. One has to consider separately the different residue classes of c modulo 4. As an example, we work out the case $c \equiv 1 \pmod 4$.

We recall (see Theorem 11.3) that $G(h, c) = \sum_{m \ (\mathrm{mod}\ c)} e^{2\pi i h m^2/c} = (\frac{h}{c}) G(1, c) = (\frac{h}{c}) \cdot \frac{1}{2}(1 + i)(1 + i^{-c}) c^{1/2}$. In the present case, $i^{-c} = i^{-1} = -i$, so that $G(1, c) = \frac{1}{2}(1 + i)(1 - i) c^{1/2} = c^{1/2}$. Consequently,

$$\begin{aligned} A_c(n) &= c^{-s} \sum_{\substack{(h,c)=1 \\ h \ (\mathrm{mod}\ c)}} G(h, c)^s e^{-2\pi i h n/c} \\ &= c^{-s} \sum_{\substack{(h,c)=1 \\ h \ (\mathrm{mod}\ c)}} \left(\frac{h}{c} \right)^s c^{s/2} e^{-2\pi i n h/c} = c^{-s/2} \sum_{h \ (\mathrm{mod}\ c)} \left(\frac{h}{c} \right)^s e^{-2\pi i h n/c}. \end{aligned}$$

The condition $(h, c) = 1$ need not be restated, because $\left(\frac{h}{c}\right) = 0$ in the case $(h, c) > 1$. Next, by Theorem 11.7, for $c \equiv 1 \pmod 4$, $\bar{v}_s(M_{c,h}) = \left(\frac{h}{c}\right)^s e^{\pi i c s/4}$, so that

$$B_c(n) = c^{-s/2} \sum_{\substack{0 \leqslant h < 2c \\ h \text{ even}}} \left(\frac{h}{c}\right)^s e^{\pi i c s/4} e^{\pi i n h/c} = c^{-s/2} e^{\pi i c s/4} \sum_{0 \leqslant 2y < 2c} \left(\frac{2y}{c}\right)^s e^{2\pi i n y/c}.$$

Hence,

$$e^{-\pi i s/4} B_c(n) = c^{-s/2} e^{\pi i s(c-1)/4} \left(\frac{2}{c}\right)^2 \sum_{0 \leqslant y < c} \left(\frac{y}{c}\right)^s e^{2\pi i n y/c}.$$

We observe that from $c \equiv 1 \pmod 4$ it follows that $\sum_{0 \leqslant y < c} \left(\frac{y}{c}\right)^s e^{2\pi i n y/c} = \sum_{0 \leqslant y < c} \left(\frac{-y}{c}\right)^s e^{-2\pi i n y/c} = \sum_{0 \leqslant y < c} \left(\frac{y}{c}\right)^s e^{-2\pi i n y/c}$. Consequently, in order to complete the proof of (13.17), it suffices to verify that $e^{\pi i s(c-1)/4} \left(\frac{2}{c}\right)^s = 1$. However,

$$\begin{aligned} e^{s\pi i(c-1)/4} \left(\tfrac{2}{c}\right)^s &= (-1)^{s(c-1)/4} (-1)^{(c^2-1)s/8} = (-1)^{s((c-1)/4)(1+(c+1)/2)} \\ &= (-1)^{s(c-1)(c+3)/8}. \end{aligned}$$

Here $c - 1 = 4g$, with $g \in \mathbb{Z}$, $c + 3 = 4(g+1)$, so that $(-1)^{s(c-1)(c+3)/8} = (-1)^{2sg(g+1)} = 1$, as we wanted to show. The other residue classes of c modulo 4 are handled in similar fashion (see Problem 17), and this completes the proof of (13.17), and hence of (13.16). Theorem 4 is proved.

§7. Modular Functions and the Number of Representations by Quadratic Forms

In §§2–6 we have seen how the method of modular functions can be used to determine the number $r_s(n)$ of representations of the integer n as a sum of s squares. It is possible to generalize this approach and to obtain not only $r_s(n)$, but in fact even $r_Q(n)$, where $Q = Q(x_1, \ldots, x_s)$ is an arbitrary positive definite quadratic form in s variables. We recall (see Chapter 1) that $r_Q(n)$ stands for the number of representations of the integer n by the quadratic form Q, i.e., for the number of solution vectors $\mathbf{x} = (x_1, \ldots, x_s)$ of the equation

$$Q(\mathbf{x}) = n, \qquad \mathbf{x} \in \mathbb{Z}^s.$$

A complete, self-contained presentation of this method, which starts with the fundamental theory of modular functions and forms (but assumes an acquaintance with the work of Hecke and of Petersson) and goes all the way up to the numerical determination (in a finite number of steps) of the $r_Q(n)$, for specific, given Q, forms the contents of the book *Modulfunktionen und quadra-*

tische Formen by H. Petersson [210]. This book appears to be the only complete presentation of the subject matter and is bound to become a classic. Its appearance (end of 1982) only after the manuscript of the present book was complete prevents a fuller discussion of its contents. It is highly recommended to the interested reader.

§8. Problems

1. Verify that $k\lambda(h/k) = G(h, k)$ $((h, k) = 1)$ and $\Gamma(s/2)c(s) = \pi^{s/2}$.
2. Determine conditions under which the series $\sum_r \lambda^s(r)(ir - i\tau)^{-s/2}$, taken over all rationals r, converges in the upper half plane $\mathscr{H} = \{\tau \mid \operatorname{Im} \tau > 0\}$.
3. Let $f(\tau)$, holomorphic in $\mathscr{H}$, and $\alpha > 0$ be given. Show that, if $f(\tau)$ admits a comparison function of dimension $-\alpha$, then this comparison function is unique.
4. Let $\phi(\tau)$ be the c.f. $-\alpha$ of $f(\tau)$, and let a be a constant. Show that $a\phi(\tau)$, $\phi(\tau + 1)$, and $(-i\tau)^{-\alpha}\phi(-1/\tau)$ are the c.f. $-\alpha$ of $af(\tau)$, of $f(\tau + 1)$, and of $(-i\tau)^{-\alpha}f(-1/\tau)$, respectively.
5. Show that the set $\mathscr{A} = \{0\} \cup \{|z| < 1, |\log z| \geqslant 2\pi\}$ is compact.
6. Verify formally the identity $\sum_r f(r) = \sum_{0<|r|\leqslant 2} \sum_{q=-\infty}^{\infty} f(r + 2q)$.
7. Show that, for $q = 1$, 2, or 3, $G_q(z)$ is real on bd $\mathscr{A} = \{|\log z| = 2\pi, |z| < 1\}$.

(Hint: For $\tau = i$, $F_q(\tau) = F_q(-\tau^{-1}) = F_q(-\bar{\tau}) = \overline{F_q(\tau)}$.)

8. Identify $r_8(n)$, as given in Theorem 2, with the expression previously obtained in Chapter 9 (see (9.19)).
9. Show that $\lambda(0) = 1$, $\lambda(1) = 0$, and, for $r \neq 0$, $\lambda^2(-r^{-1}) = (ir)^{-1}\lambda^2(r)$; also, that for $r = h/k$, $(h, k) = 1$, $|\lambda(r)| \leqslant k^{-1/2}$.
10. Give formal proofs of Lemma 1 and of the Proposition.
11. Use the results of Problem 10 to evaluate $\lambda^4(r)$ for $r = h/k$, $(h, k) = 1$.
12. Justify (13.7) and (13.8).

*13. Compute the constants d_0 and d_1, defined in §2, and verify that $d_0 d_1 \neq 0$.

14. Justify the transformation of (13.10) used in the proof of Theorem 5.
15. Use Lipschitz's formula (Theorem 8.2) to show that

$$\sum_{m=-\infty}^{\infty} (c\tau + h + 2mc)^{-s/2} = \{c^{s/2}\Gamma(s/2)e^{\pi is/4}\}^{-1}\pi^{s/2} \sum_{n=1}^{\infty} n^{(s/2)-1}e^{\pi in(\tau+h/c)}.$$

16. Fill in the gaps in the proof of Theorem 6.

17. Complete the proof of Theorem 4, by justifying (13.17) for $c \not\equiv 1 \pmod{4}$.

*18. Use the methods of the present chapter to compute $r_6(n)$ and $r_7(n)$.

*19. Recast the methods of Chapters 12 and 13, in an attempt to streamline the procedures and to reduce as much as possible the case distinctions that now appear indispensable.

Chapter 14

Recent Work

§1. Introduction

In the preceding chapters we have discussed the representation of natural integers as sums of squares of integers, and only occasionally (e.g., in Chapters 4 and 5) did particular cases of representations by more general quadratic forms occur.

Many generalizations of these results are possible. In the first place, one may consider the representation problems that were solved in the past chapters for natural integers and attempt to answer them for integers in number fields. Next, instead of algebraic extensions of the rational field $\mathbb{Q}$, one may consider transcendental extensions. This leads to problems about the representations of polynomials in a fixed, but arbitrary number of indeterminates, by squares of similar polynomials, or, more generally, of rational functions in those indeterminates. On the other hand, instead of transcendental extensions of the rational field, one may consider transcendental extensions of fields of finite characteristic, or of p-adic fields, etc.

Finally, the problems considered in the preceding chapters, as well as their extensions here outlined, may be treated in the broader context of general (not necessarily diagonal) forms.

Sometimes the same problem may be considered from more than one single point of view. For example, the representation of an element of a ring **S**, contained in the field $\mathbb{K} = \mathbb{Q}(x_1, x_2, \ldots, x_n)$ (x_i transcendental over the rational field $\mathbb{Q}$ for $i = 1, 2, \ldots, n$), by sums of squares in **S**, may be considered as the generalization of the representation of an "integer," belonging to a ring more general than $\mathbb{Z}$, by squares of elements in the same ring. The same problem may also be considered as a problem of representation of functions of n variables as sums of squares of other such functions.

One can also combine generalizations and consider, e.g., representations of functions of n variables over some very general rings as quadratic forms of functions of those n variables over the same ring.

For these reasons, a perfect classification of the topics to be discussed in the present chapter can hardly be expected.

Like much of the best mathematics of the twentieth century, a considerable amount of the progress made on the mentioned and related topics can be fitted under the headings of Hilbert problems. There are at least two of those problems that have relevance to our subject matter. In the 11th problem, Hilbert (1862–1943) asks for the study of the theory of quadratic forms in an arbitrary number of variables, over an arbitrary algebraic number field. In the 17th problem, he asks for the representation of a positive definite form (i.e., one that is never negative) in any number of variables over the reals, as a sum of squares of *rational functions*. The asymmetry in this question is due to the fact that Hilbert himself had proved (see [109]) that the more natural question of the representation of a positive definite form as a sum of squares of forms has a *negative* answer, even in the case of only two variables. For example, although $f(x_1, x_2) = 1 + x_1^2 x_2^4 + x_1^4 x_2^2 - 3x_1^2 x_2^2$ is positive definite, it is not the sum of squares of polynomials in x_1 and x_2 (see [191]).

The symmetry of the problem can be restored, by asking for the representation of an element of $\mathbb{K}_1 = \mathbb{K}(x_1, x_2, \ldots, x_n)$ ($\mathbb{K}$ a field, perhaps subject to certain restrictions; $x_i (i = 1, 2, \ldots, n)$ transcendental over $\mathbb{K}$) by a sum of squares of elements of $\mathbb{K}_1$.

In the sections that follow, we shall first recall some needed definitions, and then consider the progress made under three headings: (a) extensions of results on the representation of *natural* integers as sums of squares to *algebraic* integers; (b) progress on Hilbert's 17th problem; and (c) progress on Hilbert's 11th problem.

We recall that, as already mentioned in Chapter 1, the present chapter has a character rather different from that of the previous ones. An attempt is made to define the new concepts and to indicate clearly the nature of the problems, the direction of the investigation, and the results obtained so far. However, only sketchy indications of proofs can be given here. For complete proofs the interested reader is directed to the original papers listed in the bibliography.

§2. Notation and Definitions

A field $\mathbb{K}$, algebraic and of finite degree over the rational field $\mathbb{Q}$, is said to be *totally real* if all its conjugates are real. The field $\mathbb{K}$ is said to be *real closed* if it is real, but such that *any* algebraic extension of $\mathbb{K}$ is no longer real. A field $\mathbb{K}$ is said to be *formally real* if -1 is not a sum of squares in $\mathbb{K}$. If, for a field $\mathbb{K}$, there exists an integer n such that $1 + 1 \cdots + 1$ (n summands, each equal to the multiplicative identity of $\mathbb{K}$) equals zero in $\mathbb{K}$, then there exists a smallest such integer. That integer can be shown to be a prime p, called the *characteristic* of $\mathbb{K}$, and we write char $\mathbb{K} = p$. If no such integer exists, we set char $\mathbb{K} = \infty$. The field $\mathbb{R}$ of reals is an ordered field and admits a single ordering; this applies also to a real, closed field.

If $\mathbb{K}$ is real and $\mu \in \mathbb{K}$, then $\mu = \sum_{v_i \in \mathbb{K}} v_i^2 \neq 0$ only if $\mu > 0$. If $\mathbb{K}^{(j)}$ is a real conjugate field of $\mathbb{K} = \mathbb{K}^{(1)}$, then, by the isomorphism $\mathbb{K} \simeq \mathbb{K}^{(j)}$, it follows that $\mu^{(j)} = \sum_{v_i^{(j)} \in \mathbb{K}^{(j)}} (v_i^{(j)})^2 > 0$. It follows that the problem of representing μ by a sum of squares of elements in $\mathbb{K}$ is meaningful only if not only μ, but also all its real conjugates, are positive. An element $\mu \in \mathbb{K}$ such that all its real conjugates are positive is said to be *totally positive*, in symbols $\mu \succ 0$. Alternatively, a totally positive element may be defined as one that is positive under all possible orderings of $\mathbb{K}$.

An element $\alpha \in \mathbb{K}$ is an *algebraic integer* if it is the root of a monic polynomial with integer coefficients, i.e., if $\alpha^n + \sum_{m=1}^{n} a_m \alpha^{n-m} = 0$, $a_m \in \mathbb{Z}$. The integers of $\mathbb{K}$ form a ring, which we shall denote by $\mathbf{O}_{\mathbb{K}}$. In conclusion we may state that, in an algebraic number field, the problem of representation of $\mu \in \mathbf{O}_{\mathbb{K}}$ by sums of squares in $\mathbf{O}_{\mathbb{K}}$, is meaningful only if $\mu \succ 0$. In a field that is not real, there exists no ordering, hence no positivity, and it is convenient to consider all elements of such fields as totally positive.

Let $\mathbb{L}$ be an algebraic extension of degree n of the field $\mathbb{K}$; then, if $\mu \in \mathbb{L}$, it has n algebraic conjugates $\mu = \mu^{(1)}, \mu^{(2)}, \ldots, \mu^{(n)}$, not necessarily all distinct. The sum $S(\mu) = \sum_{j=1}^{n} \mu^{(j)}$ is called the *trace* of μ over $\mathbb{K}$, and $N(\mu) = \prod_{j=1}^{n} \mu^{(j)}$ is called the *norm* of μ over $\mathbb{K}$. If $\mathbb{K} = \mathbb{Q}$, one speaks of the *absolute* trace or norm of μ; otherwise, of the *relative* trace or norm.

A *unit* of a number field $\mathbb{K}$, of degree $[\mathbb{K} : \mathbb{Q}] = n$ over $\mathbb{Q}$, is an integer $\mu \in \mathbf{O}_{\mathbb{K}}$, such that also $\mu^{-1} \in \mathbf{O}_{\mathbb{K}}$. If $\mathbb{K}$ has r_1 real conjugate fields and r_2 pairs of complex conjugate fields, then there exist $r = r_1 + r_2 - 1$ *independent units*, i.e., units that cannot be represented as products of rational powers of the other units, or of their inverses. If $\mathbb{K} \subset \mathbb{R}$ with $r = 1$, then the smallest such unit larger than one is called the *fundamental* unit.

If $\mu \in \mathbb{K}$ is such that, for some $m \in \mathbb{Z}$, $\mu^m = 1$, then μ is called a *root of unity* of $\mathbb{K}$; obviously, $\mu \in \mathbf{O}_{\mathbb{K}}$.

The integers of $\mathbb{K}$ form an Abelian group under addition, and if $[\mathbb{K} : \mathbb{Q}] = n$, this group has n generators, say $\omega_1, \omega_2, \ldots, \omega_n$. Such a set of generators $\{\omega_i\}$ is called a *base* of $\mathbb{K}$ over $\mathbb{Q}$. A base is said to be an *integral base* if every $\mu \in \mathbf{O}_{\mathbb{K}}$ may be represented by a sum $\mu = \sum_{j=1}^{n} a_i \omega_i$ with $a_i \in \mathbb{Z}$. If $\omega_i^{(j)}$ $(j = 1, \ldots, n)$ are the n conjugates of ω_i, then the determinant $\| \omega_i^{(j)} \|$ has the property that its square Δ is in $\mathbb{Q}$ and is independent of the particular choice of the integral base $\{\omega_i\}$; it is called the *discriminant* of $\mathbb{K}$ over $\mathbb{Q}$.

An (integral) *ideal* $\mathfrak{a}$ of $\mathbf{O}_{\mathbb{K}}$ is a set of integers of $\mathbf{O}_{\mathbb{K}}$ that form an Abelian group under addition and such that if $\alpha \in \mathfrak{a}$ and $\mu \in \mathbf{O}_{\mathbb{K}}$, then $\mu\alpha \in \mathfrak{a}$. If $\alpha \in \mathbf{O}_{\mathbb{K}}$, $\beta \in \mathbf{O}_{\mathbb{K}}$, and $\alpha - \beta \in \mathfrak{a}$, we say that α and β are *congruent* to each other modulo $\mathfrak{a}$, in symbols $\alpha \equiv \beta \pmod{\mathfrak{a}}$. This congruence is an equivalence relation, and the number of equivalence classes is called the *norm* of $\mathfrak{a}$, written $N\mathfrak{a}$. The multiples of an integer $\alpha \in \mathbf{O}_{\mathbb{K}}$ form an ideal, denoted by (α). Ideals of the form (α) are called *principal ideals*, and one has the relation $N((\alpha)) = |N\alpha|$. Given two ideals $\mathfrak{a}$ and $\mathfrak{b}$, we say that $\mathfrak{a}$ *divides* $\mathfrak{b}$ if $\mathfrak{a} \supset \mathfrak{b}$; we write this as $\mathfrak{a} \mid \mathfrak{b}$.

In analogy with the definition of Riemann's zeta function $\zeta(s) = \sum_{n=1}^{\infty} n^{-s}$ $(\operatorname{Re} s > 1)$, one defines, for any algebraic number field $\mathbb{K}$, the *Dedekind zeta function* $\zeta_{\mathbb{K}}(s) = \sum (N\mathfrak{a})^{-s}$ $(\operatorname{Re} s > 1)$, where $\mathfrak{a}$ runs through all nonzero integral ideals of $\mathbf{O}_{\mathbb{K}}$.

One can define a multiplication of ideals and prove that

$$N(\mathfrak{a}\mathfrak{b}) = N\mathfrak{a} \cdot N\mathfrak{b}. \tag{14.1}$$

In general, $\mathbf{O}_{\mathbb{K}}$ does not have the property of uniqueness of factorization into indecomposable integers of $\mathbf{O}_{\mathbb{K}}$. In particular, it is *not* the case that for all algebraic number fields $\mathbb{K}$ there exists a set $\{\pi\}$ of integers, each of them not further decomposable into products of other integers of smaller norms and with the property that from $\pi \mid \alpha \cdot \beta$ and $\pi \nmid \alpha$ it follows that $\pi \mid \beta$. The corresponding property holds, however, for ideals. An ideal $\mathfrak{p}$ such that from $\mathfrak{p} \mid \mathfrak{ab}$ and $\mathfrak{p} \nmid \mathfrak{a}$ it follows that $\mathfrak{p} \mid \mathfrak{b}$ is called a *prime ideal*. The decomposition of ideals into prime ideals is (up to order) unique. The norm of a prime ideal is a power of a prime; its exponent is called the *degree of the prime ideal*.

For completeness we recall that the concept of an ideal can be enlarged to contain also *fractional ideals*, for which (14.1) remains valid. The fractional ideals form a group under multiplication.

In analogy to the completion of the rational field $\mathbb{Q}$ under the metric induced by the absolute value $|r/s|$ of a rational number, one can complete $\mathbb{Q}$ also under *p-adic valuations* for each prime p. The p-adic value of the rational $p^a r/s$ ($a, r, s \in \mathbb{Z}$, $(r, s) = 1$, $p \nmid rs$) is set equal to p^{-a}. The set $\sum_{j=m}^{\infty} a_j p^j$ ($a_j \in \mathbb{Z}$, $0 \leqslant a_j < p$, $m > -\infty$) is a field under the obviously defined operations of addition and multiplication, the field $\mathbb{Q}_p$ of *p-adic numbers*. The field $\mathbb{Q}_p$ contains the integral domain $\mathbb{Z}_p = \{\sum_{j \geqslant 0} a_j p^j,\ a_j \in \mathbb{Z},\ 0 \leqslant a_j < p\}$ of *p-adic integers*. A p-adic integer whose multiplicative inverse is also a p-adic integer is called a p-adic *unit*. A necessary and sufficient condition for the p-adic integer $\sum_{j \geqslant 0} a_j p^j$ to be a p-adic unit is that $a_0 \neq 0$.

Obviously, $\mathbb{Q}_p \supset \mathbb{Q}$ and $\mathbb{Q}_p \supset \mathbb{Z}_p$. One verifies that the p-adic valuation denoted $|\alpha|_p$ is a *norm*, in the sense that it satisfies the *norm axioms*: For α, $\beta \in \mathbb{Q}_p$, we have $|\alpha|_p \geqslant 0$, with equality only for $\alpha = 0$; $|\alpha \cdot \beta|_p = |\alpha|_p \cdot |\beta|_p$; and $|\alpha + \beta|_p \leqslant |\alpha|_p + |\beta|_p$, the "triangle inequality."

As a matter of fact, $|\alpha + \beta|_p \leqslant \max\{|\alpha|_p, |\beta|_p\}$. On account of this stronger inequality, the p-adic norm is said to be *non-Archimedean*, while a norm that satisfies only the triangle inequality is called *Archimedean*. The p-adic norm defines a distance function $d(\alpha, \beta) = |\alpha - \beta|_p$, and one verifies that under this distance function $\mathbb{Q}_p$ is a *complete* field (i.e., every bounded, infinite sequence of elements of $\mathbb{Q}_p$ has at least one limit point in $\mathbb{Q}_p$). It is usually convenient to treat the real field $\mathbb{R}$ as a completion of $\mathbb{Q}$ under the "infinite prime," and no distinction is made between the symbols $\mathbb{R}$ and $\mathbb{Q}_\infty$.

The *Hilbert symbol* $\left(\frac{\alpha, \beta}{p}\right)$ is defined in $\mathbb{Q}_p$ to be equal to $+1$ if $\alpha x^2 + \beta y^2 = 1$ has a solution in $\mathbb{Q}_p$, and -1 otherwise. From the definition follows that, for $p = \infty$, $\left(\frac{\alpha, \beta}{\infty}\right) = 1$ unless $\alpha < 0$ and $\beta < 0$, when $\left(\frac{\alpha, \beta}{\infty}\right) = -1$. In case $a, b \in \mathbb{Q}$, the product formula

$$\prod_p \left(\frac{a, b}{p}\right) = 1$$

holds, when p runs through all primes (including $p = \infty$).

Let A be the matrix of the quadratic form $Q = Q(x)$. Recall that A is symmetric (see Chapter 4). Here we use the same notation as in Chapter 4, i.e., $\mathbf{x}$ stands for the column vector of entries $(x_1, \ldots, x_n)$ and $\mathbf{x}'$ stands for the transpose of $\mathbf{x}$, i.e., the row vector of the same entries. Also $(\mathbf{x}, \mathbf{y}) = \mathbf{x}' \cdot \mathbf{y}$ stands for the inner product $x_1 y_1 + \cdots + x_n y_n$. We have $Q(\mathbf{x}) = (\mathbf{x}, A\mathbf{x}) = \mathbf{x}' \cdot A\mathbf{x}$, which we shall write, less formally, as $\mathbf{x}'A\mathbf{x}$. We may think of Q as defined over $\mathbb{Q}_p$, $\mathbb{Q}$, or $\mathbb{Z}_p$, etc. If $Q = \sum_{i=1}^{n} \alpha_i x_i^2$ is in diagonal form ($\alpha_i \in \mathbb{Q}_p$), we define the *Hasse symbol* $c_p(Q)$ by

$$c_p(Q) = \left(\frac{-1, -1}{p}\right) \prod_{1 \leqslant i \leqslant j \leqslant n} \left(\frac{\alpha_i, \alpha_j}{p}\right). \tag{14.2}$$

If Q is a (nonsingular) quadratic form, not in diagonal form, it is equivalent (see §4 of Chapter 4; also Theorem 23 below) over $\mathbb{Q}_p$ to a diagonal form Q_1, and then we set $c_p(Q) = c_p(Q_1)$, with $c_p(Q_1)$ given by (14.2). Alternatively, $c_p(Q)$ can be computed directly, as follows. Let $d_0 = 1, d_1, \ldots, d_n = d(Q)$ be the successive principal minors of A; then

$$c_p(Q) = \left(\frac{-1, -d_n}{p}\right) \prod_{j=1}^{n-1} \left(\frac{d_j, -d_{j+1}}{p}\right).$$

For more details on the p-adic field and on Hilbert and Hasse symbols, see [107], [28], [125], or [135].

In the present chapter, positive definite forms will play an important role and will have to be mentioned frequently; also, no negative definite forms will occur. For this reason, we shall (in this chapter only) simplify the expression "positive definite form" to "definite form."

§3. The Representation of Totally Positive Algebraic Integers as Sums of Squares

In 1928 F. Götzky proved the following rather surprising theorem [84]:

Theorem 1. *The field $\mathbb{K} = \mathbb{Q}(\sqrt{5})$ is the only real quadratic field in which Lagrange's Theorem 3.2 holds, i.e., in which every totally positive integer is the sum of four integral squares in $\mathbb{K}$.*

Götzky's method is a generalization of the methods of Jacobi and Mordell and uses theta and modular functions. For the present problem, Götzky constructs a theta function of two complex variables; takes its fourth power, so that the successive coefficients (see below for $r_3(\mu)$) equal $r_4(\mu)$; and shows that this function transforms appropriately under modular transformations. He then uses this result to obtain a formula for $r_4(\mu)$, from which it follows that for $\mu \succ 0$ one has $r_4(\mu) > 0$.

Götzky's result was improved in 1940 by H. Maass, who proved, among others (see [174])

Theorem 2. *Contrary to what happens in $\mathbb{Q}$, three integral squares suffice to represent every totally positive integer in $\mathbf{O}_{\mathbb{K}}$ if $\mathbb{K} = \mathbb{Q}(\sqrt{5})$; and, in strong analogy with $r_3(n)$ for $n \in \mathbb{Z} \subset \mathbb{Q}$, the number $r_3(\mu)$ of representations of the totally positive integer $\mu \in \mathbf{O}_{\mathbb{K}}$ ($\mathbb{K} = \mathbb{Q}(\sqrt{5})$) as sum of three squares of $\mathbf{O}_{\mathbb{K}}$ is proportional to the class number h_0 of the biquadratic field $\mathbb{K}_1 = \mathbb{K}(\sqrt{-\mu}) = \mathbb{Q}(\sqrt{5}, \sqrt{-\mu})$.*

Maass's method is related to that of Götzky and to that used in Chapter 9. He defines the analogue of a theta function over $\mathbb{K} = \mathbb{Q}(\sqrt{5})$ by $\theta_0(\tau) = \sum e^{2\pi i S(\nu^2\tau/\sqrt{5})}$ ($S(\alpha)$ = trace of α). The sum that defines $\theta_0(\tau)$ runs over the integers ν of $\mathbf{O}_{\mathbb{K}}$, so that $\theta(\tau) = \theta_0^3(\tau) = 1 + \sum r_3(\mu) \exp\{\pi i S(\mu\tau/\sqrt{5})\}$. First, it is shown that in $\mathbb{K}$ there exists only one class of definite ternary quadratic forms of determinent $\varepsilon^{2\nu}$ ($\varepsilon = (1 + \sqrt{5})/2$, the fundamental unit of $\mathbb{K}$), so that any such form is equivalent to $\sum_{i=1}^3 x_i^2$. Next, use is made of Siegel's theory [243] of quadratic forms. Finally, analytic methods (among others, the evaluation of residues at poles of Dedekind zeta functions) are used to show that

$$r_3(\mu) = \frac{120}{w} U(0, \mu) \left| N\left(\frac{\mu}{\Delta}\right) \right|^{1/2} h_0.$$

Here w is the number of roots of unity in $\mathbb{K}$, Δ is the relative discriminant of $\mathbb{K}_1$ over $\mathbb{K}$, N stands for the norm, and h_0 is the class number of $\mathbb{K}_1$. As for $U(0, \mu)$, this is a function of μ, comparable to the singular series of Chapter 12; it is too complicated to be defined here, but it has the property that for $u \succ 0$, $U(0, \mu) > 0$.

Continuing this line of investigation, Siegel proved in 1945 (see [246]) the following, rather startling result.

Theorem 3. *The only totally real fields in which all totally positive integers can be represented by sums of squares of integers of those fields are $\mathbb{Q}$ and $\mathbb{Q}(\sqrt{5})$.*

We recall that, by Lagrange's Theorem 3.2, four squares suffice in $\mathbb{Q}$, and by Maass's Theorem 2, three squares suffice in $\mathbb{Q}(\sqrt{5})$. On the other hand, in *any* other totally real field, there exist totally positive integers that are not sums of *any* number of integral squares. Furthermore, Siegel proves

Theorem 4. *If a field $\mathbb{K}$ is not totally real, then all totally positive integers of $\mathbb{K}$ are sums of integral squares of $\mathbb{K}$ if and only if the discriminant of $\mathbb{K}$ is odd.*

Theorem 5. *If $\mu \in \mathbf{O}_{\mathbb{K}} \subset \mathbb{K}$, $[\mathbb{K} : \mathbb{Q}] < \infty$, $\mu \succ 0$ and μ is representable as a sum of integral squares, then five such squares suffice.*

Theorem 4 is obtained by relatively elementary considerations, but the proof of Theorem 5 uses a generalization of the circle method (see Chapter 12) due to Siegel [242].

In view of the theorems of Lagrange and Maass, it was natural to investigate if, perhaps, fewer than five squares may suffice. Siegel shows that there exist infinitely many (even quadratic) fields in which three squares do not suffice. This is the case, e.g., in any imaginary quadratic field with discriminant $d \equiv 1 \pmod 8$, such as $\mathbb{Q}(\sqrt{-7})$, and, more generally, if 2 is divisible in $\mathbb{K}$ by a prime ideal of first degree. Whether four squares are sufficient to represent every algebraic integer that is a sum of integral squares, or whether there are such integers that are sums of five but not of fewer squares, is left open in Siegel's paper. He expresses the opinion that four squares may suffice. If that were the case, then Lagrange's theorem could be formulated to read, strongly generalized, as follows: "All algebraic (including the rational) integers that are sums of integral squares of their ring of integers, are already sums of four such squares."

The matter was decided with assistance from computers. In 1959, H. Cohn, apparently unaware of Siegel's paper [246], answered this question [44] by actually exhibiting a large number of algebraic integers that are sums of five but not of four integral squares of their field. He also shows that $6 + \sqrt{6}$ is the smallest algebraic integer that is not sum of any number of integral squares of its field.

The main result of [44] may be formulated as

Theorem 6. *Let $\mu \succ 0$ be an algebraic integer and $\mathbb{K} = \mathbb{Q}(\mu)$. Then, if μ is a sum of integral squares in $\mathbb{K}$, it is a sum of five squares. There are integers $\mu \succ 0$ that are sums of five but not of four squares of $\mathbf{O}_{\mathbb{K}}$.*

Siegel's paper [246] contains many other interesting results, but, as these are related mainly to generalizations of Waring's problem, we quote here only a single, particularly important one:

Theorem 7. *Let $\mathbb{K}$ be a totally real algebraic number field and m a rational integer, $m > 2$; then, if all totally positive integers of $\mathbb{K}$ are sums of totally positive integral mth powers, it follows that $\mathbb{K} = \mathbb{Q}$.*

§4. Some Special Results

In parallel with the very general investigations of Maass and Siegel, particular cases were also studied.

In 1940, I. Niven showed [198] that in the imaginary fields $\mathbb{Q}(\sqrt{-m})$, every integer of the form $a + 2b\sqrt{-m}$ $(a, b \in \mathbb{Z})$ can be represented either by $\alpha^2 + \beta^2 + \gamma^2$ or by $\alpha^2 - \beta^2 - \gamma^2$ (α, β, γ integers in $\mathbb{Q}(\sqrt{-m})$).*

---

*The author's claim is stronger, but see the counterexamples in [66] and [246]; see also the review of [66] by E. G. Straus in *Math Rev.* 15,401d.

A simple proof of Niven's theorem in the case $m = 1$ of Gaussian integers has been found by W. J. Leahy [153], and another one is due to K. S. Williams [280].

N. Eljoseph has studied [65] in particular the Gaussian field $\mathbb{Q}(\sqrt{-1})$. The author shows that a necessary and sufficient condition for $\alpha \in \mathbb{Z}[i]$ to be a sum of squares in that ring is (as already shown by Niven in [198]) that $\alpha = a + 2bi$ ($a, b \in \mathbb{Z}$). In that case two squares also suffice, unless $a = 2a_1$, with $2 \nmid ba_1$, when three squares are needed. In either case, if $\alpha = a + 2bi$, there are infinitely many representations as a sum of three squares, while the number of representations of α as a sum of two squares is always finite.

In another paper [66], Eljoseph has studied the representations of rational, positive integers by sums of squares of quadratic integers $a + b\sqrt{D}$ ($a, b, D \in \mathbb{Z}$, D not a square). His main result may be formulated as follows: For such a representation two squares suffice if and only if $D = -1$ (i.e., only for Gaussian integers); three squares suffice if $D = 2, 3, 5, 6$, but not if $D \geqslant 8$, with the case $D = 7$ left open.

The *number* of representations of $\alpha = a + 2b\rho$, where $\rho^2 = -1$ or 2, or $\rho^2 + \rho + 1 = 0$, was obtained by G. Pall in 1951 (see [207]), by a method based on the study of the representation of a quadratic form as sum of squares of two linear forms (a particular case of Hilbert's 17th problem). The case $\rho^2 = -1$ is again that of Gaussian integers, already discussed by Niven and Eljoseph, but the other cases are new. For this particular case, K. S. Williams gave a simple proof. In [280], [281], and [282], he (i) indicates new expressions for $r_2(\mu)$, $\mu \in \mathbb{Z}[i]$, and (ii) shows the equivalence (which is not obvious) of his results with those of Pall [207].

In J. Hardy [99], one finds an extension of the work of Niven and of Pall to quadratic fields $\mathbb{Q}(\sqrt{m})$ for $m = 3, 7, 13, 37$ and $m = -p$ (p a rational prime). Some of these results are already found in the earlier work of T. Nagell, to be discussed in what follows. Other related, somewhat older work is due to Mordell [185–187], H. Braun [29], and C. Ko [134]. H. Lenz shows [159] that if $\mathbb{Q} \subset \mathbb{G} \subset \mathbb{K}$, with $[\mathbb{K} : \mathbb{G}] = 2$, then, if $\gamma \in \mathbb{K}$, $\gamma \succ 0$ (γ not necessarily an integer), one has $\gamma = \alpha^2 + \sum_{i=1}^{m} a_i^2$, with $\alpha \in \mathbb{K}$, $a_i \in \mathbb{G}$ ($i = 1, 2, \ldots, m$). It follows from Siegel's results that one may take $m = 4$.

The case of the representation of algebraic integers as sums of *two* integral squares is taken up in detail by T. Nagell. He calls such integers that are sums of two squares *A-numbers* (see [192]). For $m = 1, 2, 3, 7, 11, 19, 43, 67$, and 163, all A-numbers of the quadratic fields $\mathbb{Q}(\sqrt{\pm m})$ are identified ($m = -1$ leads back to the often studied Gaussian integers); all fields here considered have class number $h = 1$, and there are no other imaginary quadratic fields with $h = 1$. Some more general results on A-numbers in general quadratic fields are also obtained. In later papers [193], [195]. Nagell extends some of these results to $m = \pm 5$, ± 13 and ± 37. In [194] also the question of finiteness versus infinity for the number of representations of A-numbers in arbitrary algebraic number fields is studied.

The relevance of the number of classes in the principal genus, i.e., in the genus of $\sum_{j=1}^{k} x_j^2$, appeared in Chapter 4 and then again in the present one (see the work of Maass). In particular, only when this number is one can simple approaches yield the number of representations $r_k(\mu)$ for $\mu \succ 0$. J. Dzewas investigated the real quadratic fields with one class in the principal genus. He showed in [61] that this is never the case if $k \geqslant 9$ (not even for $\mathbb{K} = \mathbb{Q}$, as Eisenstein seems to have known already). For $k = 8$, this is the case only if $\mathbb{K} = \mathbb{Q}$. If $2 \leqslant k \leqslant 7$, then for each k there exist only finitely many totally real fields $\mathbb{K}$ with a single class in the principal genus. None of their discriminants exceeds 62,122,500. Once more, A-numbers (see the discussion of the work of T. Nagell) receive special attention. For the fields $\mathbb{K}$ with a single class in the principal genus, the values of $r_k(\mu)$ for $\mu \in \mathbf{O}_{\mathbb{K}}$, $\mu \succ 0$, are determined.

For $\mathbb{Q}(\sqrt{3})$, R. Salamon [233] shows that the genus of $\sum_{i=1}^{4} x_j^2$ has three classes, explicitly given, and the genus of $\sum_{i=1}^{3} x_i^2$ has two classes.

Some of the empirical results of [44] are proven by Cohn in two papers, one [45] on representations by sums of four integral squares in $\mathbb{Q}(\sqrt{2})$ and $\mathbb{Q}(\sqrt{3})$, and the other [46] on representations by sums of three integral squares in $\mathbb{Q}(\sqrt{m})$, $m = 2, 3$, or 5. In both papers, Cohn uses essentially Götzky's method. Let us denote the number of divisors of μ (up to units) of even and odd norm by G and H, respectively. Then the main results may be formulated as follows:

In $\mathbb{Q}(\sqrt{2})$, for $\mu = a + 2b\sqrt{2} \succ 0$,

$$r_4(\mu) = \begin{cases} 8\,G & \text{if } N\mu \equiv 1 \pmod 2 \\ 32\,G & \text{if } N\mu \equiv 4 \pmod 8 \\ 48\,G + 6\,H & \text{if } N\mu \equiv 0 \pmod 8. \end{cases}$$

In $\mathbb{Q}(\sqrt{3})$ a similar result holds for $\mu = \mathrm{a} + \mathrm{b}\sqrt{3} \succ 0$, but the formulae are no longer exact; instead, they contain a "small" auxiliary term. Consequently, $r_3(\mu) > 0$ follows only for sufficiently large $N\mu$. If we denote as before by h_0 the class number of the biquadratic field $\mathbb{K}_1 = \mathbb{Q}(\sqrt{m}, \sqrt{-\mu})$, $m = 2$, 3, or 5, $\mu \in \mathbf{O}_{\mathbb{K}}$, $\mathbb{K} = \mathbb{Q}(\sqrt{m})$, $\mu \succ 0$, then $r_3(v) = h_0 g + E(v)$, where $E(v) = o(r_3(v))$ and g depends on the prime ideal factorization of the ideal (2) in $\mathbb{K}_1$; also, in general, $v = \mu$. However, if $v \neq a + b\sqrt{d}$ (d = discriminant of $\mathbb{K}$),* then $v = 2\mu$ if $d = 8$, and $v = (1 + \sqrt{3})^2\mu$ if $d = 12$. The numerical evidence of [44] suggests that, at least for $(\mu, 6) = 1$, $E(\mu) = 0$.

§5. The Circle Problem in Algebraic Number Fields

Generalizations of many different kinds of the results of previous chapters are possible. In §7 of Chapter 2 we considered the *circle problem*, i.e., the determination of the number $A(t) = \sum_{n \leqslant t} r_2(n)$ of lattice points inside the circle of radius

---

*This may happen for $d = 8$ and $d = 12$.

$\sqrt{t}$. While it was easy to show that $\lim_{t\to\infty} A(t)/\pi t = 1$, the determination of $E(t) = A(t) - \pi t$ is a far from trivial problem, one in fact that is still not solved satisfactorily. In the case of a real quadratic field $\mathbb{K} = \mathbb{Q}(\sqrt{m})$, the integers are $(a + b\sqrt{m})/2$, with $a \equiv b \pmod 2$ (a, b odd is possible only if $m \equiv 1 \pmod 4$)). If $\mu = a + b\sqrt{m}$ and is totally positive, then its conjugate $\mu' = a - b\sqrt{m}$ is also positive, so that $\mu\mu' = (a + b\sqrt{m})(a - b\sqrt{m}) = a^2 - b^2 m > 0$. We then find, in direct generalization of the results in $\mathbb{Q}$, that if G stands for the region defined by $0 \leqslant \mu \leqslant t$, $0 \leqslant \mu' \leqslant t_1$, and if, as before, we let $r_2(\mu)$ be the number of representations of the integer $\mu \in \mathbf{O}_{\mathbb{K}}$, $\mu \succ 0$, as a sum of two squares of integers of $\mathbf{O}_{\mathbb{K}}$, then $\lim_{tt_1\to\infty} \{\sum_{\mu\in G} r_2(\mu)/(\mu\mu')\} = \pi^2/d$, where d is the discriminant of $\mathbb{K}$ ($d = m$ if $m \equiv 1 \pmod 4$), $d = 4m$ if $m \equiv 2, 3 \pmod 4$)). Once more the determination of $\sum_{\mu\in G} r_2(\mu) - \pi^2(\mu\mu')/d = E(t, t_1)$ say, is difficult. In 1962, W. Schaal obtained in [236] the following results: (i) For every $\delta > 0$ and $tt_1 \to \infty$, $E(tt_1) = O((tt_1)^{(2/3)+\delta})$; (ii) $\lim_{tt_1\to\infty} \sup|E(t, t_1)|(tt_1)^{-1/4} > 0$, i.e., $E(t, t_1) = \Omega((tt_1)^{1/4})$.

The estimates for both bounds use methods of C. L. Siegel (see [241] and [244], respectively). The result has been extended by Schaal [237] to general totally real algebraic number fields $\mathbb{K}$, of discriminant d, even under the restriction $\mu \in \mathfrak{a} \subset \mathbb{K}$ ($\mathfrak{a}$ an arbitrary ideal of $\mathbb{K}$). Specifically, let $t_1, t_2, \ldots, t_n$ be positive, $0 < \mu^{(j)} \leqslant t_j$, and $\mathfrak{a} \mid \mu$; then

$$\sum_{\substack{0<\mu^{(j)}\leqslant t_j \\ \mathfrak{a}\mid\mu}} r_2(\mu) = \frac{\pi^n}{dN\mathfrak{a}^2}(t_1 t_2 \cdots t_n) + E(t_1, \ldots, t_n)$$

holds with $E(t_1, t_2, \ldots, t_n) = O((t_1 \cdots t_n)^{n/(n+1)+\delta})$ for every $\delta > 0$ and $t_1 t_2 \cdots t_n \to \infty$. For $n = 2$ this reduces to the previous O-statement.

§6. Hilbert's 17th Problem

In Chapter 4 (see especially §13) we sketched Gauss's method of proof in his determination of representations of an integer n as a sum of three squares. The essential idea was the representation of a form in two variables by a form in three variables, both quadratic. One may ask, more generally, for the possibility of representing a form, or a more general function in m variables, by a form in n ($\geqslant m$) variables. In particular, one may ask for the representation of a function in m variables as a sum of n squares. This is the essence of Hilbert's 17th problem. In his famous address of 1900 to the International Congress of Mathematicians in Paris, he stated it as follows*:

> The square of any form is evidently always a definite form. But since, as I have shown [109], not every definite form can be obtained by the addition of squares of forms, the question arises (I have answered it affirmatively for ternary forms),

* This is a slightly paraphrased translation; for the original German text, see [111] or [112]. The first complete English translation appears in [113]; see also [3].

whether any definite form may not be expressed as a quotient of sums of squares of forms.

As already observed in §1, there is a certain lack of symmetry in the formulation, due to the quoted negative result of Hilbert. For that reason, the problem has often been reformulated to ask for the representation of a definite rational function over a field as a sum of squares of rational functions over the same field. In view of Hilbert's result, the best one can hope for, even for definite *polynomials* in n variables, is to represent them by sums of squares of rational functions in those variables. The existence of such a representation for $n = 2$ was proved by Hilbert himself [110]. His method shows in fact that, in analogy with Lagrange's Theorem 3.2, four squares suffice (at least for $n = 2$), a result actually due to Landau [147]. Another improvement is also due to Landau, who, in the same paper, considered the case of a rational function in a single variable over the rationals and showed that, if it is definite, then it is the sum of at most eight squares of rational functions of one variable with rational coefficients. For a recent improvement see Theorem 21 or [219].

§7. The Work of Artin

The decisive progress on Hilbert's 17th problem is due to Artin, who considered the problem in a fairly general setting. Artin and Schreier had given [11] an abstract characterization of the field $\mathbb{R}$ of reals. Based on this, Artin showed (see [10]) that in a field $\mathbb{K}$ with char $\mathbb{K} \neq 2$ and that is *not* formally real, i.e., in which -1 is the sum of, say, n squares, every element α is sum of $n + 1$ squares. Indeed, if $-1 = \sum_{j=1}^{n} \xi_j^2$, $\xi_j \in \mathbb{K}$, and $\alpha \in \mathbb{K}$, then, if char $\mathbb{K} \neq 2$,

$$\alpha = \left(\frac{1+\alpha}{2}\right)^2 - \left(\frac{1-\alpha}{2}\right)^2 = \left(\frac{1+\alpha}{2}\right)^2 + \left(\sum_{j=1}^{n} \xi_j^2\right)\left(\frac{1-\alpha}{2}\right)^2$$

$$= \left(\frac{1+\alpha}{2}\right)^2 + \sum_{j=1}^{n} \left(\xi_j \frac{1-\alpha}{2}\right)^2.$$

If, on the other hand, char $\mathbb{K} = 2$, then the elements that are sums of squares are the squares themselves, because $(\sum_{j=1}^{n} \xi_j)^2 = \sum_{j=1}^{n} \xi_j^2$.

For future use, we reformulate this statement (compare with Theorem 24) as

Theorem 8. *If the quadratic form* $Q(\mathbf{x}) = \sum_{j=1}^{n+1} x_j^2$ *represents zero nontrivially in* $\mathbb{K}$ (char $\mathbb{K} \neq 2$), *then Q represents, in fact, every element* $\alpha \in \mathbb{K}$.

As already observed, only totally positive elements are candidates to be sums of squares of elements of $\mathbb{K}$, and Artin proves that also the converse holds.*

---

* We recall that in a field that is not formally real, with char $\mathbb{K} \neq 2$, every element is totally real.

Theorem 9. *The totally positive elements of a field $\mathbb{K}$, and only those, are sums of squares of elements of $\mathbb{K}$.*

For a more general version of this statement, see [10, §1].

In the particular case of algebraic number fields, Theorem 9 had already been obtained by Hilbert and by Landau [151].

To avoid any possible misunderstanding, one should perhaps observe at this point that the Hilbert–Landau theorem (i.e., Theorem 9 as applied to algebraic number fields), while stating that every totally positive element of $\mathbb{K}$ is a sum of squares in $\mathbb{K}$ and conversely, in no way conflicts with Siegel's Theorems 3 and 4, because the latter refer to representations by sums of *integral* squares and not of squares of arbitrary field elements.

At first sight, it may appear that Hilbert's 17th problem has now been completely solved by Theorem 9. Indeed, given an arbitrary field $\mathbb{K}$, one may consider a transcendental extension $\mathbb{L} = \mathbb{K}(x_1, \ldots, x_n)$. Then, if $f(x_1, \ldots, x_n) \in \mathbb{L}$ and is definite, it is tempting to assume that f is a totally positive element of $\mathbb{L}$, so that Theorem 9 applies. This statement, while true under certain restrictions on the field $\mathbb{K}$, turns out to be anything but easy to prove. The difficulty is due to a large extent to the fact that several different orderings of $\mathbb{L}$ may coincide on $\mathbb{K}$ with a given ordering of $\mathbb{K}$. By using the previously mentioned joint work with Schreier, Artin succeeds in proving that the definite rational functions of n variables over a field $\mathbb{K}$ are, in fact, totally positive—and hence, by Theorem 9, sums of squares in $\mathbb{L}$—under the additional condition that $\mathbb{K}$ is either a real closed field or a field with a single Archimedean ordering. We state this result formally as

Theorem 10. *If $\mathbb{K}$ is a real closed field, or has a single Archimedean ordering, then the definite rational functions in any number of variables $x_1, x_2, \ldots, x_n$ are sums of squares of rational functions in $x_1, x_2, \ldots, x_n$ over $\mathbb{K}$.*

As a corollary we obtain (nontrivially; see Satz 4 of [10]):

Theorem 11. *If $\mathbb{K} = \mathbb{Q}$, the field of rationals, then every definite, rational function in n variables with coefficients in $\mathbb{Q}$ is a sum of squares of rational functions with coefficients in $\mathbb{Q}$.*

Artin's proof has as its principal ingredient the idea of *specialization.* A property E of a finite set of functions $f_j(t, \mathbf{x})$ $(j = 1, 2, \ldots, k;\ \mathbf{x}' = (x_1, \ldots, x_n))$ admits specialization if the following holds: There exists a set of rational functions $\phi_i(\mathbf{x})$, such that, for all $\mathbf{a}' = (a_1, \ldots, a_n) \in \mathbb{K}^n$ for which all $\phi_i(\mathbf{a})$ are defined and have (as elements of $\mathbb{K}$) the same sign as the $\phi_i(\mathbf{x})$ (as elements of $\mathbb{L} = \mathbb{K}(\mathbf{x})$), property E holds also for the set $f_j(t, \mathbf{a})$.

Let $\mathbb{P} \supset \mathbb{K}$, where $\mathbb{P}$ is a real closed field, with an ordering induced by that of $\mathbb{K}$; then it is shown by induction on n that the following two properties admit specialization:

(i) $f(t, \mathbf{x})$ has, as a function of t, exactly r real zeros in $\mathbb{P}$; and
(ii) each of the finitely many functions of the sequence $f_j(t, \mathbf{x})$ has a real zero $\alpha_j \in \mathbb{P}$, with $\alpha_i < \alpha_j$ for $i < j$.

The proof of these two properties makes essential use of Sturm's theorem on the separation of real roots, and this theorem, in turn, is valid only in Archimedean or in real closed fields.

By using (i) and (ii) it is possible to prove

Theorem 12. *Let $\mathbb{K} \subset \mathbb{R}$ and $\mathbb{L} = \mathbb{K}(\mathbf{x})$ ($\mathbf{x} = (x_1, \ldots, x_n)$), with $\mathbb{L}$ equipped with an arbitrary, but fixed ordering induced by that of $\mathbb{K}$. If $\phi_i(\mathbf{x})$ is an arbitrary, finite set of rational functions, i.e., $\phi_i(\mathbf{x}) \in \mathbb{L}$ for $i = 1, 2, \ldots, m$, say, then there exist n rational numbers $a_1, \ldots, a_n$ ($=\mathbf{a}$, for short) such that all $\phi_i(\mathbf{x})$ are defined for $\mathbf{x} = \mathbf{a}$ and the $\phi_i(\mathbf{x})$ (as elements of $\mathbb{L}$) have the same sign as $\phi_i(\mathbf{a})$ (as elements of $\mathbb{K}$) for $i = 1, 2, \ldots, m$.*

This theorem permits, under the stated restrictions on $\mathbb{K}$, to link the sign of $f(\mathbf{x}) \in \mathbb{L}$ to the sign of $f(\mathbf{a}) \in \mathbb{K}$ and to complete the proof that if $f(\mathbf{x})$ is a definite rational function over $\mathbb{K}$, then it is a totally positive element of $\mathbb{L}$, i.e., it remains positive under all orderings of $\mathbb{L}$ compatible with the fixed ordering of $\mathbb{K}$. On account of Theorem 9, this completes the proof of Theorem 10, and hence that of Theorem 11.

The validity of these results can be extended further. Let $\mathbb{K}$ be an arbitrary field of algebraic numbers, and let $f(\mathbf{x}) \in \mathbb{K}(x_1, \ldots, x_n)$. Let us say that f is *totally definite* if it takes only totally positive values at all $\mathbf{x} \in \mathbb{K}^n$ where it is defined. We obtain from Theorem 10 that the following theorem holds:

Theorem 13. *If $\mathbb{K}$ is an algebraic number field, every totally definite rational function with coefficients in $\mathbb{K}$ is a sum of squares of rational functions in $\mathbb{K}(\mathbf{x})$.*

At first view it may seem that these results can be vastly enlarged by the remark that in [11] (see also [9]), Artin and Schreier have given an abstract axiomatic characterization of the real field, and it is clear that any field that satisfies the axioms of the real field needed in the preceding proofs can be substituted for $\mathbb{R}$ in the statements of the results. Unfortunately, this generalization is illusory. Indeed, a theorem (Satz 8a) of [11] states that, up to isomorphisms, there exists one and only one field that is real closed and whose elements are algebraic over $\mathbb{Q}$, namely the field of all real algebraic numbers.

Of more interest are the applications of Theorems 8 to 13 to the particular case of integral elements, i.e., to polynomials. In generalization of a result of Landau [147], one obtains, by induction on the degree of the polynomials, that the following theorem holds.

Theorem 14. *Let $\mathbb{K} \subset \mathbb{R}$, and set $\mathbb{L} = \mathbb{K}(x)$, the field of rational functions of a single variable over $\mathbb{K}$. If the polynomial $F(x) \in \mathbb{K}[x]$ is a sum of squares in $\mathbb{K}(x)$, then it is also a sum of squares of polynomials in x.*

Theorem 14 is essentially the case $n = 1$ of the following, more general theorem.

Theorem 15. *In the representation of definite polynomials as sums of squares of rational functions in n variables* (*guaranteed to exist by Theorem* 10), *one can select those rational functions in such a way that all are actually polynomials in one arbitrarily chosen variable.*

For further results, in particular on the representation of algebraic definite functions, we must refer the reader to Artin's original paper.

§8. From Artin to Pfister

In spite of its brilliance, Artin's work appeared incomplete on at least two counts. Firstly, it did not apply to all fields, and secondly, it did not give any bounds for the number of squares needed in any given representation. Also, its use of Sturm's theorem made some mathematicians unhappy. This was in part because its analytic character appeared somewhat incongruous for handling the problem under consideration. More important, however, was the fact that precisely because of the use of Sturm's theorem, the nature of the fields $\mathbb{K}$ to which the conclusions applied appeared severely restricted.

Several mathematicians attempted to extend the validity of the proofs to more general fields, by avoiding Sturm's theorem—and the essentially equivalent ones of Bézout.

By strengthening the assumption of definiteness to *strict definiteness* (i.e., $f(\mathbf{x}) \geqslant \varepsilon$, $\varepsilon > 0$ fixed, instead of $f(\mathbf{x}) \geqslant 0$), Habicht gave [91] a surprisingly short (6 pages) proof of representability as sum of squares, but still only over real closed fields.

Several proofs of Artin's main result were given by use of concepts of mathematical logic (see G. Kreisel [142] and A. Robinson [232]). However, all attempts to extend Artin's theorems to unrestricted fields failed. The reason became clear only rather recently (1967), when Dubois [59] constructed a field $\mathbb{K}$ with a single (but, naturally, in view of Theorem 10, non-Archimedean) ordering, in which an explicitly given, definite rational function—in fact, a polynomial—is not sum of squares of rational functions over $\mathbb{K}$. It follows that, while it may still be possible to reduce the restrictions on $\mathbb{K}$ in Artin's Theorem 10, these cannot be totally eliminated.

Another proof of some of Artin's results that avoids Sturm's theorem is due to Gross and Hafner [85]; it uses weaker analytic tools (especially the mean value theorem), but it still refers only to real closed fields.

A rather recent proof by Knebusch [130] uses no deep analytic tools and relies instead on the theory of Witt rings.

A beautiful set of results, valid in general fields and destined to play an important role in later research, is due to J. W. S. Cassels and is proved on barely 4 pages in [35]. We formulate the findings here as

Theorem 16.

(i) *Any polynomial in a single variable x that is the sum of squares of rational functions over a field $\mathbb{K}$ is the sum of the same number of squares of polynomials in x.*

(ii) *If $d \in \mathbb{K}$ and* char $\mathbb{K} \neq 2$, *then $x^2 + d$ is a sum of $n > 1$ squares in $\mathbb{K}(x)$ if and only if either -1 or d can be written as a sum of $n - 1$ squares in $\mathbb{K}$.*

(iii) *In the particular case $\mathbb{K} = \mathbb{R}$, let $x_1, \ldots, x_n$ be indeterminates over $\mathbb{R}$; then the form $\sum_{j=1}^{n} x_j^2$ is not the sum of $n - 1$ squares in $\mathbb{R}(x_1, \ldots, x_n)$.*

We observe that (i) is a considerable strengthening of Theorem 14 and of the corresponding result of Landau [147]. From (iii) it follows, in particular, that in $\mathbb{K}(x_1, \ldots, x_n)$, at least $n + 1$ squares of rational functions are needed for the representation of a definite rational function $f(x_1, x_2, \ldots, x_n) \in \mathbb{K}(x_1, \ldots, x_n)$, because $1 + x_1^2 + \cdots x_n^2$ cannot be represented as sum of n squares. Even today this is the best known *lower* bound for the number of squares needed in the representation of a definite rational function of n variables over an arbitrary field $\mathbb{K}$.

We call attention to the condition $-1 = \sum_{j=1}^{n-1} f_j^2(x)$, $f_j(x) \in \mathbb{K}(x)$ (see (1) in [10] with $\xi_j = f_j(x) \in \mathbb{K}(x)$), which cannot hold if $\mathbb{K} \subset \mathbb{R}$; it will play an important role in later work by Pfister.

Cassels's work had been preceded by a paper of Davenport [50]. In answering a particular case of Hilbert's 17th problem (formulated by N. Fine), Davenport showed that for no triple of rational functions F_j $(j = 1, 2, 3)$ of x_1, x_2, x_3, x_4 with real coefficients can an equation $\sum_{i=1}^{4} x_i^2 = \sum_{j=1}^{3} F_j^2$ hold identically.

Except for Landau's older results on four and on eight squares, Cassels's results seem to be the first ones that mention the *number* of squares in a given representation, thus addressing themselves to the second lacuna in Artin's theorems.

The decisive breakthrough on this problem is due to A. Pfister and will be discussed in the next section. It is based in part on the mentioned work of Cassels and also on the following theorem of C. Tsen [268], rediscovered independently by S. Lang [152].

Theorem 17. *Let $\mathbb{K}$ be an algebraically closed field and $\mathbb{L}$ a field of transcendency degree n over $\mathbb{K}$. Let f be a form of degree d in more than d^n variables over $\mathbb{L}$. Then f has a nontrivial zero in $\mathbb{L}$.*

For $d = 2$ we obtain the following

Corollary. *Every quadratic form in more than 2^n variables, with coefficients in a field $\mathbb{L}$ of transcendency degree n over an algebraically closed field $\mathbb{K}$, has a nontrivial zero in $\mathbb{L}$.*

It is by no means obvious that the existence of such a nontrivial zero is closely connected to the representability by sums of squares. This fact may have been recognized only later (see O'Meara [200]).

§9. The Work of Pfister and Related Work

In what precedes, the question of the representation of -1 in a given field $\mathbb{K}$ occurred several times. In Cassels's work the minimal number $s = s(\mathbb{K})$ of squares needed to represent -1 in a given field occurs explicitly. This number, which has been called *Stufe* in German, will be called here the *level* of the field. The first question to be considered is the set of possible values of s.

In an ordered field and, in particular in subfields of the real field $\mathbb{R}$, no such representation of -1 as the sum of a finite number of squares is possible, and s is not finite.

In the complex field $\mathbb{C}$, $-1 = (\sqrt{-1})^2$ and a single square suffices, $s(\mathbb{C}) = 1$. The same is true in a finite field of p elements if p is a prime and $p \equiv 1 \pmod 4$, because then -1 is a quadratic residue modulo p. However, if $p \equiv 3 \pmod 4$, then *two* squares are needed and $s = 2$. So, e.g., $1^2 + 1^2 \equiv -1 \pmod 3$. For a proof in the general case, see, e.g., [98, Theorem 87]. The necessary and sufficient conditions on an algebraic number field $\mathbb{K}$ in order to have $s(\mathbb{K}) = 2$ were determined by Hasse and by B. Fein, B. Gordon, and J. H. Smith [78]; the case of cyclotomic fields $\mathbb{Q}(e^{2\pi i/p})$ had been settled earlier by P. Chowla [40] (see also [41]).

On the other hand, no field can have level 3. Indeed, if we assume that in some field $\mathbb{K}$, $-1 = \sum_{j=1}^{3} x_j^2$, while no smaller number of squares will suffice, then (i) the equation $\sum_{j=0}^{4} x_j^2 = 0$ has a nontrivial solution in $\mathbb{K}$, and (ii) the solution can be selected so that $x_2^2 + x_3^2 \neq 0$ (otherwise, with $x_0 = 1$, $-1 = x_1^2$, contrary to the minimality of three squares). Then, however, by use of (2.2),

$$0 = (x_2^2 + x_3^2) \sum_{j=0}^{4} x_j^2 = (x_2^2 + x_3^2)(x_0^2 + x_1^2) + (x_2^2 + x_3^2)^2$$

$$= (x_0 x_2 - x_1 x_3)^2 + (x_0 x_3 + x_1 x_2)^2 + (x_2^2 + x_3^2)^2,$$

or, recalling that $x_2^2 + x_3^2 \neq 0$, we have $-1 = x^2 + y^2$ with $x = (x_0 x_2 - x_1 x_3)/(x_2^2 + x_3^2)$, $y = (x_0 x_3 + x_1 x_2)/(x_2^2 + x_3^2)$, which again contradicts the minimality of three squares.

Similarly, by using (3.5) instead of (2.2), one shows that 5, 6, or 7 cannot occur as levels, while 4 or 8 cannot be excluded as levels by this method. This suggests that perhaps only powers of 2, including 1, 2, and possibly 4, 8, etc., can be values of levels. This question had been asked as a problem in the *Jahresbericht der DMV* [273] in 1932 by van der Waerden. A partial answer was given soon (1934) by H. Kneser [131], who proved that the only possible levels are 1, 2, 4, 8, and $16m$ ($m \in \mathbb{Z}$), and that the levels 1, 2, and 4 actually occur.

It was A. Pfister who answered the question completely, some 30 years later [213]. His proofs are surprisingly elementary, particularly in view of their generality. The main result of [213] can be formulated as follows.

Theorem 18.

(a) *Only powers of 2 can be levels; and*
(b) *every power of 2 actually occurs as level of some field.*

Part (b) is an immediate consequence of

Theorem 19. *Let $\mathbb{K}$ be a formally real field, let $x_1, x_2, \ldots, x_n$ be indeterminates, and denote by y a solution of $y^2 + \sum_{j=1}^n x_j^2 = 0$. If $2^m \leqslant n < 2^{m+1}$, then the field $\mathbb{K}_n = \mathbb{K}(x_1, \ldots, x_n)$, y is of level 2^m.*

In addition to Cassels's quoted results, the proofs use only three simple lemmas. For $\mathbb{K}$ an arbitrary field, define $G_n = G_n(\mathbb{K}) = \{\alpha \in \mathbb{K} \mid \alpha = \sum_{i=1}^n \beta_j^2 \neq 0,\ \beta_j \in \mathbb{K}\}$, and denote by $G_n G_m$ the set $\{w = uv \mid u \in G_n,\ v \in G_m\}$. Also, denote by $m \circ n$ the smallest integer k such that $G_n G_m \subset G_k$ holds for *all* fields. With these notations, the three lemmas can be formulated as follows:

Lemma 1. *If G_n is a group, then $G_n G_{n+1} = G_{2n}$.*

Lemma 2. *Let $n = 2^m$; then G_n is a group. More specifically, if u_i, v_i $(i = 1, 2, \ldots, n)$ are any $2n$ elements of $\mathbb{K}$, then there exist elements w_i $(i = 1, 2, \ldots, n)$, such that $\sum_{i=1}^n u_i^2 \sum_{i=1}^n v_i^2 = \sum_{i=1}^n w_i^2$; moreover, one may always take $w_1 = \sum_{i=1}^n u_i v_i$.*

Lemma 3. *If $m \leqslant n$, then $m \circ n$ can be computed by induction on m; $k = m \circ n \leqslant n + m - 1$ and $G_m G_n \subset G_k$.*

To give some of the flavor of this work, we sketch here some of the proofs.

Proof of Lemma 1. If $\sum_{i=1}^n u_i^2 = 0$, then $\sum_{i=1}^{2n} u_i^2 = \sum_{i=n+1}^{2n} u_i^2 \in G_n \subset G_n G_{n+1}$; otherwise, $\sum_{i=1}^{2n} u_i^2 = (\sum_{i=1}^n u_i^2)\{1 + (\sum_{i=1}^n u_i^2)(\sum_{i=n+1}^{2n} u_i^2)/(\sum_{i=1}^n u_i^2)^2\} \in G_n G_{n+1}$, by use of the assumed group property. This proves $G_{2n} \subset G_n G_{n+1}$. Also,

$$\sum_{i=1}^{n} u_i^2 \sum_{j=1}^{n+1} v_j^2 = \sum_{i=1}^{n} u_i^2 \sum_{j=1}^{n} v_j^2 + \left(\sum_{i=1}^{n} u_i^2\right) v_{n+1}^2$$

$$= \sum_{i=1}^{n} w_i^2 + \sum_{i=1}^{n} (u_i v_{n+1})^2 \in G_{2n},$$

by use of the group property; hence, $G_n G_{n+1} \subset G_{2n}$. $\square$

Proof of Lemma 2. For $m = 0$, $u_1^2 v_1^2 = (u_1 v_1)^2$. This covers, in particular, the case of char $\mathbb{K} = 2$, when every sum of squares is a square. From here on we may—and shall—assume that char $\mathbb{K} \neq 2$. For $m = 1$, $(u_1^2 + u_2^2)(v_1^2 + v_2^2) = (u_1 v_1 + u_2 v_2)^2 + (u_1 v_2 - u_2 v_1)^2$ by (2.2), so that Lemma 2 holds. For $m > 1$, one uses induction on m. With $n = 2^m$, one considers separately the cases $-1 = \sum_{j=2}^{2n} e_j^2$, a sum of $2n - 1$ squares, and -1 *not* a sum of $2n - 1$ squares. In the first case, $0 = (1 + e_2^2)(1 + \sum_{j=2}^{2n} e_j^2) = (1 + e_2^2)^2 + \sum_{j=2}^{n} (1 + e_2^2)(e_{2j-1}^2 + e_{2j}^2)$. By (2.2), each product is the sum of two squares; hence, $0 = \sum_{j=1}^{2n-1} v_j^2$. As we now assume char $\mathbb{K} \neq 2$, by Theorem 8 the form $\sum_{i=1}^{2n-1} x_i^2$, which represents zero nontrivially, represents every $\alpha \in \mathbb{K}$. In particular,

$$\alpha = \sum_{i=1}^{2n} u_i^2 \sum_{i=1}^{2n} v_i^2 - \left(\sum_{i=1}^{2n} u_i v_i \right)^2 = \sum_{i=2}^{2n} w_i^2 \quad \text{and} \quad \sum_{i=1}^{2n} u_i^2 \sum_{i=1}^{2n} v_i^2 = \sum_{i=1}^{2n} w_i^2,$$

with $w_1 = \sum_{i=1}^{2n} u_i v_i$. As $n = 2^m$ by the induction hypothesis, this completes the proof for the first case. The proof for the second case is longer. The original, also rather lengthy proof that $w_1 = \sum_{i=1}^{n} u_i v_i$ is an admissible choice has been simplified by Cassels. We omit both these proofs. □

Proof of Lemma 3. For $m = 1$, $1 \circ n = n$, $G_1 G_n = G_n$. For $m > 1$, define r by $2^{r-1} < m \leqslant 2^r$; set $n = k \cdot 2^r + j$ with $k \geqslant 0$, $1 \leqslant j \leqslant 2^r$; assume the lemma for all (p, q) with $p \leqslant q$, $p < m$; and, assuming the value of $p \circ q$ already known in this case, proceed by induction on m. □

Remarks.

1. For $\mathbb{K} = \mathbb{Q}(x_1, \ldots, x_m; y_1, \ldots, y_n)$, Lemma 3 asserts that $(\sum_{i=1}^{m} x_i^2) \times (\sum_{j=1}^{n} y_j^2) = \sum_{r=1}^{k} z_r^2$ with $k = m \circ n$, which is a sweeping generalization of (2.2) and (3.5).
2. Obviously, $k = m \circ n$, which holds for *all* fields, is an upper bound for the integers t for which $G_m G_n \subset G_t$ may hold in any specific field. If char $\mathbb{K} \neq 2$ and $s(\mathbb{K})$ is finite, then obviously $G_j = G_{j+1}$ for all $j \geqslant s + 1$, because, by Artin's Theorem 8, for every $\alpha \in \mathbb{K}$, $\alpha = \sum_{i=1}^{s+1} \alpha_i^2$. If char $\mathbb{K} = 2$, then for every j, $G_j = G_1$.

Proof of Theorem 18(a). Assume that the level s satisfies $2^n < s < 2^{n+1}$. Then $-1 = \sum_{j=1}^{s} x_j^2$, $0 = 1 + \sum_{j=1}^{s} x_j^2$, and the equation $\sum_{j=1}^{s+1} x_j^2 = 0$ has a nontrivial solution. As $s < 2^{n+1}$, we have $s + 1 \leqslant 2^{n+1}$ and, for some u_j, with $u_1 \neq 0$, $\sum_{j=1}^{2^{n+1}} u_j^2 = 0$. Then $0 = \sum_{j=1}^{2^n} u_j^2 \sum_{j=1}^{2^{n+1}} u_j^2 = (\sum_{j=1}^{2^n} u_j^2)^2 + \sum_{j=1}^{2^n} u_j^2 \sum_{j=2^n+1}^{2^{n+1}} u_j^2$. By Lemma 2, the second product is in G_{2^n} (because G_{2^n} is a group; the reasoning would fail in general for G_m, $m \neq 2^n$); hence, $\sum_{i=1}^{2^n+1} w_i^2 = 0$, where $w_1 = \sum_{j=1}^{2^n} u_j^2 \neq 0$. It follows that $-w_1^2 = \sum_{i=2}^{2^n+1} w_i^2$, or with $z_i = w_{i+1}/w_1$, that $-1 = \sum_{i=1}^{2^n} z_i^2$ with $2^n < s$, which contradicts the definition of s. □

Proof of Theorem 19. From the definition of y, we have $-1 = \sum_{i=1}^{n} (x_i/y)^2$; hence, $s = s(\mathbb{K}_n) \leqslant n < 2^{m+1}$, and by Theorem 18(a), $s \leqslant 2^m$. If one assumes strict inequality, $-1 = \sum_{i=1}^{s} u_i^2$ with $s < 2^m$, where $u_i = p_i + q_i y$, p_i, q_i polynomials in $x_1, x_2, \ldots, x_n$, because $y^2 = -\sum_{i=1}^{n} x_i^2$ and similarly for higher powers of y. If we substitute these values of the u_i, we obtain, using also the assumed formal reality of $\mathbb{K}$, that

$$\sum_{i=1}^{2^m} p_i^2 = -\sum_{i=1}^{2^m} q_i^2 y^2 = \sum_{i=1}^{2^m} q_i^2 \sum_{j=1}^{n} x_j^2, \qquad \sum_{i=1}^{2^m} p_i q_i = 0, \qquad \sum_{i=1}^{2^m} q_i^2 \neq 0.$$

By Lemma 2, $(\sum_{i=1}^{2^m} p_i^2)(\sum_{j=1}^{2^m} q_j^2)$ is a sum of 2^m squares. However, the first may be taken as $(\sum_{i=1}^{2^m} p_i q_i)^2$ and thus vanishes. Hence, $(\sum_{i=1}^{2^m} p_i^2)(\sum_{j=1}^{2^m} q_j^2) = (\sum_{j=1}^{2^m} q_j^2)^2 \sum_{i=1}^{n} x_i^2 = \sum_{r=1}^{2^m-1} w_r^2$, and, dividing by $(\sum_{j=1}^{2^m} q_j^2)^2$, we have $\sum_{i=1}^{n} x_i^2 = \sum_{r=1}^{2^m-1} z_r^2$ with $z_r = w_r/(\sum_{j=1}^{2^m} q_j^2)$. However, $n \geqslant 2^m$, so that we have a contradiction to Cassels's result (iii). The proof is complete. □

Numerical values for $s(\mathbb{K})$ have been computed for a variety of local and global fields by C. Moser [190], and an expository account of Pfister's and related work is due to Ph. Revoy [230].

The main result of Pfister, related to Hilbert's 17th problem, is contained in the following theorem, published in 1967 (see [215]):

Theorem 20. *Let $\mathbb{K}$ be a real closed field, and let $\mathbf{x}$ stand for the vector $x_1, x_2, \ldots, x_n$; also, let $f(\mathbf{x})$ be a definite rational function, with coefficients in $\mathbb{K}$. Then $f(\mathbf{x})$ is a sum of at most 2^n squares in $\mathbb{K}(\mathbf{x})$.*

The particular case $n = 3$ of Theorem 20 had been proved and the correct bound 2^n for $n > 3$ had been conjectured by J. Ax in an unpublished manuscript.

The gap between this upper bound of 2^n and the mentioned lower bound $n + 1$ that follows from Cassels's Theorem 16(iii) is large, but this may well be in the nature of the problem. At least for $n = 2$, the upper bound is attained, as follows from Motzkin's polynomial. Indeed, Cassels, Ellison, and Pfister have shown in [36] that $f(x_1, x_2) = 1 + x_1^2 x_2^4 + x_1^4 x_2^2 - 3x_1^2 x_2^2$ (which, as we know from Section 1, is not the sum of any number of squares of polynomials over $\mathbb{Z}$) is not the sum of less than four squares of rational functions. The proof uses fairly deep considerations concerning the Tate-Šafarevici group of a certain elliptic curve over $\mathbb{C}(x)$ and cannot be presented here. From Theorem 20, on the other hand, it follows that $f(x_1, x_2)$ is a sum of four squares of rational functions in x_1, x_2 with real coefficients.

It is not possible to give here the complete proof of Theorem 20, mainly because it would require the study of still another paper of Pfister [214] that relies heavily on the theory of the Witt group of a field. A skeleton of the proof, however, may be sketched.

One first proves

Lemma 4. *Let $\mathbb{L}$ be a nonreal field. If in $\mathbb{L}(i)$ every quadratic form of more than 2^n indeterminates has a nontrivial zero, then -1 is a sum of 2^n squares in $\mathbb{L}$.*

By Theorem 17 of Tsen and Lang and its corollary, it follows that in the intended application, the hypothesis of Lemma 4 is verified, in the sense that -1 is, in fact, a sum of 2^{n-1} squares. Indeed, we are interested in a field $\mathbb{L}$ that is a nonreal algebraic extension of $\mathbb{K}(x_1, \ldots, x_{n-1})$, with $\mathbb{K}$ a real, closed field. It follows that in such a field $\mathbb{L}$, -1 is a sum of 2^{n-1} squares.

One also has

Lemma 5 (A. Pfister). *Let $\mathbb{K}_1$ be a formally real field, and let $f(x) \in \mathbb{K}_1(x)$; assume also that -1 is a sum of 2^{n-1} squares in every nonreal algebraic extension of $\mathbb{K}_1$. Then, if $f(x)$ can be represented as a sum of squares in $\mathbb{K}_1(x)$, it is possible to write it as a sum of at most 2^n squares in $\mathbb{K}_1(x)$.*

If we take $\mathbb{K}_1 = \mathbb{K}(x_1, \ldots, x_{n-1})$ and $x = x_n$, it follows from Lemma 5, with $\mathbb{L} = \mathbb{K}_1(x) = \mathbb{K}(x_1, \ldots, x_{n-1}, x_n)$, on account of Lemma 4, that $f(x)$ is a sum of 2^n squares in $\mathbb{K}(x_1, x_2, \ldots, x_{n-1}, x_n)$, provided that it is a sum of any finite number of squares in $\mathbb{K}_1(x)$. This is, however, indeed the case, on account of Artin's Theorem 10. This finishes the proof of Theorem 20.

§10. Some Comments and Additions

Pfister's work yields the strongest known results on Hilbert's 17th problem. It gives the very usable upper bound 2^n for the number of squares needed to represent definite rational functions of n variables over certain fields $\mathbb{K}$. Together with Cassels's lower bound of $n + 1$ squares, the correct number of needed squares is quite effectively bracketed. However, while Cassels's result is valid over any field $\mathbb{K}$, Pfister's, like Artin's, is restricted to real closed fields. To realize how severe this restriction is, let us observe that we don't even know an upper bound for the number of squares needed to represent a rational definite function of n variables over the rationals Q, by squares of such rational functions over $\mathbb{Q}$.

In the case of $n = 1$ variable, Landau's result was improved as follows by Pourchet [219] in 1971.

Theorem 21.

(a) *Let $\mathbb{K}$ be a field of algebraic numbers and $f(x) \in \mathbb{K}(x)$ be a definite rational function with coefficients in $\mathbb{K}$. Then $f(x)$ is a sum of five squares of functions in $\mathbb{K}(x)$, and this number is, in general, the best possible.*

(b) *If f is a polynomial over $\mathbb{Q}$, $f \succ 0$, then f is sum of five squares of polynomials in $\mathbb{Q}[x]$, and this is the best possible number.*

(c) *The polynomials $f \in \mathbb{Q}[x]$, $f \succ 0$, that are sums of four squares of polynomials are precisely characterized.*

Theorem 21 is a consequence of more general results of Pourchet, too complicated to be reproduced here.

It may well be a pure coincidence that the formula $2^n + 3$ gives the exact bound of the squares needed over $\mathbb{Q}$, as well in the case $n = 0$, i.e., that of integral constants (by Lagrange's Theorem 3.2), as also in the second known case $n = 1$, of rational integral functions (i.e., polynomials in one variable), as follows from Pourchet's Theorem 21. The conjecture that $2^n + 3$ is the correct bound for the number of squares of $\mathbb{Q}(x_1, \ldots, x_n)$ needed to represent every definite $f(x_1, \ldots, x_n) \in \mathbb{Q}(x_1, \ldots, x_n)$ is formulated in [115].

We finish this section with a recent result that answers, albeit in a very particular case, a very general and very challenging question: Given an integral domain $\mathbf{A}$ and its field of quotients $\mathbb{K}$, suppose that an element of $\mathbf{A}$ is a sum of n squares in $\mathbb{K}$; under what conditions is it also a sum of n squares of $\mathbf{A}$? or of any (finite) number of squares in $\mathbf{A}$?

An integer that is a sum of squares in $\mathbb{Q}$ is also a sum of the same number of squares in $\mathbb{Z}$. As we know from Hilbert and Landau, a polynomial in one variable is a sum of squares of polynomials whenever it is definite, i.e., whenever it is a sum of squares of rational functions. On the other hand, already for $n = 2$, we know that a polynomial in n variables is a sum of at most 2^n rational functions (at least over real closed fields), while it need not be a sum of any number of squares of polynomials (Pfister and Hilbert, respectively).

In [39] Choi et al. prove the following theorem, whose somewhat restrictive character appears to invite generalizations; these are unlikely to be easy.

Theorem 22. *Let $\mathbf{A}$ be a unique factorization domain, with $\mathbb{K}$ as its field of quotients. Assume also that $x^2 + 1 = 0$ has no solution in $\mathbb{K}$, but that, if one adjoins such a solution, say i, to $\mathbf{A}$, then the integral domain $\mathbf{A}[i] = \{a_1 + ia_2 \mid a_1, a_2 \in \mathbf{A}\}$ is also a domain with uniqueness of factorization. If $a \in \mathbf{A}$ is a sum of two squares in $\mathbb{K}$, then it is also a sum of two squares in $\mathbf{A}$.*

§11. Hilbert's 11th Problem

In his address of 1900 to the International Congress of Mathematicians in Paris, Hilbert outlines the problem at hand, in part, as follows: "... to attack successfully the theory of quadratic forms, with any number of variables and with any algebraic numerical coefficients. ..."

Hilbert's 11th problem encompasses a broad range of research and its investigation has led to many beautiful and deep results in a variety of directions (see, e.g., [200]). Here, however, we shall discuss only the developments that are connected with the problem of representations by sums of squares and its natural generalizations.

§12. The Classification Problem and Related Topics

This problem and the interest attached to it may be explained as follows. Let us suppose that a specific type of quadratic forms is considered, such as forms over the integers, forms with discriminant different from zero, forms that represent zero nontrivially, or others. Then we try to establish an equivalence relation among the forms considered; this relation defines a certain number of classes. Next, one determines a number of parameters that are identical for all forms of a given class and that are sufficient to characterize a given class completely, the so-called *class invariants*. Finally, one defines in each class a unique, distinguished, particularly simple form, the so-called *canonical form.*

If this program has been completed successfully, one may take advantage of it as follows: Let us suppose that we want to solve a problem that involves a specific quadratic form of the type just considered. We first compute its class invariants; these determine the class, and hence its canonical form. Next, we transform the problem so that the given form becomes the equivalent canonical one. In this simplified form, the problem is, in general, more manageable than the original one. If we succeed in solving it, we then return to the original setting and interpret the results obtained for the canonical form in terms of the original formulation. This is, basically, the same procedure that we use in elementary geometry, when a problem on, say, a general quadric is reduced by a change of axes to one involving the quadric in canonical (often called *standard*) form, with its geometrical axes parallel to the coordinate axes. In this position, the quadric has a simple equation, so that the given problem can be solved more easily, and then one refers to the original quadric, by returning to the original axes.

The classification problem of quadratic forms depends strongly on the setting in which it is considered. It is easiest to handle over a field, because that allows the greatest freedom in the selection of the matrices denoted by C and D in Chapter 4.

A considerable amount of this classification work had been done already before Hilbert, or was done by Hilbert himself and his contemporaries, especially by H. Minkowski and C. Jordan (1838–1922). For completeness, we start by recalling here a few classical results (see also Chapter 4).

Two quadratic forms $Q(\mathbf{x})$ and $Q_1(\mathbf{y})$ are said to be equivalent over $\mathbb{Z}$ (in symbols, $Q \simeq Q_1$) if there exist nonsingular matrices C and D with integral entries, inverse to each other, so that $CD = I$, the identity matrix, and such that if $\mathbf{x} = C\mathbf{y}$, then $Q(\mathbf{x}) = Q_1(\mathbf{y})$, and if $\mathbf{y} = D\mathbf{x}$, then $Q_1(\mathbf{y}) = Q(\mathbf{x})$. The condition $CD = I$ forces the determinants of C and of D to be equal and either $+1$ or -1. Alternatively, the equivalence of Q and Q_1 may also be defined by the condition that both matrix equations $D'AD = B$ and $C'BC = A$ (A the matrix of Q and B that of Q_1; X' = transpose of X) should have solutions in integral matrices D and C, respectively. This definition implies, in particular, that the two matrix congruences $D'AD \equiv B \pmod{m}$ and $C'BC \equiv A \pmod{m}$ have solutions in integral matrices D and C, respectively, for all integers m. The converse, however, does not hold. If two quadratic forms Q and Q_1 have matrices A and B, respectively, such that for every integer m the above congruences have solutions in integral matrices D and C, respectively, then Q and Q_1 are said to belong to the same *genus*. It may be shown that this definition of the genus is consistent with our previous use of this term. We recall that each genus consists of a certain number of classes, because all forms of the same class belong to a given genus.

One says that the form $Q(\mathbf{x})$ in n variables *represents* the form $Q_1(\mathbf{y})$ in m variables, with $m \leqslant n$, if $Q(\mathbf{x}) = Q(C\mathbf{y}) = Q_1(\mathbf{y})$, where C is an $n \times m$ matrix. This definition generalizes the representation of binary forms by ternary forms used in Chapter 4.

Over $\mathbb{Z}$ (as stated by Eisenstein and proven by Hermite), for every definite quadratic form Q in n variables with $1 \leqslant n < 8$, if $d(Q) = 1$, one can find a matrix C with integral entries such that if we set $\mathbf{x} = C\mathbf{y}$, then $Q(\mathbf{x}) = Q_1(\mathbf{y}) = \sum_{i=1}^{n} y_i^2$. The reason for that is that in these cases, there exists only one class of definite forms Q with $d(Q) = 1$. This statement is false for $n = 8$, in which case there are two classes with $d(Q) = 1$, and incidentally, also two genera, each with one class in each genus. We took advantage of the fortunate situation for $n < 8$ in Chapter 4, when we had $n = 2$ and $n = 3$.

Returning to the solvability of the matrix equations $D'AD = B$ in integral matrices D, a special difficulty arises when one is interested not only in the existence, but also in the number of distict solutions D. Indeed, there may exist matrices $U \neq I$ such that $U'AU = A$. Each such matrix, sometimes called an *automorph* of A, leads to solutions $D'(U'AU)D = D'U'AUD = (UD)'A(UD) = B$, and two solutions that differ only by an automorph of A, say $D_1 = UD_2$, are usually not counted as essentially distinct solutions.

Over the rational field $\mathbb{Q}$, Lemmas 2 and 4 of Chapter 4 on binary and ternary forms, respectively, generalize as follows in the case of n variables: Over the rational field $\mathbb{Q}$, in each class of definite forms in n variables and $d(Q) \neq 0$, there exists a form $Q(\mathbf{x}) = \sum_{i,j=1}^{n} a_{ij}x_ix_j$ such that $0 < |a_{11}| \leqslant (\frac{4}{3})^{(n-1)/2}\sqrt[n]{|d|}$, $|a_{11}| \geqslant 2|a_{1i}|$ for $i > 1$, and $a_{11}Q = (\sum_{i=1}^{n} a_{1i}x_i)^2 + Q_1$, where Q_1 depends only on $x_2, x_3, \ldots, x_n$, satisfies (*mutatis mutandis*) the same conditions as Q, and has a discriminant $d(Q_1) = a_{11}^{n-2}d(Q)$. We recall that $d(Q)$ is the determinant of A. If we set $n = 2$ or $n = 3$ in the above formulas, we recover Lemmas 4.2 and 4.4, respectively. As already mention in Chapter 4, this is Hermite's reduction of quadratic forms.

Over the real field, for each symmetric matrix A that corresponds to a quadratic form Q in n variables, we can find (see Theorem 23) a nonsingular real matrix D such that $B = D'AD$ has on its main diagonal r entries $+1$, s entries -1, and $n - (r + s) \geqslant 0$ entries zero, with all nondiagonal entries also zero. Consequently, the quadratic form $Q = \sum_{i,j=1}^{n} a_{ij}x_ix_j$ has been transformed by a substitution $\mathbf{x} = C\mathbf{y}$ into the quadratic form $Q_1(\mathbf{y}) = \sum_{i=1}^{r} y_i^2 - \sum_{i=r+1}^{r+s} y_i^2$. If $n > r + s$, so that there are effectively zeros on the main diagonal, then the determinant of $Q = d(Q) = d(Q_1) = 0$ and Q is said to be *singular*. It is obvious that a singular quadratic form such as Q_1 (hence, also Q) represents zero nontrivially, by taking, e.g., $y_i = 0$ for $1 \leqslant i \leqslant r + s$, $y_i = 1$ for $r + s + 1 \leqslant i \leqslant n$. If $r + s = n$ but $s > 0$, then, as we recall from Chapter 4, the form is called indefinite. The determinant $d(Q_1) = (-1)^s \neq 0$; nevertheless, the form represents zero nontrivially, by taking, e.g., $y_1 = y_{r+1} = 1$, $y_i = 0$ for $i \neq 1$, $r + 1$. Such a form, with $d(Q) = d(Q_1) \neq 0$ but representing zero nontrivially, is called a *zero form*. We have discussed a particular case of such forms in Chapter 5. If, on the other hand, the form Q, with coefficients in a field $\mathbb{K}$, represents all nonzero elements of $\mathbb{K}$, it is called a *universal form*. If $r = n, s = 0$, then $Q_1 = \sum_{i=1}^{n} y_i^2$, and then Q_1 and (hence) Q are definite forms over the reals.

The form Q_1 is considered the canonical form, and given n, then either r and s or, equivalently (and customarily), the *rank* $r + s$ and the *signature* $r - s$ are the class invariants. This classical statement (see, e.g., [108] for proofs) is known as *Sylvester's law of inertia*.

More generally, let $A = (\alpha_{ij})$ be the nonsingular, symmetric matrix of the quadratic form $Q(\mathbf{x})$ (as before, $\mathbf{x}' = (x_1, x_2, \ldots, x_n)$), with $\alpha_{ij} \in \mathbb{K}$, an arbitrary field with char $\mathbb{K} \neq 2$. Then the change of variables $\mathbf{x} = C\mathbf{y}$ leads to $Q(\mathbf{x}) = Q(C\mathbf{y})$, and the quadratic form $Q(\mathbf{x}) = (\mathbf{x}, A\mathbf{x}) = \mathbf{x}'A\mathbf{x}$ becomes $\mathbf{y}'C'\ A\ C\mathbf{y} = \mathbf{y}'B\mathbf{y} = (\mathbf{y}, B\mathbf{y}) = Q_1(\mathbf{y})$, say, with B the matrix of the form Q_1. By proper choice of C, one can obtain B in diagonal form, so that $Q_1(\mathbf{y})$ contains no mixed terms $y_i y_j$ with $i \neq j$. We state this result formally as

Theorem 23. *Over a field, every quadratic form is equivalent to a diagonal form.*

If, in particular, $\mathbb{K}$ is a finite field with $q = p^m$ elements, then every quadratic form Q in n variables over $\mathbb{K}$ with $d(Q) = d$ is equivalent to the form $dx_0^2 + \sum_{i=1}^{n-1} x_i^2$.

The most frequently used canonical form over a field is the *Jordan form*. This has, however, little relevance to representations by sums of squares, and unless the field $\mathbb{K}$ over which the form is defined is algebraically closed, it requires the consideration of algebraic extensions of $\mathbb{K}$. For that reason, we shall not discuss it further here, but direct the interested reader to [108].

For further results on class numbers of definite quadratic forms, see M. Kneser [132]. For methods applicable to quadratic forms with $d(Q) = 1$ in many variables, see H.-V. Niemeier [197].

Of perhaps unexpected importance are the zero forms. Their definition, previously given over $\mathbb{R}$, extends unchanged to any field $\mathbb{K}$. Their interest becomes obvious in view of the following theorem, which is a generalization of Theorem 8 to forms that are not necessarily diagonal forms.

Theorem 24. *Let $\mathbb{K}$ be any field with* char $\mathbb{K} \neq 2$ *(or a p-adic field for $p \neq 2$, with certain restrictions on Q and the solutions of $Q = 0$). Then the quadratic form $Q(x_1, \ldots, x_n)$ represents any element of $\mathbb{K}$, provided that Q represents zero nontrivially in $\mathbb{K}$.*

Remark. It is obvious that if $Q(x_1, \ldots, x_n)$ represents $\alpha \in \mathbb{K}$, then $Q_1 = Q(x_1, \ldots, x_n) - \alpha x_{n+1}^2$ is a zero form (just take $x_{n+1} = 1$), and the converse is almost equally obvious, provided that, in the representation of zero by Q_1, we have $x_{n+1} \neq 0$. Indeed, if $Q(x_1, \ldots, x_n) = \alpha x_{n+1}^2$, $x_{n+1} \neq 0$, then $Q(y_1, \ldots, y_n) = \alpha$, provided that $y_i = x_i/x_{n+1}$ $(i = 1, 2, \ldots, n)$. In order to complete the proof of the theorem, we have to show that the statement holds also if Q_1 represents zero nontrivially only with $x_{n+1} = 0$; the converse, of course, is clear, as then $Q(x_1, \ldots, x_n) = \alpha x_{n+1} = 0$ nontrivially.

Theorem 24 is nontrivial; some particular cases were proved by A. Meyer [177]. The following simple proof, valid under very general conditions, is due to C. L. Siegel [245].

Proof of Theorem 24. First, we use Theorem 23 to reduce Q to diagonal form, say $Q \simeq \sum_{i=1}^n a_i y_i^2$. Then let $y_i = r_i$ be a nontrivial solution of $\sum_{i=1}^n a_i r_i^2 = 0$, where, without loss of generality, we may assume $a_1 r_1 \neq 0$. Set $\tau = \alpha/4a_1 r_1^2$, and verify that $z_1 = (1 + \tau) r_1$, $z_i = (1 - \tau) r_i$ for $i = 2, 3, \ldots, n$ leads to

$$\begin{aligned}
\sum_{i=1}^n a_i z_i^2 &= a_1(1+\tau)^2 r_1^2 + \sum_{i=2}^n a_i (1-\tau)^2 r_i^2 \\
&= (1+\tau^2) \sum_{i=1}^n a_i r_i^2 + 2\tau\left(a_1 r_1^2 - \sum_{i=2}^n a_i r_i^2\right) = 2\tau\left(2a_1 r_1^2 - \sum_{i=1}^n a_i r_i^2\right) \\
&= 4a_1 r_1^2 \tau = \alpha. \quad \square
\end{aligned}$$

Also the following theorem holds.

Theorem 25. *If Q is a nonsingular quadratic form with coefficients in a field $\mathbb{K}$, then $Q(\mathbf{x})$ represents $\alpha \in \mathbb{K}$ $(\alpha \neq 0)$, if and only if $Q(\mathbf{x}) - \alpha x_{n+1}^2$ represents zero nontrivially.*

Proof. If $Q(\mathbf{x}) = \alpha$ for $\mathbf{x} = \mathbf{r}$, then $Q(\mathbf{r}) - \alpha x_{n+1}^2$ represents 0 for $x_{n+1} = 1$. Conversely, if $Q(\mathbf{x}) - \alpha x_{n+1}^2$ represents zero for $\mathbf{x} = \mathbf{r}$ and $x_{n+1} \neq 0$, then $Q(\mathbf{r}/x_{n+1}) = \alpha$; otherwise, if $Q(\mathbf{x}) - \alpha x_{n+1}^2$ represents zero for $\mathbf{x} = \mathbf{r}$ and $x_{n+1} = 0$, then $Q(\mathbf{r}) = 0$, and by Theorem 24, $Q(\mathbf{x})$ represents all nonzero elements $\alpha \in \mathbb{K}$. $\square$

§13. Quadratic Forms over $\mathbb{Q}_p$

Much of the progress on quadratic forms made since Hilbert is due to the introduction of p-adic numbers by Hensel (1861–1941). We recall from Chapter 5 that a whole sequence of congruences modulo prime powers can be replaced by a single equation in a p-adic field. In his original work, Minkowski presented (somewhat sketchily) conditions for equivalence of quadratic forms [179] as sets of congruences; these can be recast as equations in $\mathbb{Q}_p$. This work was performed by Hasse, a former student of Hensel's. In what follows, we mention some of the more important results of this theory. Most of them—those of the present section, as well as those of §14—are due to H. Hasse, and the original proofs may be found in his five papers [100–104]. Occasionally, simpler proofs, obtained later, will be quoted.

By Theorem 23, every nonsingular quadratic form over $\mathbb{Q}_p$ is equivalent to a diagonal one, say $\sum_{i=1}^{n} \alpha_i y_i^2$. If in the term $\alpha_i y_i^2$, with $\alpha_i = p^{2r_i+\delta_i}\varepsilon_i$ (ε_i a p-adic unit, $\delta_i = 0$, or 1) one sets $x_i = p_i^{r_i} y_i$, then the term becomes $\varepsilon_i x_i^2$ or $\varepsilon_i p x_i^2$. This proves

Theorem 26. *Over $\mathbb{Q}_p$, every nonsingular quadratic form is equivalent to*

$$Q(x) = \sum_{i=1}^{r} \varepsilon_i x_i^2 + p \sum_{i=r+1}^{n} \varepsilon_i x_i^2 = f_1(\mathbf{x}) + pf_2(\mathbf{y}), \qquad \textit{say},$$

with $\mathbf{x}' = (x_1, \ldots, x_r)$ and $\mathbf{y}' = (x_{r+1}, \ldots, x_n)$.

Furthermore, the following theorem holds.

Theorem 27. *A nonsingular quadratic form over $\mathbb{Q}_p$ ($p \neq 2$), with $0 < r < n$, represents zero nontrivially if and only if, in Theorem 26, either f_1 or f_2 or both represent zero nontrivially.*

The sufficiency is obvious (take $\mathbf{x} = 0$ if $f_2(\mathbf{y})$ represents zero nontrivially, and $\mathbf{y} = 0$ if f_1 represents zero nontrivially). For the long proof of the necessity, see [28, p. 50].

By using a well-known theorem of number theory (see, e.g., [86, Chapter 4, Theorem 25]) we obtain

Theorem 28. *With the previous notation, $\sum_{i=1}^{n} \varepsilon_i x_i^2 = 0$ has a nontrivial solution in $\mathbb{Q}_p$ if and only if $\sum_{i=1}^{n} \varepsilon_i x_i^2 \equiv 0 \pmod{p}$ has a nontrivial solution in p-adic integers.*

On the basis of a general theorem of Chevalley (see [28]) that insures the existence of a nontrivial solution to the congruence $f(x_1, \ldots, x_n) \equiv 0 \pmod{p}$ for any form f of degree $m < n$ with integer coefficients, it follows that if Q is a quadratic form over $\mathbb{Z}$ in at least three variables, then the congruence $Q(x_1, x_2, x_3, \ldots) \equiv 0 \pmod{p}$ has a nontrivial solution. From this and Theorem 28 follows

Theorem 29. *If $p \neq 2$, $n \geqslant 3$, and ε_i are p-adic units, then the form $\sum_{i=1}^{n} \varepsilon_i x_1^2$ represents zero in $\mathbb{Q}_p$.*

If we combine Theorem 29 with Theorem 26, a simple consideration leads to

Theorem 30. *For $p \neq 2$, the form $Q(\mathbf{x})$ of Theorem 26 represents zero nontrivially if and only if $Q(\mathbf{x}) \equiv 0 \pmod{p^2}$ has a solution $x_1, \ldots, x_n$ in $\mathbb{Q}_p$ with at least one x_j $(j = 1, 2, \ldots, n)$ a p-adic unit.*

For completeness we add that, for $p = 2$, the form $Q(\mathbf{x})$ of Theorem 26 represents zero if and only if $Q(\mathbf{x}) \equiv 0 \pmod{16}$ has a solution with at least one x_j odd.

From the preceding results we finally obtain

Theorem 31. *A quadratic form $Q(x)$ over $\mathbb{Q}_p$ in $n \geqslant 5$ variables represents zero in $\mathbb{Q}_p$.*

Proof. In Theorem 26 we may take, without loss of generality, $r \geqslant n/2$. For $n \geqslant 5$ we have $r \geqslant 3$, and for $p \neq 2$ the result follows from Theorems 29 and 27. We suppress here the special considerations needed to prove the theorem for $p = 2$. □

From Theorems 31 and 25 follows

Theorem 32. *A nonsingular quadratic form in $n \geqslant 4$ variables over $\mathbb{Q}_p$ represents all p-adic numbers.*

From Theorem 32 immediately follows, as a corollary,

Theorem 33. *Let $Q(\mathbf{x})$ be a nonsingular quadratic form over $\mathbb{Z}$ in n variables. If $n \geqslant 5$ and $m \in \mathbb{Z}$, then $Q(\mathbf{x}) \equiv 0 \pmod{m}$ has a nontrivial solution.*

We quote without proofs (these may be found in [125] or [28]) the corresponding statements in the case of fewer variables. We recall that the Hasse symbol $c_p(Q)$ has been defined by (14.2).

Theorem 34. *Let $Q(\mathbf{x})$ be a quadratic form over $\mathbb{Q}_p$ (p finite) in n variables and with $d(Q) = d$. Then $Q(\mathbf{x})$ is a zero form in $\mathbb{Q}_p$ under the following conditions: For $n = 2$, if and only if $-d$ is a square in $\mathbb{Q}_p$; for $n = 3$, if and only if $c_p(Q) = 1$; for $n = 4$, if and only if $c_p(Q) = 1$ whenever d is a square.*

For $n \geqslant 5$, we recall of course that, by Theorem 31, $Q(\mathbf{x})$ is always a zero form.

As a consequence of Theorems 34 and 25, we obtain

Theorem 35. *Let $Q(\mathbf{x})$ be a quadratic form over $\mathbb{Q}_p$, $p < \infty$, in n variables, and let $d(Q) = d$. Then Q is universal over $\mathbb{Q}_p$ (i.e., $Q(\mathbf{x}) = \alpha$ is solvable in $\mathbb{Q}_p$ for $\alpha \in \mathbb{Q}_p$, $\alpha \neq 0$) under the following conditions: for $n = 1$, if and only if $d\alpha$ is a square in $\mathbb{Q}_p$; for $n = 2$, if and only if*

$$c_p(Q) = \left(\frac{-d, -\alpha}{p}\right);$$

for $n = 3$, if and only if

$$c_p(Q) = \left(\frac{-d, -\alpha}{p}\right)$$

in case $-d\alpha$ *is a square in* $\mathbb{Q}_p$ (otherwise unconditionally).

By Theorem 32, Q is always universal if $n \geqslant 4$.

Corollary 1. *For p finite or infinite, Q is universal if and only if it is a zero form, except for $n = 4$, p finite, $c_p(Q) = -1$, and $d(Q)$ a square, when Q may be universal, but not a zero form.*

We collect a few complementary results in

Theorem 36. *If $n \geqslant 2$, $\alpha \in Z_p$, Q has coefficients in Z_p, and $d(Q) = d$, then $Q(\mathbf{x}) = \alpha$ is solvable in $\mathbb{Q}_p$ provided that $p \nmid 2d\alpha$. For $n \geqslant 3$ and $p \neq 2$, Q is a zero form in $\mathbb{Q}_p$ provided that $p \nmid d$. A form that represents all nonzero elements in $\mathbb{Q}_p$ also represents zero nontrivially in $\mathbb{Q}_p$ (both for finite p and for $\mathbb{Q}_\infty = \mathbb{R}$), except if $n = 4$, p is finite, $c_p(Q) = -1$, and d is a square.*

For $\mathbb{Z}$ we have only the following, weaker statement:

Theorem 37. *A quadratic form $Q(\mathbf{x})$ over $\mathbb{Z}$ in n variables, $n \neq 4$, that represents all integers $n \neq 0$ also represents zero.*

It is worth remarking that the converse of Theorem 37 does *not* hold.

The problem of the equivalence of quadratic forms over $\mathbb{Q}_p$ is settled by the following theorem.

Theorem 38. *Let Q and Q_1 be quadratic forms of the same number n of variables, with coefficients in $\mathbb{Q}_p$ and with $d(Q) = d$, $d(Q_1) = d_1$. Then $Q \simeq Q_1$ over $\mathbb{Q}_p$, for p finite, if and only if $c_p(Q) = c_p(Q_1)$ and d/d_1 is a square in $\mathbb{Q}_p$. If $p = \infty$, then $Q \simeq Q_1$ implies $c_p(Q) = c_p(Q_1)$ and $d/d_1 = \alpha^2$, but not conversely.*

For the representation over $\mathbb{Q}_p$ of a form $Q(\mathbf{x})$ of n variables by a form $Q_1(\mathbf{y})$ of m variables, $n < m$, the following theorem holds:

Theorem 39. *In general, $Q_1(\mathbf{y})$ represents $Q(\mathbf{x})$ over $\mathbb{Q}_p$. However, if $m = n + 1$, then this occurs only if*

$$c_p(Q)c_p(Q_1) = \left(\frac{-d(Q), d(Q_1)}{d}\right),$$

and if $m = n + 2$ and $-d(Q)/d(Q_1)$ is a square, only if

$$c_p(Q)c_p(Q_1) = \left(\frac{-1, -d(Q)}{p}\right).$$

§14. The Hasse Principle

In Theorem 5.1 we met for the first time a quadratic form that, while nonsingular, is a zero form. We proved then that the conditions of Legendre for the form to be, in fact, a zero form are equivalent to the solvability of the corresponding equation $Q(x, y, z) = 0$ in *all* p-adic fields (i.e., for all finite p, as well as for $p = \infty$, i.e., in $\mathbb{R}$). The validity of similar conclusions in a more general context was sketched (in the language of congruences modulo prime powers) by Minkowski [179]. As already mentioned in Chapter 5, the complete proofs of these theorems, now formulated as equations in the p-adic fields $\mathbb{Q}_p$ and in their algebraic extensions, are due to Hasse. The main result may be formulated as

Theorem 40 (Minkowski, Hasse). *Let $Q(\mathbf{x})$ and $Q_1(\mathbf{x})$ ($\mathbf{x}' = (x_1, \ldots, x_n)$) be two quadratic forms with rational coefficients. Then Q and Q_1 are equivalent over $\mathbb{Q}$ if and only if they are equivalent over all $\mathbb{Q}_p$ (including $p = \infty$).*

The "only if" part is trivial, and the interest of the theorem resides in the "if" part. This follows from the theorems of §13, especially Theorem 38 and the remark that $\mathbb{Q} \subset \mathbb{Q}_p$ for every p. For a proof, see [101].

For the application of this principle, we have to be able to compute the class invariants of a given quadratic form over each $\mathbb{Q}_p$. We recall from §13 that these are the number of variables, the rank, and the signature for $p = \infty$ (by Sylvester's law of inertia), and the p-adic discriminant and the Hasse symbol for $p < \infty$. One may remark that these can all be obtained with a finite amount of computation.

Theorem 40 can be generalized. Let $\mathbb{K}$ be a finite, algebraic extension of $\mathbb{Q}$. Next, consider all completions of $\mathbb{K}$. These are of two kinds, namely those corresponding to the non-Archimedean valuations of $\mathbb{K}$ (these parallel closely the p-adic completions of $\mathbb{Q}$, but now are related to the prime ideals of $\mathbb{K}$), and those that correspond to the Archimedean valuations, such as the completion of $\mathbb{Q}$ to $\mathbb{R}$ under the absolute value. One has to observe, however, that there are, in general, several inequivalent Archimedean valuations in $\mathbb{K}$ (corresponding roughly to the different conjugate fields). Each of the Archimedean valuations corresponds, in the terminology already used, to a distinct "infinite prime" and leads to a completion isomorphic to either $\mathbb{R}$ or $\mathbb{C}$. Over $\mathbb{C}$, however, all quadratic forms with a given number of variables are equivalent, so that no new conditions arise. In completions isomorphic to $\mathbb{R}$, the invariants are again those of Sylvester's law of inertia. As for the completions corresponding to the prime ideals $\mathfrak{p}$ of $\mathbb{K}$, the invariants are, as before, the number of variables, the $\mathfrak{p}$-adic determinants, and the $\mathfrak{p}$-adic Hasse symbol. The latter is defined exactly like that over $\mathbb{Q}$, except that now one obtains it as a product of Hilbert symbols of the type $(\frac{\alpha, \beta}{\mathfrak{p}})$; this, in turn, is defined as $+1$ if $\alpha\xi^2 + \beta\eta^2 = 1$ is solvable in the completion $\mathbb{K}_\mathfrak{p}$ of $\mathbb{K}$, and as -1 otherwise. With these definitions the following theorem holds.

Theorem 41. *Two quadratic forms defined in $\mathbb{K}$ are equivalent over $\mathbb{K}$ if and only if they are equivalent over each $\mathbb{K}_{\mathfrak{p}}$ ($\mathfrak{p}$ any prime ideal of $\mathbb{K}$, finite or infinite).*

In particular, we have

Theorem 42. *$Q(\mathbf{x})$ is a zero form over $\mathbb{K}$ if and only if it is a zero form over each $\mathbb{K}_{\mathfrak{p}}$.*

If we take $n = 3$ and $K = Q$ in Theorem 42, we obtain the following

Corollary. *If $Q = ax_1^2 + bx_2^2 + cx_3^2$, a, b, c squarefree, $(a, b) = (b, c) = (c, a) = 1$, then Q is a zero form in $\mathbb{Z}$ if and only if*

$$\left(\frac{-ab}{c}\right) = \left(\frac{-bc}{a}\right) = \left(\frac{-ca}{b}\right) = +1$$

and not all coefficients have the same sign.

The corollary is precisely Legendre's theorem 5.1. As for its proof in the present context, the first conditions insure the solvability of $Q(\mathbf{x}) = 0$ in all fields $\mathbb{Q}_p$ for $p \mid abc$, while the last one is the condition that $Q(\mathbf{x}) = 0$ should have solution in $\mathbb{Q}_\infty = \mathbb{R}$. It is easy to verify (see Problem 3 in Chapter 5) that for $p \nmid abc$, $Q(\mathbf{x})$ is a zero form in $\mathbb{Q}_p$. We conclude by Theorem 42 that $Q = 0$ has solutions in $\mathbb{Q}$, and hence (multiply by a common denominator) in $\mathbb{Z}$.

The Hasse principle is valid also in the problem of representation of a quadratic form $Q(\mathbf{x})$ of n variables by a quadratic form $Q_1(\mathbf{y})$ of m variables, $m \geqslant n$. Indeed, the following theorem holds.

Theorem 43. *The quadratic form $Q_1(\mathbf{y})$ represents $Q(\mathbf{x})$ over $\mathbb{K}$ if and only if it represents it over each $\mathbb{K}_{\mathfrak{p}}$.*

As an immediate corollary of Theorem 43, we have

Theorem 44. *If Q is a quadratic form in n variables with rational coefficients and $m \in \mathbb{Q}$, then $Q(\mathbf{x}) = m$ is solvable in $\mathbb{Q}$ if and only if it is solvable in each $\mathbb{Q}_p$ (including $\mathbb{Q}_\infty = \mathbb{R}$).*

While Theorem 44 is only a very particular instance of this broad theorem, it is sometimes called "the" Hasse principle.

As already stated in Chapter 5, the precise extent of the validity of the Hasse principle is not known. Many of previous theorems hold if $\mathbb{K}$, instead of being an algebraic extension of $\mathbb{Q}$, is a finite algebraic extension of a field of functions of a single variable over a finite field of constants. It does *not* hold unrestrictedly, however, even for such simple cases as the equivalence of quadratic forms over $\mathbb{Z}$ (see, e.g., Theorem 37, for which the converse fails to hold), or the equivalence of forms of degree higher than two over $\mathbb{Q}$.

The study of the applicability of the Hasse principle to quadratic forms over $\mathbb{Z}$ led Siegel to his definition of forms as belonging to the same genus provided that they are equivalent over all $\mathbb{Z}_p$ and also over $\mathbb{Z}_\infty$ $(=\mathbb{Q}_\infty = \mathbb{R})$. He counts the number of representations of one form by another one modulo prime powers. The result of these investigations is Siegel's definition of the density of solutions and of representations that we already know from §5 of Chapter 12. It should be mentioned, however, that while in Chapter 12 all forms were definite ones, Siegel's theory is much more general and covers also the highly nontrivial case of indefinite forms [243, II, Chapter 1].

Appendix: Open Problems

Some currently unsolved problems were marked with asterisks at the end of several of the preceding chapters. Most of them seem to be at least approachable. No such problems are listed at the end of Chapter 14. Indeed, for the problems left open in Chapter 14 it is hardly possible at the present time to distinguish among those that may be solved with moderate effort and those that are completely hopeless. Instead, it appears preferable to list here a few such problems whose solution appears important, without any suggestion concerning the depth of the problem, or its degree of difficulty.

1. Let us formulate the Hasse principle as follows: "For a certain class $\mathscr{C}$ of polynomial equations in n variables over the rational field, solvability in $\mathbb{Q}$ is equivalent to solvability in all $\mathbb{Q}_p$, for $p \leqslant \infty$." The problem is to define as accurately as possible the extent of the class $\mathscr{C}$.

2. As seen, Artin and Pfister have solved Hilbert's 17th problem, except for certain restrictions on the field $\mathbb{K}$. Dubois has shown that the theorems of Artin and Pfister are not valid in all fields $\mathbb{K}$. Is it possible (by, perhaps, increasing Pfister's bound on the number of needed squares) to relax the present limitations on $\mathbb{K}$? How far? In particular, what can be said in the case $\mathbb{K} = \mathbb{Q}$? Is the conjectured bound $2^n + 3$ valid? Does there exist, perhaps for a certain class of fields (say, subfields of the reals that are not real closed) a bound of the type $2^n + O(1)$, or $2^n + O(n)$, or $2^n(1 + o(1))$? Here the implied constants may be absolute, or depend on the field.

3. Hilbert showed that, if the function f to be represented by a sum of squares belongs to a ring $\mathbf{T}$, it is, in general necessary to accept members of the field of fractions of $\mathbf{T}$ as elements to be squared. On the other hand, in certain cases (see [147] and [39]) this is not necessary and $f = \sum_i g_i^2$, with $g_i \in \mathbf{T}$, and with the number of needed squares unchanged. Are there any other such cases? Are there cases in which it is possible to select $g_i \in \mathbf{T}$ by increasing the number of squares?

4. Let $\mathbb{K}$ be a field, $\mathbb{K}_1 = \mathbb{K}(x_1, \ldots, x_n)$ and $Q(x_1, \ldots, x_n) \in \mathbb{K}_1$; under what conditions is $Q(\mathbf{x}) = \alpha$ solvable for every $0 \neq \alpha \in \mathbb{K}$? Is it sufficient to have Q a form of $m \geqslant 2^n$ variables? The natural setting for this problem seems to be $\mathbb{K}$ real closed; is that in fact the case?

5. As seen, the genus of $\sum_{i=1}^{n} x_i^2$ (over $\mathbb{Z}$) contains a single class for $1 \leqslant n < 8$; is that true in a more general setting? In particular, what can be said about the genera of a preassigned number of classes, when arbitrary algebraic number fields are considered? This problem is partly solved (see [211], [212], [188], and [233]), and it appears that, at least for $n \geqslant 3$, only in finitely many algebraic number fields does the genus contain only one class.

6. In more elaborate presentations of the circle method, one computes separately the contributions to the main integral of the *major* and of the *minor* arcs. The major arcs consist of those points of the unit circle that are "close" to points $e^{2\pi ih/k}$ with "small" k. The complements of the major arcs are the minor arcs. The integral along the major arcs leads to the principal term; that along the minor arcs leads to an error term. For precise definitions see [71] or [271]. A really good estimate for the contributions of the minor arcs would be an important achievement.

References

* Books are marked by an asterisk.

*1. M. Abramowitz and I. A. Stegun. *Handbook of Mathematical Functions*. 7th ed., New York: Dover Publications Inc., 1968.
*2. L. Ahlfors. *Complex Analysis*. 2nd ed. New York: McGraw-Hill, 1966.
*3. *Amer. Math. Soc. Proc. Symposia Pure Math*. Vol. 28. Providence, RI, 1976.
4. K. Ananda Rau. On the representation of a number as the sum of an even number of squares, J. Madras Univ. Part B 24 (1954), 61–89.
*5. G. Andrews. *The Theory of Partitions*. Encyclopedia of Mathematics and Its Applications, Vol. 2. Reading, MA: Addison-Wesley Publishing Company, 1976.
6. G. Andrews. Applications of basic hypergeometric functions, SIAM Rev. 16 (1974), 441–484.
7. N. C. Ankeny. Representations of primes by quadratic forms, Amer. J. Math. 74 (1952), 913–919.
8. N. C. Ankeny. Sums of 3 squares, Proc. Amer. Math. Soc. 8 (1957), 316–319.
9. E. Artin. Kennzeichnung des Körpers der reelen algebraischen Zahlen, Abh. Sem. Hamburg. Univ. 3 (1924), 319–323.
10. E. Artin. Über die Zerlegung definiter Funktionen in Quadrate, Abh. Sem. Hamburg. Univ. 5 (1926), 100–115.
11. E. Artin and O. Schreier. Algebraische Konstruktion reeller Körper, Abh. Seminar Hamburg. Univ. 5 (1926), 85–99.
*12. P. Bachmann. *Die Arithmetik der Quadratischen Formen*. Leipzig: Teubner, 1898.
13. A. Baker. On the class number of imaginary quadratic fields, Bull. Amer. Math. Soc. 77 (1971), 678–684.
14. A. Baker. Imaginary quadratic fields with class-number two, Ann. of Math. (2) 94 (1971), 139–152.
15. A. Baltes, P. K. J. Draxl, and E. R. Hilf. Quadratsummen und gewisse Randwertprobleme der Mathematischen Physik, J. Reine Angew. Math. 268/269 (1974), 410–417.
16. R. P. Bambah and S. Chowla. On numbers that can be expressed as a sum of two squares, Proc. Nat. Inst. Sci. India 13 (1947), 101–103.
17. D. P. Banerjee. On the application of the congruence property of Ramanujan's Function to certain quaternary forms, Bull. Calcutta Math. Soc. 37 (1945), 24–26.
18. P. T. Bateman. On the representation of a number as the sum of three squares, Trans. Amer. Math. Soc. 71 (1951), 70–101.
19. P. T. Bateman. Problem E 2051, Amer. Math. Monthly. Solution by the proposer, Amer. Math Monthly 76 (1969) 190–191.
20. P. T. Bateman and E. Grosswald. Positive integers expressible as a sum of three squares in essentially only one way, J. Number Theory 19 (1984), 301–308.
*21. R. Bellman. *A Brief Introduction to Theta Functions*. New York: Holt, Rinehart &Winston, 1961.
*22. H. Behnke and F. Sommer. *Theorie der analytischen Funktionen einer komplexen Veränderlichen*. Grundlehren Math. Wiss. 74. Berlin: Springer-Verlag, 1955.
23. G. Benneton. Sur la représentation des nombres entiers par la somme de 2^m carrés et sa mise en facteurs, *C. R. Acad Sci. Paris* 212 (1941), 591–593, 637–639.
24. B. Berndt. *The evaluation of character series by contour integration*, Publ. Elektrotechn.

Fac. Ser. Mat. i Fiz., No. 381–409 (1972), 25–29.
*25. G. Birkhof and S. MacLane. *A Survey of Modern Algebra.* New York: Macmillan Co., 1947.
26. M. N. Bleicher and M. I. Knopp. Lattice points in a sphere, Acta Arithmetica 10 (1965), 369–376.
27. F. van der Blij. On the theory of simultaneous linear and quadratic representations, Parts I, II, III, IV, V, Nederl. Akad. Wetensch. Proc. 50, 31–40; 41–48, 166–172, 298–306, 390–396; Indag. Math. 9 (1947), 16–25, 26–33, 129–135, 188–196, 248–254.
*28. Z. I. Borevich and I. R. Šafarevich. *Number Theory.* New York: Academic Press, 1966.
29. H. Braun. Über die Zerlegung quadratischer Formen in Quadrate. J. Reine Angew. Math. 178 (1938), 34–64.
30. P. Bronkhorst. *On the Number of Solutions of the System of Diophantine Equations* $x_1^2 + \cdots + x_s^2 = n$, $x_1 + \cdots + x_s = m$ *for* $s = 6$ *and* 8 (in Dutch). Thesis, Univ. of Groningen, 1943. Amsterdam: North Holland Publishing Co., No. 6, 1943.
31. D. A. Buell. Small class numbers and extreme values of L-functions of quadratic fields, Math. Comp. 31 (1977), 786–796.
32. L. Carlitz. On the representation of an integer as the sum of 24 squares, Nederl. Akad. Wetensch. Indag. Math 17 (1955), 504–506.
33. L. Carlitz. Some partition formulas related to sums of squares, Nieuw Arch. Wisk. (3) 3 (1955), 129–133.
34. L. Carlitz. Note on sums of 4 and 6 squares, Proc. Amer. Math. Soc. 8 (1957), 120–124.
35. J. W. S. Cassels. On the representation of rational functions as sums of squares, Acta Arith. 9 (1964), 79–82.
36. J. W. S. Cassels, W. J. Ellison, and A. Pfister. On sums of squares and on elliptic curves over function fields, J. Number Theory 3 (1971), 125–149.
37. M. C. Chakrabarti. On the limit points of a function connected with the 3-square problem, Bull. Calcutta Math. Soc. 32 (1940), 1–6.
38. Chen, Jin-run. The lattice points in a circle, Sci. Sinica 12 (1963), 633–649.
39. D. M. Choi, T. Y. Lam, B. Resnick, and A. Rosenberg. Sums of squares in some integral domains, J. Algebra 65 (1980), 234–256.
40. P. Chowla. On the representation of -1 as a sum of two squares of cyclotomic integers, Norske Vid. Selsk. Forh. (Trondheim) 42 (1969), 51–52. See also J. Number Theory 1 (1968), 208–210.
41. P. Chowla and S. Chowla. Determination of the *Stufe* of certain cyclotomic fields, J. Number Theory 2 (1970), 271–272.
42. S. Chowla. An extension of Heilbronn's class-number theorem, Quart. J. Math. Oxford 5 (1934), 304–307.
*43. H. Cohn. *Advanced Number Theory.* New York: Dover Publications, 1980.
44. H. Cohn. Numerical study of the representation of a totally positive quadratic integer as the sum of quadratic integral squares, *Numer. Math.* 1 (1959), 121–134.
45. H. Cohn. Decomposition into 4 integral squares in the fields of $2^{1/2}$ and $3^{1/3}$, Amer. J. Math. 82 (1960), 301–322.
46. H. Cohn. Calculation of class numbers by decomposition into 3 integral squares in the field of $2^{1/2}$ and $3^{1/2}$, Amer. J. Math. 83 (1961), 33–56.
47. J. G. van der Corput. Neue Zahlentheoretische Abschätzungen, Math. Ann. 89 (1923), 215–254.
48. K. Corradi and I. Katai. A comment on K. S. Gangadharan's paper entitled "Two classical lattice point problems", Magyar Tud. Akad, Mat. Fiz. Oszt. Közl. 17 (1967), 89–97.
*49. H. Davenport. *Analytical Methods for Diophantine Equalities and Diophantine Inequalities.* Ann Arbor, Univ. of Michigan: Campus Publishers, 1962.
50. H. Davenport. A problematic identity, Mathematika 10 (1963), 10–12; *Collected Works* 4, No. 140, 1745–1747.
51. P. Deligne. *La Conjecture de Weil I.* Institut de Hautes Etudes Scient., Publ. Math. No. 43, 1974, 273–307. See also review by N. M. Katz in Math. Rev. 49 (1975), 5013.
52. B. Derasimovič. A Fermat theorem, Mat. Vesnik 6 (21) (1969), 423–424.
*53. R. Descartes. Oeuvres (Adam and Tannery, editors), 11 volumes. Paris, 1898. (See especially letters to Mersenne from July 27 and Aug. 23, 1638 in vol. II, pp. 256 and 337–338, respectively.)
*54. L. E. Dickson. *History of the Theory of Numbers*; (especially vol. II). Washington: Carnegie Inst. 1920; reprinted, New York: Chelsea Publishing Co., 1971.

*55. P. G. L. Dirichlet. *Vorlesungen über Zahlentheorie* (R. Dedekind, editor), 4th edition. Braunschweig: Vieweg, 1894.

56. P. G. L. Dirichlet. Über die Zerlegbarkeit der Zahlen in 3 Quadrate, J. Reine Angew. Math. 40 (1850), 228–232; Werke 2, 1897, 91–95.

57. J. D. Dixon. Another proof of Lagrange's 4-square theorem, Amer. Math. Monthly 71 (1964), 286–288.

58. J. Drach. Sur quelques points de théorie des nombres at sur la théorie générale des courbes algébriques, C. R. Acad. Sci. Paris 221 (1945), 729–732.

59. D. W. Dubois. Note on Artin's solution of Hilbert's 17th problem, Bull. Amer. Math. Soc. 73 (1967), 540–541.

60. E. Dubouis, (Solution of a problem of J. Tannery), Intermédiaire Math. 18 (1911), 55–56.

61. J. Dzewas. Quadratsummen in reellquadratischen Zahlkörpern, Math. Nachr. 21 (1960), 233–284.

62. I. Ebel. Analytische Bestimmung der Darstellungsanzahlen natürlicher Zahlen durch spezielle ternäre quadratische Formen mit Kongruenzbedingungen, Math. Z. 64 (1956), 216–228.

*63. H. M. Edwards. *Fermat's Last Theorem*. New York: Springer-Verlag, 1977.

64. G. Eisenstein. Note sur la représentation par la somme de cinq carrés, J. Reine Angew. Math. 35 (1847), 368.

65. N. Eljoseph. Notes on a theorem of Lagrange, Riveon Lemat. 5 (1962), 74–79.

66. N. Eljoseph. On the representation of a number as a sum of squares, Riveon Lemat. 7 (1954), 38–42.

67. O. Emersleben. Anwendung zahlentheoretischer Abschätzungen bei numerischen Rechnungen, Z. Angew. Math. Mech. 33 (1953), 265–268.

68. O. Emersleben. Ueber Funktionalgleichungen zwischen Epsteinschen Zetafunktionen gleichen Argumentes—Anhang: Summen ungerader Quadrate, Math. Nachr. 44 (1970), 205–230.

69. O. Emersleben. Ueber die Reihe $\sum_{k=1}^{\infty} k/(k^2 + a^2)^2$, Math. Ann. 125 (1952), 165–171.

*70. *Encyclopedic Dictionary of Mathematics* (Iyanaga and Kawada, editors). Cambridge, Mass: MIT Press, 1977.

*71. Th. Estermann. *Introduction to Modern Prime Number Theory*. Cambridge Tract. Cambridge, England: Cambridge University Press, 1952.

72. Th. Estermann. On the representations of a number as a sum of squares, Acta Arith. 2 (1936), 47–79.

73. Th. Estermann. On the representations of a number as a sum of 3 squares, Proc. London Math. Soc. (3) 9 (1959), 575–594.

*74. L. Euler *Vollständige Anleitung zur Algebra*. St. Petersburg, 1770. Also *Opera* (*1*), vol. 1.

75. L. Euler. Novi Comment. Ac. Petropolitensis 5 (1754/5), 3 et seq.

76. L. Euler. Novi Comment. Ac. Petropolitensis 15 (1770), 75.

77. L. Euler. Corresp. Math.-Phys. (Fuss, editor), vol. 1. Petersburg, 1843.

78. B. Fein, B. Gordon, and J. H. Smith. On the representation of -1 as a sum of two squares in an algebric number field, J. Number Theory 3 (1971), 310–315.

*79. P. de Fermat. *Oeuvres*, 3 volumes. Paris: Gauthier-Villars, 1841, 1894, 1896.

80. K. S. Gangadharan. Two classical lattice point problems, Proc. Cambridge Philos. Soc. 57 (1961), 699–721.

*81. C. F. Gauss. *Disquistiones Arithmeticae*. Leipzig, 1801. English translation by A. A. Clarke, S. J., New Haven, Conn.: Yale University Press, 1965.

82. C. F. Gauss. *Werke*, 12 volumes. Göttingen: Gesellschaft der Wissenschaft, 1863–1930.

83. M. L. Glasser and I. J. Zucker. Lattice sums, in *Theoretical Chemistry, Advances and Perspectives*, vol. 5, 1980, 68–96.

84. F. Götzky. Über eine zahlentheoretische Anwendung von Modulfunktionen einer Veränderlichen, Math. Ann. 100 (1928), 411–437.

85. H. Gross and P. Hafner. Über die Eindeutigkeit des reellen Abschlusses eines angeordneten Körpers, Comment. Math. Helv. 44 (1969), 491–494.

*86. E. Grosswald. *Topics from the Theory of Numbers*, 2nd edition. Boston: Birkhäuser Boston, Inc., 1984.

87. E. Grosswald. Negative discriminants of binary quadratic forms with one class in each genus, Acta Arith. 8 (1963). 295–306.

88. E. Grosswald, A. Calloway, and J. Calloway. The representation of integers by three positive squares, Proc. Amer. Math. Soc. 10 (1959), 451–455.

*89. R. C. Gunning. *Lectures on Modular Forms* (Notes by A. Brumer). Annals of Mathematics Studies 48. Princeton, N.J.: Princeton University Press, 1962.
90. H. Gupta. On numbers of the form $4^a(8b + 7)$, *J. Indian Math. Soc.* (*N.S.*) 5 (1941), 172–202.
91. W. Habicht. Über die Zerlegung strikte definiter Formen in Quadrate, Comment. Math. Helv. 12 (1940), 317–322.
92. J. Hafner. New Ω-theorems for two classical lattice point problems, Invent. Math. 63 (1981), 181–186.
93. F. Halter-Koch. Darstellung natürlicher Zahlen als Summe von Quadraten, Acta Arith. 42 (1982), 11–20.
94. G. H. Hardy. On the expression of a number as the sum of two squares, Quart. J. Math. 46 (1915), 263–283.
95. G. H. Hardy. On Dirichlet's divisor problem, Proc. London Math. Soc. (2) 15 (1916), 1–25.
96. G. H. Hardy. On the representation of a number as a sum of any number of squares and in particular of 5, Trans. Amer. Math. Soc. 17 (1920), 255–284.
*97. G. H. Hardy. *Ramanujan*. 12 lectures. Cambridge: Cambridge University Press, 1940: reprinted, New York: Chelsea Publishing Company, 1959. (See especially Lecture IX, pp. 132–160.)
*98. G. H. Hardy and E. M. Wright. *An Introduction to the Theory of Numbers*, 3rd. edition. Oxford, England: Clarendon Press, 1954.
99. J. Hardy. Sums of two squares in a quadratic ring, Acta Arith. 14 (1967/8), 357–369.
100. H. Hasse. Über die Darstellbarkeit von Zahlen durch quadratische Formen im Körper der rationalen Zahlen, J. Reine Angew. Math. 152 (1923), 129–148.
101. H. Hasse. Über die Äquivalenz quadratischer Formen im Körper der rationalen Zahlen, J. Reine Angew. Math. 152 (1923), 205–224.
102. H. Hasse. Symmetrische Matrizen im Körper der rationalen Zahlen, J. Reine Angew. Math. 153 (1924), 12–43.
103. H. Hasse. Darstellbarkeit von Zahlen durch quadratische Formen in einem beliebigen algebraischen Zahlkörper, J. Reine Angew. Math. 153 (1924), 113–130.
104. H. Hasse. Äquivalenz quadratischer Formen in einem beliebigen Zahlkorper, J. Reine Angew. Math 153 (1924), 158–162.
*105. E. Hecke. *Vorlesungen über die Theorie der algebraischen Zahlen*. Leipzig: Akad. Verlagsgesellschaft, 1923. Reprinted, New York: Chelsea Publishing Co., 1948.
106. E. Hecke. Theorie der Eisensteinschen Reihen höherer Stufe und ihre Anwendungen, Abh. Math. Sem. Univ. Hamburg. 5 (1927), 199–224; *Mathematische Werke*, 461–486.
*107. K. Hensel. *Theorie der Algebraischen Zahlen*. Leipzig: Teubner, 1908.
*108. I. N. Herstein. *Topics in Algebra*. Waltham, Mass.: Blaisdell Publishing Co., 1964.
109. D. Hilbert. Über die Darstellung definiter Formen als Summe von Formenquadraten, Math. Ann. 32 (1888), 342–350; *Gesammelte Abhandlungen* 2, 154–161.
110. D. Hilbert. Über ternäre definite Formen, Acta Math. 17 (1893), 169–197; *Gesammelte Abhandlungen* 2, 345–366.
111. D. Hilbert. Mathematische Probleme, Göttinger Nachrichten 1900, 253–297.
112. D. Hilbert. Mathematische Probleme, Arch. Math. Phys. (3) 1, 44–63, 213–237.
113. D. Hilbert. English translation of [111], Bull. Amer. Math. Soc. 8 (1902), 437–479; reprinted in [3].
*114. E. Hille. *Analytic Function Theory*, 2 volumes. Boston, New York: Ginn & Co., 1959.
115. J. S. Hsia and R. P. Johnson. On the representation in sums of squares for definite functions in one variable over an algebraic number field, to appear.
116. L. K. Hua. The lattice-points in a circle, Quart. J. Math. Oxford 13 (1942), 18–29.
117. G. Humbert. Sur la représentation d'un entier par une somme de 10 ou 12 carrés, C. R. Acad Sci. Paris 144 (1907), 874–878.
118. A. Hurwitz. Sur la décomposition des nombres en cinq carrés, C. R. Acad. Sci. Paris 98 (1884), 504–507; *Werke*, vol. 2, 5–7.
119. A. Hurwitz. Intérmediaire Math. 14 (1907), 107; *Werke*, vol. 2, 751.
*120. D. Jackson. *Fourier Series and Orthogonal polynomials*. Carus Monograph No. 6. Oberlin, Ohio: Math Assoc. Amer., 1941.
*121. C. G. J. Jacobi. *Fundamenta Nova Theoriae Functionum Ellipticarum*. 1829.
122. C. G. J. Jacobi. Note sur la décomposition d'un nombre donné en quatre carrés, J. Reine Angew. Math. 3 (1828), 191; *Werke*, vol. I, 247.
123. C. G. J. Jacobi. De compositione numerorum e quator quadratis, J. Reine Angew. Math. 12 (1834), 167–172; *Werke*, vol. VI, 1891, 245–251.

124. C. G. J. Jacobi. Correspondence mathématique entre Legendre et Jacobi, J. Reine Angew. Math. 80 (1875), 205–279 (letter of September 9, 1828, 240–243).
*125. B. W. Jones. *The Arithmetic Theory of Quadratic Forms*. Carus Monograph No. 10. Math. Assoc. Amer., 1950.
126. T. Kano. On the number of integers representable as the sum of two squares, J. Fac. Sci. Shinsu Univ. 4 (1969), 57–65.
127. A. A. Kiselev and I. S. Slavutskii. Some congruences for the number of representations as sums of an odd number of squares (in Russian), Dokl. Akad. Nauk SSSR 143 (1962), 272–274.
*128. F. Klein and R. Fricke. *Theorie der Elliptischen Modulfunctionen* (2 volumes). Leipzig: Teubner, 1890, 1892.
129. H. D. Kloostermann. Simultane Darstellung zweier ganzer Zahlen als eine Summe von ganzen Zahlen und deren Quadratsumme, Math. Ann. 118 (1942), 319–364.
130. M. Knebusch. On the uniqueness of real closures and the existence of real places, Comment. Math. Helv. 47 (1972), 260–269.
131. H. Kneser. Verschwindende Quadratsummen in Körpern, Jahresber. Deutsch. Math.-Verein 44 (1934), 143–146.
132. M. Kneser. Klassenzahl definiter quadratischer Formen, Arch. Math. 8 (1957), 241–250.
*133. M. Knopp. *Modular Functions in Analytic Number Theory*. Markham Mathematics Series. Chicago: Markham Publishing Co., 1970.
134. C. Ko. On the representation of a quadratic form as a sum of squares of linear forms. Quart. J. Math. Oxford 8 (1937), 81–98. See also Proc. London Math. Soc. 42 (1936), 171–185.
*135. N. Koblitz. *p-adic Numbers, p-adic Analysis and Zeta-Functions*. New York, Berlin: Springer-Verlag, 1977.
136. G. A. Kolesnik. On the order of $\zeta(\frac{1}{2} + it)$ and $\Delta(R)$. Pacific J. Math. 98 (1982), 107–122.
137. G. A. Kolesnik. Private communication, December 1982.
138. E. Krätzel. Über die Anzahl der Darstellungen von natürlichen Zahlen als Summe von $4k$ Quadraten, Wiss. Z. Friedrich-Schiller-Univ. Jena 10 (1960/61), 33–37.
139. E. Krätzel. Über die Anzahl Der Darstellungen von natürlichen Zahlen als Summe von $4k + 2$ Quadraten. Wiss. Z. Friedrich-Schiller-Univ. Jena 11 (1962/63), 115–120.
*140. E. Krätzel. *Zahlentheorie*. Studienbücherei, No. 19. Berlin: Deutscher Verlag der Wissenschaften, 1981.
141. A. Krazer, *Lehrbuch der Thetafunctionen*. Leipzig: B. G. Teubner, 1903.
142. G. Kreisel. *Hilbert's 17th Problem—Summer Institute of Symbolic Logic*. Cornell Univ., 1957.
143. L. Kronecker. Ueber die Anzahl der verschiedenen Classen quadratischer Formen von negativer Determinante, J. Reine Angew. Math. 57 (1860), 248–255.
144. I. P. Kubilyus and Yu. V. Linnik. On the decomposition of the product of 3 numbers into the sum of 2 squares (in Russian), *Trudy Mat. Inst. Steklov*, vol. 38. Moscow: Izdat. Akad. Nauk SSSR, 1951, 170–172.
145. J. L. Lagrange. Nouveau Mém. Acad. Roy. Sci. Berlin (1772), 123–133; *Oeuvres*, vol. 3, 1869, 189–201.
*146. E. Landau. *Vorlesungen über Zahlentheorie* (3 volumes). Leipzig: Hirzel, 1927.
147. E. Landau. Über die Darstellung definiter Funktionen durch Quadrate, Math. Ann. 62 (1906), 272–285.
148. E. Landau. Über die Einteilung der ... Zahlen in 4 Klassen ..., Arch. Math. Phys. (3) 13 (1908), 305–312.
149. E. Landau. Über die Anzahl der Gitterpunkte in gewissen Bereichen, Göttinger Nachr., 1912, 687–771; see especially 691–692, 765–766.
150. E. Landau. Ausgewählte Abhandlungen zur Gitterpunktlehre (A. Walfisz, editor). Berlin: VEB Deutscher Verlag der Wissenschaften, 1962.
151. E. Landau. Über die Zerlegung total positiver Zahlen in Quadrate, Göttinger Nachr., 1919, 392–396.
152. S. Lang. On quasi-algebraic closure. Ann. of Math. 55 (1952), 373–390.
153. W. J. Leahy. A note on a theorem of I. Niven, Proc. Amer. Math. Soc. 16 (1965), 1130–1131.
154. A.-M. Legendre. *Théorie des Nombres*. 1798. Reprinted, Paris: Blanchard, 1955.
155. A.-M. Legendre. Mémoire Acad. Sci. Paris, 1785.
156. D. H. Lehmer. On the partition of numbers into squares, Amer. Math. Monthly 55 (1948), 476–481.

*157. J. Lehner. *Discontinuous Groups and Automorphic Functions*. Providence, R. I.: Amer. Math. Soc., 1964.

*158. J. Lehner. *A Short Course in Automorphic Functions*. New York: Holt, Rinehart and Winston, 1966.

159. H. Lenz. Zur Quadratsummendarstellung in relativ-quadratischen Zahlkörpern. Sitzungsber. Math.-Nat. Klasse, Bayer. Akad. Wiss., (1953/4), 283–288.

160. W. LeVeque. *Reviews in Number Theory*. Providence, R.I.: Amer. Math. Soc., 1974.

161. Yu. V. Linnik. The asymptotic distribution of lattice points on a sphere (in Russian), Dokl. Akad. Nauk SSSR (N.S.) 96 (1954), 909–912.

162. Yu. V. Linnik. Asymptotic-geometric and ergodic properties of sets of lattice points on a sphere (in Russian), Mat. Sbornik (N.S.) 43 (85) (1957), 257–276.

163. Yu. V. Linnik. Application of methods of D. Burgess to the investigation of integer points on large spheres, *Symposia Math*., vol. IV. (Rome: INDAM, 1968/1969), 99–112; London: Academic Press, 1970.

164. J. H. van Lint. Über einige Dirichletsche Reihen. Nederl. Akad. Wetensch., Indag. Math. 20 (1958), 56–60.

165. J. Liouville. Math. Pures Appl. (2) 3 (1858), 143–152, 193–200, 201–208, 241–250, 273–288, 325–336; 4 (1859), 1–8, 72–80, 111–120, 195–204, 281–304; 5 (1860), 1–8; 9 (1864), 249–256, 281–288, 321–336, 389–400; 10 (1865), 135–144, 169–176.

166. J. Liouville. Extrait d'une lettre à M. Besge, J. Math. Pures Appl. (2) 9 (1864), 296–298.

167. R. Lipschitz. Untersuchungen der Eigenschaften einer Gattung von unendlichen Reihen, J. Reine Angew. Math. 105 (1889), 127–156.

168. G. A. Lomadze. On the representation of numbers by sums of squares (in Russian; Georgian summary), Akad. Nauk Gruzin. SSR Trudy Tbiliss. Mat. Inst. Razmadze 16 (1948), 231–275.

169. G. A. Lomadze. On the representations of numbers by sums of an odd number of squares (in Georgian; Russian summary), Akad. Nauk Gruzin. SSR Trudy Tbiliss, Mat. Inst. Razmadze 17 (1949), 281–314.

170. G. A. Lomadze. On the simultaneous representation of two whole numbers by sums of whole numbers and their squares (in Russian; Georgian summary), Akad. Nauk Gruzin. SSR Trudy Tbiliss, Mat. Inst. Razmadze 18 (1951), 153–181.

171. G. A. Lomadze. On the summation of the singular series I and II (in Russian; Georgian summary), Akad. Nauk Gruzin. SSR Trudy Tbiliss., Mat. Inst. Razmadze 19 (1953), 61–77 and 20 (1954), 21–45.

172. G. A. Lomadze. On the representation of numbers as sums of squares (in Russian), Akad. Nauk Gruzin. SSR Trudy Tbiliss., Mat. Inst. Razmadze 20 (1954), 47–87.

173. Gino Loria. Sulla scomposizione di un entero nella somma di numeri poligonali, Accad. Naz. Lincei, Atti Cl. Sci. Fis. Mat. Nat. Rendiconti (8) 1 (1946), 7–15.

174. H. Maass. Über die Darstellung total positiver Zahlen des Körpers $R(\sqrt{-5})$ als Summe von drei Quadraten, Abh. Math. Sem. Hamburg. Univ. 14 (1941), 185–191.

175. A. V. Malyshev. The distribution of integer points on a four-dimensional sphere (in Russian), Dokl. Akad. Nauk SSSR (N.S.) 114 (1954), 25–28.

176. Ju. I. Manin. On Hilbert's 17th problem (in Russian), *Hilbert's Problems*, Moscow: Izdatebśtvo "Nauka" 1969, 196–199.

177. A. Meyer. Über die Kriterien für die Auflösbarkeit der Gleichung $ax^2 + by^2 + cz^2 + du^2 = 0$ in ganzen Zahlen, Vierteljahrsschrift Naturforsch. Ges. Zürich 29 (1884), 209–222.

178. H. Minkowski. Mémoire présenté à l' Adad. Sci., Inst. France (2) 29 (1884) No. 2; *Gesammelte Abhandlungen*, vol. 1, 1911, 118–119, 133–134.

179. H. Minkowski. Über die Bedingungen unter welchen zwei quadratische Formen mit rationalen Koeffizienten ineinander rational transformiert werden können, J. Reine Angew. Math. 106 (1890), 5–26.

*180. L. J. Mordell. *Diophantine Equations*. New York: Academic Press, 1969.

181. L. J. Mordell. Note on class relations, Messenger of Math. 45 (1915), 76–80.

182. L. J. Mordell. On the representations of numbers as sums of 2r squares, Quart. J. Math. Oxford 48 (1917), 93–104.

183. L. J. Mordell. On M. Ramanujan's empirical expansions of modular functions, Proc. Cambridge Philos. Soc. 19 (1917), 117–124.

184. L. J. Mordell. On the representations of a number as a sum of an odd number of squares, Trans. Cambridge Philos. Soc. 22 (1919), 361–372.

185. L. J. Mordell. On binary quadratic forms as a sum of 3 linear squares with integer coefficients, J. Reine Angew. Math. 167 (1931), 12–19.
186. L. J. Mordell. On the representation of a binary quadratic as a sum of squares of linear forms, Math. Z. 35 (1932), 1–15.
187. L. J. Mordell. An application of quaternions to the representation of a binary quadratic form as a sum of four linear squares, Quart. J. Math. Oxford 8 (1937), 58–61.
188. L. J. Mordell. The definite quadratic forms in 8 variables with determinant unity, J. Math. Pures Appl. 17 (1938), 41–46.
189. L. J. Mordell. On the representations of a number as a sum of 3 squares, Rev. Math. Pures Appl. 3 (1958), 25–27.
190. C. Moser. Représentation de -1 par une somme de carrés dans certains corps locaux et globaux et dans certains annaux d'entiers algébriques, C. R. Acad. Sci. Paris Ser. A-B 271 (1970), A1200–1203.
191. T. S. Motzkin. The arithmetic-geometric inequality, *Inequalities* (Oved Shisha, editor). New York: Academic Press, 1967, 205–224.
192 T. Nagell. On the representations of integers as the sum of two integral squares in algebraic, mainly quadratic fields, Nova Acta Soc. Sci. Upsala (4) 15 (1953), No. 11 (73pp.).
193. T. Nagell. On the sum of two integral squares in certain quadratic fields, Ark. Math. 4 (1961), 267–286.
194. T. Nagell. On the number of representations of an A-number in an algebraic field, Ark. Math. 4 (1962), 467–478.
195. T. Nagell. On the A-numbers in the quadratic fields $K(\sqrt{\pm 37})$, Ark. Math. 4 (1963), 511–521.
196. M. Newman. Subgroups of the modular group and sums of squares, Amer. J. Math. 82 (1960), 761–778.
197. H.-V. Niemeier. Definite quadratische Formen der Dimension 24 und Diskriminante 1, J. Number Theory 5 (1963), 142–178.
198. I. Niven. Integers in quadratic fields as sums of squares, Trans. Amer. Math. Soc. 48 (1940), 405–417.
199. I. Niven and H. S. Zuckerman. Variations on sums of squares, Pi Mu Epsilon J. 4 (1968), 407–410.
*200. O. T. O'Meara. *Introduction to Quadratic Forms.* Grundlehren d. Math. Wiss. No. 117. Berlin: Springer-Verlag, 1963.
*201. A. Ogg. *Modular Forms and Dirichlet Series.* New York, Amsterdam: W. A. Benjamin, 1969.
202. C. D. Olds. On the representations $N_3(n^2)$ Bull. Amer. Math. Soc. 47 (1941), 499–503.
203. C. D. Olds. On the representations $N_7(n^2)$, Bull Amer. Math. Soc. 47 (1941), 624–628.
204. G. Pall. On sums of squares, Amer. Math. Monthly 40 (1933), 10–18.
205. G. Pall. On the rational automorphs of $x_1^2 + x_2^2 + x_3^2$, Ann. of Math. (2) 41 (1940), 754–766.
206. G. Pall. On the arithemtic of quaternions, Trans. Amer. Math. Soc. 47 (1940), 487–500.
207. G. Pall. Sums of two squares in a quadratic field, Duke Math. J. 18 (1951), 399–409.
208. G. Pall and O. Taussky. Applications of quaternions to the representations of a binary quadratic form as a sum of four squares, Proc. Roy. Irish Acad. Sect. A 58 (1957), 23–28.
209. H. Petersson. Über die Zerlegung des Kreisteilungspolynoms von Primzahlordnung, Math. Nachr. 14 (1956), 361–375.
*210. H. Petersson. *Modulfunktionen und Quadratische Formen.* Ergebnisse der Math. u. ihrer Grenzgebiete, vol. 100. Berlin: Springer-Verlag, 1982.
211. H. Pfeuffer. Quadratsummen in totalreellen algebr. Zahlkörpern, J. Reine Angew. Math. 249 (1971), 208–216.
212. H. Pfeuffer. Einklassige Geschlechter totalpositiver quadratischer Formen in totalreellen algebraischen Zahlkörpern, J. Number Theory 3 (1971), 371–411.
213. A. Pfister. Zur Darstellung von -1 als summe von Quadraten in einem Körper, J. London Math. Soc. 40 (1965), 150–165.
214. A. Pfister. Multiplikative quadratische Formen, Arch. Math. 16 (1965), 363–370.
215. A. Pfister. Zur Darstellung definiter Funktionen als Summen von Quadraten, Invent. Math. 4 (1967), 229–237.
216. B. van der Pol. The representation of numbers as sums of 8, 16, and 24 squares, Nederl. Akad. Wetensch. Indag. Math. 16 (1954), 349–361.
217. G. Pólya. Elementarer Beweis einer Thetaformel, Sitzungsber. Akad. Wiss. Berlin. Phys.-Math. Kl (1927) 158–161.

218. A. G. Postnikov. Additive problems with a growing number of terms (in Russian), Dokl. Akad. Nauk SSSR (N.S.) 108 (1956), 392.
219. Y. Pourchet. Sur la représentation en somme de carrés des polynômes à une indéterminée sur un corps de nombres algébriques, Acta Arith. 19 (1971), 89–104.
220. D. Pumplün. Über die Darstellungsanzahlen von quadratfreien Zahlen durch Quadratsummen, Math. Ann. 170 (1967), 253–264.
*221. H. Rademacher. *Topics in Analytic Number Theory*. Grundlehren d. Math. Wiss. No. 169. Berlin, New York: Springer-Verlag, 1970.
222. S. Ramanujan. On certain arithmetical functions, Trans. Cambridge Philos. Soc. 22 (1916), 159–184; *Collected Papers*, No. 18.
223. S. Ramanujan. On certain trigonometrical sums and their applications in the theory of numbers, Trans. Cambridge Philos. Soc. 22 (1918), 259–275; *Collected Papers*, No. 21.
224. B. Randol. A lattice point problem, I, II, Trans. Amer. Math. Soc. 121 (1966), 257–268; 125 (1966), 101–113.
225. R. A. Rankin. On the representations of a number as a sum of squares and certain related identities, Proc. Cambridge Philos. Soc. 41 (1945), 1–11.
226 R. A. Rankin. Representations of a number as the sum of a large number of squares, Proc. Roy. Soc. Edinburgh Sect. A 65 (1960/61), 318–331.
227. R. A. Rankin. On the representation of a number as the sum of any number of squares and in particular of 20, Acta Arith. 7 (1961/62), 399–407.
228. R. A. Rankin. Sums of squares and cusp forms, Amer. J. Math. 87 (1965), 857–860.
229. L. Reitan. On the solution of the number theoretical equation $x^2 + y^2 + z^2 + v^2 = t$ for a given t (in Norwegian), Norsk. Mat. Tidsskr. 28 (1945), 21–23.
230. Ph. Revoy. Formes quadratiques et représentation de -1 comme somme de carrés dans un corps, *Mélanges d'Algebre Pure et Appliquée. Montpellier: Univ. de Montpellier*, 1970, 27–41.
231. U. Richards. Risoluzione elementare dell'equazione indeterminata $u^2 + v^2 = p$, essendo p un numero primo, Atti Accad. Sci. Torino Cl. Sci. Fis. Mat. Nat. 75 (1940), 268–273.
232. A. Robinson. On ordered fields and definite functions, Math. Ann. 130 (1955), 257–271.
233. R. Salamon. Die Klassen der Geschlechten von $x_1^2 + x_2^2 + x_3^2$ und $x_1^2 + x_2^2 + x_3^2 + x_4^2$ über $Z[\sqrt{3}]$, Arch. Math. (Basel) 20 (1969), 523–530.
234. H. F. Sandham. A square as the sum of 7 squares, Quart. J. Math. Oxford Ser. (2) 4 (1953), 230–236.
235. H. F. Sandham. A square as the sum of 9, 11, and 13 squares, J. London Math. Soc. 29 (1954), 31–38.
236. W. Schaal. Übertragung des Kreisproblems auf reell-quadratische Zahlkörper, Math. Ann. 145 (1962), 273–284.
237. W. Schaal. On the expression of a number as the sum of 2 squares in totally real algebraic number fields, Proc. Amer. Math. Soc. 16 (1965), 529–537.
238. A. Schinzel. On a certain conjecture concerning partitions into sums of 3 squares (in Polish), Wiadom. Mat.(2) 1 (1955/56), 205.
239. A. Schinzel. Sur des sommes de trois carrés (Russian summary), Bull. Acad. Polon. Sci. Ser. Sci. Math. Astronom. Phys. 7 (1959), 307–310.
240. L. Seshu. On the simultaneous representation of a given pair of integers as the sum respectively of four integers and their squares, I, II, Nederl. Akad. Wetensch. Indag. Math. 23 (1961), 64–79, 80–88.
241. C. L. Siegel. Additive theorie der Zahlkörper I, Math. Ann. 87 (1922), 1–35; *Gesammelte Abhandlungen*, vol. 1, 119–153.
242. C. L. Siegel. Additive theory der Zahlkörper II, Math. Ann. 88 (1923), 184–210; *Gesammelte Abhandlungen*, vol. 2, 180–206.
243. C. L. Siegel. Über die analytische Theorie der quadratischen Formen I, II, III, Ann. of Math. 36 (1935), 527–606, 37 (1936), 230–263, 38 (1937), 212–291; *Gesammelte Abhandlungen*, vol. 1, 326–405, 410–443, 468–548.
244. C. L. Siegel. Mittelwerte arithmetischer Funktionen in Zahlkörpern, Trans. Amer. Math. Soc. 39 (1936), 219–224; *Gesammelte Abhandlungen*, vol. 1, 453–458.
245. C. L. Siegel. Equivalence of quadratic forms, Amer. J. Math. 63 (1941), 658–680; *Gesammelte Abhandlungen*, vol. 2, 217–239.
246. C. L. Siegel. Sums of m-th powers of algebraic integers, Ann. of Math. 46 (1945), 313–339; *Gesammelte Abhandlungen*, vol. 3, 12–38.
*247. W. Sierpiński. *Elementary Theory of Numbers* (Hulanicki, translator). Monograph. Mat.,

vol. 42. Warsaw, 1964
248. W. Sierpiński. Sur un problème du calcul des fonctions asymptotiques (in Polish), Prace. Mat. Fiz (Warsaw) 17 (1906), 77–118.
249. W. Sierpiński. Proc. Sci. Soc. Warsaw 2 (1909), 117–119.
250. W. Sierpiński. Sur les nombres impairs admettant une seule décomposition en une somme de 2 carrés de nombres naturels, premiers entre eux, Elem. Math. 16 (1961), 27–30.
251. Th. Skolem. On orthogonally situated lattice points on spheres (in Norwegian), Norsk. Mat. Tidsskr. 23 (1941), 54–61.
252. Th. Skolem. A relation between the congruence $x^2 + y^2 + z^2 + u^2 \equiv 0 \pmod{m}$ and the equation $x^2 + y^2 + z^2 + u^2 = m$ (In Norwegian), Norsk. Mat. Tidsskr. 25 (1943), 76–87.
253. H. J. S. Smith. On the orders and genera of quadratic forms containing more than 3 indeterminates, Proc. Roy. Soc. London 16 (1867), 197–208; *Collected Mathematical Papers*, vol. 1, 1894, 510–523.
254. H. J. S. Smith. Mémoire sur la représentation des nombres par 5 carrés, Mémoires Savants Etrangers, Acad. Sci. Paris (2) 29 (1887); *Collected Mathematical Papers*, vol. 2, 1894, No. 44, 623–680.
255. R. Spira. Polynomial representations of sums of two squares, Amer. Math. Monthly 71 (1964), 286–288.
256. R. Sprague. Über Zerlegung in ungleiche Quadratzahlen, Math. Z. 51 (1948), 289–290.
257. H. Stark. A complete determination of the complex quadratic fields with class-number one, Michigan Math. J. 14 (1967), 1–27.
258. H. Stark. On complex quadratic fields with class-number two, Math. Comp. 29 (1975), 289–302.
259. T. Stieltjes. Sur le nombre de décompositions d'un entier en cinq carrés, C. R. Acad. Sci. Paris 97 (1883), 1545–1547; *Ouevres*, vol. 1, 329–331.
260. T. Stieltjes. Sur quelques applications arithmétiques ... (from a letter to Hermite), C. R. Acad. Sci. Paris 98 (1884), 663–664; *Oeuvres*, vol. 1, 360–361.
261. M. V. Subba Rao. On the representation of numbers as sums of two squares, Math. Student 26 (1958), 161–163.
*262. J. Tannery and J. Molk. *Eléments de la Théorie des Fonctions Elliptiques* (4 volumes), Paris: Gauthier-Villars, 1893, 1896, 1898, 1902.
263. O. Taussky. Sums of squares, Amer. Math. Monthly 77 (1970), 805–830.
264. O. Taussky. *History of Sums of Squares in Algebra*. American Math. Heritage. Texas Technical University, 1981.
*265. E. C. Tichmarsh. *The Theory of Fourier Integrals*. Oxford: Clarendon Press, 1937.
*266. E. C. Titchmarsh. *The Theory of the Riemann Zeta-Function*. Oxford: Clarendon Press, 1951.
267. E. C. Titchmarsh. The lattice-points in a circle, Proc. London Math. Soc. (2) 38 (1934), 96–115.
268. C. Tsen. Zur Stufentheorie der Quasi-algebraisch Abgeschlossenheit kommutativer Körper, J. Chinese Math. Soc. 1 (1936), 81–92.
269. J. V. Uspenskii. A new proof of Jacobi's theorem (in Spanish), Math. Notae 4 (1944), 80–89.
*270. J. V. Uspenskii and M. H. Heaslet. *Elementary Number Theory*. New York, London: McGraw-Hill Book Co., 1939.
*271. R. C. Vaughan. *The Hardy–Littlewood Method*. Cambridge Tracts in Mathematics, No. 80. Cambridge: Cambridge University Press, 1981.
272. I. M. Vinogradov. On the number of integral points in a given domain (in Russian), Izv. Akad. Nauk SSSR Ser. Mat. 24 (1960), 777–786.
273. B. L. van der Waerden. Problem 144, Jahresber. Deutsch. Math. Verein, 42 (1932), 71.
274. C. Waid. Generalized automorphs of the sum of four squares, J. Number Theory 3 (1971), 468–473.
275. A. Z. Walfisz. On the representation of numbers by sums of squares; asymptotic formulas, Uspehi Mat. Nauk (N.S.) 52 (6) (1952), 91–178 (in Russian).
276. A. Z. Walfisz. English translation of [275], Amer. Math. Soc. Transl. (2) 3 (1956), 163–248.
277. A. Z. Walfisz. The additive theory of numbers XI (in Russian, Georgian summary), Akad. Nauk Gruzin. SSR Trudy Tbiliss., Mat. Inst. Razmadze 19 (1953), 13–59.
278. A. Z. Walfisz. Additive number theory XII (in Russian), Akad. Nauk Gruzin. SSR Trudy Tbiliss., Mat. Inst. Razmadze 22 (1956), 3–31.
279. P. J. Weinberger. Exponents of the class groups of complex quadratic fields, Acta Arith. 22

(1973), 117–124.
280. K. S. Williams. On a theorem of Niven, Canad. Math. Bull. 10 (1967), 573–578.
281. K. S. Williams. Addendum "On a theorem of Niven", Canad. math. Bull. 11 (1968), 145.
282. K. S. Williams. Note on a theorem of Pall, Proc. Amer. Math. Soc. 28 (1971), 315–316.
283. W.-lin Yin. The lattice points in a circle, Sci. Sinica 11 (1962), 10–15.

Bibliography

* Books are marked by an asterisk.

1. D. Allison. On square values of quadratics, Math. Colloq. Univ. Capetown 9 (1974), 135–141.
2. E. A. Anferteva and N. G. Cudakov. The minima of a normed function in imaginary quadratic fields, Dokl. Akad. Nauk SSSR 183 (1968), 255–256; Erratum, 187 (1969), No. 1–3, vi; translation, Soviet Math. Dokl. 9 (1968), 1342–1344.
3. J. K. Arason and A. Pfister. Zur theorie der quadratischen Formen über formalreellen Körpern, Math. Z. 153 (1977), 289–296.
4. G. Pall (editor). *Arithmetical Theory of Quadratic Forms*. Proceedings of Conference at Louisiana State University, Baton Rouge, La., 1972. Dedicated to L. J. Mordell. No. 5 and No. 6 of the *Journal of Number Theory* contain 18 papers and 10 abstracts of papers presented at the conference. The papers considered of interest to the readers of the present book are listed under the respective authors.
5. R. Ayoub and S. Chowla. On a theorem of Müller and Carlitz, J. Number Theory 2 (1970), 342–344.
6. R. Baeza. Ueber die Stufe eines semi-localem Ringes, Math. Ann. 215 (1975), 13–21.
7. Ph. Barkan. Sur l'equation diophantienne $X^2 - mY^2 = -2$ (English summary), C. R. Acad Sci. Paris. Sér. A.-B 282 (1976), A1215–A1218.
8. C. W. Barnes. The representation of primes of the form 4n + 1 as the sum of two squares, Enseign. Math. (2) 18 (1972), 289–299 (1973).
9. F. W. Barnes. On the *Stufe* of an algebraic number field. J. Number Theory 4 (1972), 474–476.
10. E. S. Barnes and M. J. Cohn. On Minkowski reduction of positive quaternary quadratic forms, Mathematika 23 (1976), 156–158.
11. P. Barrucand and H. Cohn. Note on primes of the type $x^2 + 32y^2$, class number and residuacy, J. Reine Angew. Math. 238 (1969), 67–70.
12. V. J. D. Baston. Extreme copositive quadratic forms, Acta Arith. 15 (1969), 319–328.
13. H. Belkner. Die Diophantische Gleichung $a^2 + kb^2 = kc^2$, Wiss. Z. Pädagog. Hochsch. Potsdam, Math. Naturwiss. Reihe 14 (1970), 187–193.
14. H. Belkner. Die Diophantische Gleichung $x_1^2 \cdots + x_n^2 = x_{n+1}^k$, Wiss. Z. Pädagog. Hochsch. Potsdam, Math. Naturwiss. Reihe 14 (1970) 181–186.
15. H. Belkner. Die Diophantische Gleichungen vom Typ $\sum_{j=1}^k a_j x_j^2 - \sum_{j=1}^m a_{k+j} x_{k+j}^2 = a_{k+m+1}$, Wiss. Z. Pädagog. Hochsch. Potsdam, Math. Naturwiss. Reihe 20 (1976), 131–135.
16. R. I. Beridze and G. A. Gogišvili. The number of representations of numbers by certain quadratic forms in 6 variables (in Russian), Sakharth. SSR Mecn. Akad. Math. Inst. Šrom. 57 (1977), 5–15.
17. D. G. Beverage. On Ramanujan's form $x^2 + y^2 + 10z^2$, Univ. Nac. Tucumán, Rev. Ser. A 19 (1969), 195–199.
18. A. Bijloma. Representations of a natural number by positive binary quadratic forms of given discriminant, Nieuw Arch. Wisk. (3) 23 (1975), 105–114.
19. P. E. Blanksby. A restricted inhomogeneous minimum for forms. J. Austral. Math. Soc. 9 (1969), 363–386.
20. F. van der Blij. Quelques remarques sur la théorie des formes quadratiques (expository), Bull. Soc. Math. Belge 20 (1968), 205–212.

21. L. Bröcker. Zur Theorie der quadratischen Formen über formal reellen Körpern, Math. Ann. 210 (1974), 233–256.
22. L. Bröcker. Ueber eine Klasse pythagoreischer Körper, Arch. Math. (Basel) 23 (1972), 405–407.
23. M. Broué. Codes et formes quadratiques, *Séminaire P. Dubreil, F. Aribaud, M.-P. Malliavin*, 1974/1975, Algèbre, Exp. 23.
24. J. Browkin. On zero forms, Bull. Acad. Pol. Sci. Ser. Sci. Math. Astron. Phys. 17 (1969), 611–616.
25. E. Brown. Representations of discriminantal divisors by binary quadratic forms, J. Number Theory 3 (1971), 213–255.
26. E. Brown. Binary quadratic forms of determinant -pq, J. Number Theory 4 (1972), 408–410.
27. E. Brown. Quadratic forms and biquadratic reciprocity, J. Reine Angew. Math. 253 (1972), 214–220.
28. E. Brown. Discriminantal divisors and binary quadratic forms, Glasgow Math. J. 13 (1972), 69–73.
29. E. Brown. The class number of $Q(\sqrt{-p})$ for $p \equiv 1 \pmod 8$ a prime, Proc. Amer. Math. Soc. 31 (1972), 381–383.
30. E. Brown. The power of 2 dividing the class number of a binary quadratic discriminant, J. Number Theory 5 (1973), 413–419.
31. E. Brown. Class numbers of complex quadratic fields, J. Number Theory 6 (1974), 185–191.
32. H. S. Butts and G. Pall. Modules and binary quadratic forms, Acta Arith. 15 (1968), 23–44.
33. G. Cantor. Zwei Sätze aus der Theorie der binären, quadratischen Formen, Z. Math. Phys. 13 (1868), 259–261.
34. L. Carlitz. Bulygin's method for sums of squares, J. Number Theory 5 (1973), 405–412.
35. J. W. S. Cassels (Kassels). Sums of squares (in Russian), Proceedings of the International Conference on Number Theory, Trudy Mat. Inst. Steklov 132 (1973), 114–117, 265.
*36. J. W. S. Cassels. *Rational Quadratic forms*. London Math. Soc. Monographs, No. 13. London, New York: Academic Press, 1978.
37. H. H. Chalk. Zeros of quadratic forms, C. R.—Math. Rep. Acad. Sci Canada 1 (1978/79), 275–278.
38. K. Chang. Diskriminanten und Signaturen gerader quadratischer Formen, Arch. Math. (Basel) 21 (1970), 59–65.
39. P. L. Chang. On the number of square classes of a field with finite *Stufe*, J. Number Theory 6 (1974), 360–368.
40. M. D. Choi. Positive, semidefinite biquadratic forms, Linear Algebra Appl. 12 (1975), 95–100.
41. P. Chowla. On the representation of -1 as a sum of squares in a cyclotomic field, J. Number Theory 1 (1969), 208–210.
42. S. Chowla and P. Hartung. The 3-squares theorem, Indian J. Pure Appl. Math. 6 (1975), 1077–1079.
43. S. Chowla and P. Chowla. "Metodo rapido" for finding real quadratic fields of class number 1, Proc. Nat. Acad. Sci. U.S.A. 70 (1973), 395.
44. M. R. Christie. Positive definite rational functions of two variables, which are not the sum of three squares, J. Number Theory 8 (1976), 224–232.
45. H. Cohen. Représentations comme sommes de carrés, *Séminaire Théorie des Nombres*, 1971–1972, Univ. Bordeaux I, Talence, Exp. No. 21.
46. H. Cohen. Sommes de carrés, fonctions *L* et formes modulaires, C.R. Acad. Sci. Paris Sér. A.-B. 277 (1973), 827–830.
47. J.-L. Colliot-Thélène. Formes quadratiques sur les anneaux sémi-locaux réguliers, Colloque sur les Formes Quadratiques (Montpellier, 1977) 2; Bull. Soc. Math. France, Mém. No. 59 (1979), 13–31.
48. I. Connell. The *Stufe* of number fields, Math. Z. 124 (1972), 20–22.
49. J. H. Conway. Invariants for quadratic forms, J. Number Theory 5 (1973), 390–404.
50. R. J. Cook. Simultaneous quadratic equations I, J. London Math. Soc. 4 (1971), 319–326.
51. R. J. Cook. Simultaneous quadratic equations II, Acta Arith. 25 (1974/5), 1–5.
52. C. M. Cordes. Representation sets for integral quadratic forms over the rationals, J. Number Theory 10 (1978), 127–134.
53. C. M. Cordes and J. R. Ramsey. Quadratic forms over fields with $u = q/2 < \infty$, Fund.

Math. 99 (1978), 1–10.
54. H. Davenport and M. Hall, Jr. On the equation $ax^2 + by^2 + cz^2 = 0$, Quart. J. Math. 19 (1948), 189–192.
55. H. Davenport and D. J. Lewis. Gaps between values of positive definite quadratic forms, Acta Arith. 22 (1972), 87–105.
56. V. A. Demjanenko. The representation of numbers by binary quadratic forms (in Russian), Mat. Sb. 80 (122) (1969), 445–452.
57. T. J. Dickson. On Voronoi reduction of positive definite quadratic forms, J. Number Theory 4 (1972), 330–341.
58. B. Diviš. On lattice points in high-dimensional ellipsoids, *Number Theory* (Colloq. J. Bolyai Math. Soc. Debrecen, 1968). Amsterdam: North-Holland, 1970, 27–30.
59. B. Diviš. Ueber Gitterpunkte in mehrdimensionalen Ellipsoiden I, II, Czechoslovak Math. J. 20 (95) (1970), 130–139, 149–159.
60. D. Z. Djokowič. Level of a field and the number of square classes, Math. Z. 135 (1973/74), 267–269.
61. P. K. J. Draxl. *La Représentation de* -1 *comme Somme de Carrés dans les Ordres d'un Corps de Nombres Algébriques*. Journées arithmétiques Françaises, Mai 1971. Marseille: Université de Provence, 1971.
62. B. J. Dubin and H. S. Butts. Composition of binary quadratic forms over integral domains, Acta Arith. 20 (1972), 223–251.
63. V. C. Dumir. Asymmetric inequalities for non-homogeneous ternary quadratic forms, J. Number Theory 1 (1969), 326–345.
64. V. C. Dumir. On positive values of indefinite binary quadratic forms. Math. Student 38 (1970), 177–182.
65. M. Eichler. Ueber ambige definite quadratische Formen von Primzahldiskriminante, Comment. Pure Appl. Math. 29 (6) (1976), 623–647. (In abbreviated form, essentially the same results appear in Represéntation moyenne de nombres par des formes quadratiques quaternaires, *Journées Arithmétiques de Caen*, Université de Caen, 1976, 199–202; Astérisque No. 41–42, Société Mathématique de France, Paris, 1977.)
66. W. J. Ellison. On sums of squares in $Q^{1/2}$ (X), etc. Séminaire Théorie des Nombres, 1970/71, Université de Bordeaux I, Talence, Exp. No. 9.
67. R. Elman, T. Y. Lam, and A. Prestel. On some Hasse principle over formally real fields, Math. Z. 134 (1973), 291–301.
68. R. Elman and T. Y. Lam. Quadratic forms over formally real fields and pythagorean fields, Amer J. Math. 94 (1972), 1155–1194.
69. R. Elman and T. Y. Lam. Classification theorems for quadratic forms over fields, Comment. Math. Helv. 49 (1974), 373–381.
70. P. Epstein. Zur Auflösbarkeit der Gleichung $x^2 - Dy^2 = -1$, J. Reine Angew. Math. 171 (1934), 243–252.
71. P. Erdös and C. Ko. On definite quadratic forms, which are *not* the sum of two definite, or semi-definite forms, Acta Arith. 3 (1939), 102–122.
72. D. Estes and G. Pall. The definite octonary quadratic forms of determinant 1, Illinois J. Math. 14 (1970), 159–163.
73. A. S. Fainleib. The limit theorem for the number of classes of primitive quadratic forms with negative determinants, Doklady Akad. Nauk SSSR 184 (1969), 1048–1049; translation, Soviet Math. Dokl. 10 (1969), 206–207.
74. T. V. Fedorova. The representation of numbers by certain quadratic forms with 8 variables (in Russian), Taškent Gos. Univ. Naučn. Trudy Vyp. 292 (1967), 44–60.
75. T. V. Fedorova. Certain quadratic forms with 6 and 8 variables (in Russian), Dokl. Akad. Nauk UzSSR, 1966, No. 3, 3–5.
76. B. Fein. A note on the two-squares theorem, Canad. Math. Bull. 20 (1977), 93–94.
77. B. Fein. Sums of squares rings, Canad. J. Math. 29 (1977), 155–160.
78. R. Finkelstein and H. London. On triangular numbers, which are sums of consecutive squares, J. Number Theory 4 (1972), 455–462.
79. O. M. Fomenko. An application of Eichler's reduction formula to the representation of numbers by certain quaternary quadratic forms (in Russian), Mat. Zametki 9 (1971), 71–76.
80. O. Fraser and B. Gordon. On representing a square as the sum of 3 squares, Amer. Math. Monthly 76 (1969), 922–923.
81. J. Friedländer and H. Iwaniec. Quadratic polynomials and quadratic forms, Acta Math.

141 (1978), 1–15.

82. D. A. Garbanati. An algorithm for the representation of zero by a quadratic form, J. Pure Appl. Algebra 13 (1978), 57–63.
83. L. J. Gerstein. The growth of class numbers of quadratic forms, Amer. J. Math. 94 (1972), 221–236.
84. L. J. Gerstein. A new proof of a theorem of Cassels and Pfister, Proc. Amer. Math. Soc. (1973), 327–328.
85. L. J. Gerstein. Unimodular quadratic forms over global function fields, J. Number Theory 11 (1979), 529–541.
86. J. R. Gillet. An alternative representation of sums of two squares, Math. Gaz. 55 (391) (1971), 59–60.
87. S. J. F. Gilman. A bound on the number of representations of quadratic forms, Acta Arth. 13 (1967/68), 363–374.
88. G. P. Gogišvili. The summation of a singular series that is connected with diagonal quadratic forms in 4 variables (in Russian; Georgian summary), Sakharth. SSR Mecn. Akad. Mat. Inst. Šrom 38 (1970), 5–30.
89. G. P. Gogišvili. The number of representaions of numbers by positive quaternary diagonal quadratic forms (in Russian; Georgian summary) Sakharth. SSR Mecn. Akad. Mat. Inst. Šrom 40 (1971), 59–105.
90. G. P. Gogišvili. A relation between the number of representations of numbers by quadratic forms and the corresponding singular series (in Russian), Sakharth. SSR Mecn. Akad. Mat. Inst.Šrom 57 (1977), 40–62.
91. L. Goldstein. Imaginary quadratic fields of class number 2, J. Number Theory 4 (1972), 286–301.
92. E. P. Golubeva. The representation of large numbers by ternary quadratic forms (in Russian), Dokl. Akad. Nauk SSSR 191 (1970), 519–521; translation, Soviet Math. Dokl. 11 (1970), 394–396.
93. D. Gondard. Sur le 17-ème problème de Hilbert dans $R(V)$, C.R. Acad. Sci. Paris, Sér. A.-B. 276 (1973), A585–A590.
94. D. Gondard. Quelques généralisations du 17-ème problème de Hilbert, Séminaire Théorie des Nombres 1973/74, Univ. Bordeaux I, Talence, Exp. 20.
95. R. S. Gongadze. The representation of numbers by certain forms of the type $x^2 + 2^{2k+1}y^2 + 32z^2 + 32t^2$ (in Russian; Georgian summary), Sakharth. SSR Mecn. Akad. Moambe 50 (1968), 519–524.
96. M. Gottschalk and K.-H. Indlekofer. Ueber binäre additive Probleme, J. Reine u. Angew. Math. 297 (1978), 65–79.
97. G. Greaves. On the representation of a number in the form $x^2 + y^2 + p^2 + q^2$, where p, q are odd primes, Acta Arith. 29 (1976), 257–274.
98. H. Gross. Untersuchungen über quadratische Formen in Körpern der Charakteristik 2, J. Reine Angew. Math. 297 (1978), 80–91.
99. B. Gruber. On the minimum of a positive definite quadratic form in 3 variables, Glasnik Mat. Ser. III, 5 (25) (1970), 1–18.
100. H. Gupta. Ramanujan's ternary quadratic form $x^2 + y^2 + 10z^2$, Res. Bull. Panjab Univ. (N.S.) 24 (1973), 57 (1977).
101. K. Györy. Représentation des nombres entiers par des formes binaires, Publ. Math. Debrecen 24 (1977), 363–375.
102. E. H. Hadlock. On the existence of a ternary quadratic form, Univ. Nac. Tucumán Rev. Ser. A 18 (1968), 81–122.
103. E. H. Hadlock and T. O. Moore. A new definition of a reduced form, Proc. Amer. Math. Soc. 25 (1970), 105–113.
104. E. H. Hadlock and T. O. Moore. A new definition of a reduced binary quadratic form, J. Natur. Sci. Math. 11 (1971), 257–263.
105. J. Hardy. A note on the representability of binary quadratic forms with Gaussian integer coefficients as sums of squares of two linear forms, Acta Arith. 15 (1968/69), 77–84.
106. E. Härtter and J. Zöllner. Darstellung natürlicher Zahlen als Summe und als Differenz von Quadraten (English summary), Norske Videnskab. Selsk. Skr. (Trondheim), 1977, No. 1.
107, K. Hashimoto. Some examples of integral definite quaternary quadratic forms with prime discriminant, Nagoya Math. J. 77 (1980), 167–175.
108. H. Hasse. Ueber die Klassenzahl der Körper $P(\sqrt{-p})$ mit einer Primzahl $p \equiv 1 \pmod{2^3}$, Aequationes Math. 3 (1969), 165–169.

109. H. Hasse. Ueber die Klassenzahl des Körpers $P(\sqrt{-2p})$ mit einer Primzahl $p \neq 2$, J. Number Theory 1 (1969), 231–234.
110. H. Hasse. Ueber die Teilbarkeit durch 2^3 der Klassenzahl imaginärquadratischer Zahlkörper mit genau zwei verschiedenen Diskriminantenprimteilern, J. Reine Angew. Math. 241 (1970), 1–6.
111. C. J. Hightower. The minimum of indefinite binary quadratic forms, J. Number Theory 2 (1970), 364–378.
112. L. Holzer. Minimal solutions of diophantine equations, Canad. J. Math. 2 (1950), 238–244.
113. C. Hooley. On the intervals between numbers that are sums of two squares I, Acta Math. 127 (1971), 279–297.
114. C. Hooley. On the intervals between numbers that are sums of two squares II, J. Number Theory 5 (1973), 215–217.
115. C. Hooley. On the intervals between numbers that are sums of two squares III, J. Reine Angew. Math. 267 (1974), 207–218.
116. J. S. Hsia. On the Hasse principle for quadratic forms, Proc. Amer. Math. Soc. 39 (1973), 468–470.
117. J. S. Hsia. On the representation of cyclotomic polynomials as sums of squares, Acta Arith. 25 (1973/74), 115–120.
118. J. S. Hsia. Representations by integral quadratic forms over algebraic number fields, *Proceedings of the Conference on Quadratic forms, 1976*. Queen's Papers in Pure and Appl. Math. No. 46. Kingston, Ontario: Queen's University, 1977, 528–537.
119. J. S. Hsia and R. P. Johnson. Round and Pfister forms over $R(t)$, Pacific J. Math. 49 (1973), 101–108.
120. J. S. Hsia and R. P. Johnson. Round and group quadratic forms over global fields, J. Number Theory 5 (1973), 356–366.
121. J. S. Hsia, Y. Kitaoka, and M. Kneser. Representations of positive definite quadratic forms, J. Reine Angew. Math. 301 (1978), 132–141.
122. J. L. Hunsucker. Primitive representation of a binary quadratic form as a sum of 4 squares, Acta Arith. 19 (1971), 321–325.
123. K.-H. Indlekofer and W. Schwarz. Ueber *B*-Zwillinge, Arch. Math. (Basel) 23 (1972), 251–256.
*124. O. Intrau. Kompositionstafeln quaternärer quadratischer Formen, Arch. Sächs. Akad. Wiss. Leipzig, Math.-Naturw. Kl. 50 (2) (1970).
125. V. A. Iskovskih. A counterexample to the Hasse principle for systems of two quadratic forms in five variables (in Russian), Mat. Zametki 10 (1971), 253–257.
126. D. Ismailov. The distribution of integral points on elliptic cones of higher order (in Russian; Tajiki summary) Doklady Akad. Nauk Tadžik SSR 13 (1970), 7–11.
127. H. Iwaniec. Primes represented by quadratic polynomials in two variables (Russian summary), Bull. Acad. Polon. Sci. Ser. Sci. Math. Astron. Phys. 20 (1972), 195–202.
128. H. Iwaniec. Primes represented by quadratic polynomials in two variables, Acta Arith. 24 (1972/73), 435–459.
129. H. Iwaniec. On indefinite quadratic forms in 4 variables, Acta Arith. 33 (1977), 209–229.
130. T. H. Jackson. Small positive values of indefinite quadratic forms, J. London Math. Soc. 2 (1969), 643–659.
131. D. G. James. On Witt's Theorem for unimodular, quadratic forms I, II, Pacific J. Math. 26 (1968), 303–316, 33 (1970), 645–652.
132. D. G. James. Representations by integral quadratic forms, J. Number Theory 4 (1972), 321–329.
133. D. G. James. The structure of orthogonal groups of 2-adic unimodular quadratic forms, J. Number Theory 5 (1973), 444–455.
134. V. Jarnik. Bemerkungen zu Landauschen Methoden in der Gitterpunktlehre, *Number Theory and Analysis* (*Papers in Honor of E. Landau*). New York: Plenum Press, 1969, 137–156.
135. T. Kano. On the number of integers representable as the sum of two squares, J. Fac. Sci. Shinsu Univ. 4 (1969), 57–65.
136. P. Kaplan. Représentation de nombres premiers par des formes quadratiques binaires de discriminant-π où $\pi \equiv 1 \pmod 4$, C. R. Acad. Sci. Paris Sér. A.-B. 276 (1973), A1535–A1537.
137. P. Kaplan. Divisibilité par 8 du nombre des classes des corps quadratiques, dont le 2-groupe de classes est cyclique et réciprocité biquadratique, J. Math. Soc. Japn 25 (1973),

596–608.

138. E. Karst. New quadratic forms with high density of primes, Elem. Math. 28 (1973), 116–118.
139. M. Kassner. Darstellungen mit Nebenbedingungen durch quadratische Formen, J. Reine Angew. Math. 331 (1982), 151–191.
140. I. Kátai and I. Környei. On the distribution of lattice points on circles, Ann. Univ. Sci. Budapest Eötvös Sect. Math. 19 (1976), 87–91 (1977).
141. J. Kelemen. On a ternary quadratic Diophantine equation, Mat. Lapok 19 (1968), 367–371.
142. Y. Kitaoka. Quaternary even positive definite quadratic forms of prime discriminant, Nagoya Math. J. 52 (1973), 147–161.
143. M. Knebusch and W. Scharlau. Quadratische Formen und quadratische Reziprozitätsgesätze über algebraischen Zahlkörpern, Math. Z. 121 (1971), 346–368.
144. M. Kneser. Kleine Lösungen der diophantischen Gleichung $ax^2 + by^2 = cz^2$, Abh. Math. Sem. Univ. Hamburg. 23 (1959), 163–173.
*145. M. Kneser. *Quadratische Formen*. Lecture Notes. Göttingen, 1973/74.
146. L. A. Kogan. Theory of modular forms and the problem of finding formulas for the number of representations of numbers by positive quadratic forms, Dokl Akad. Nauk SSSR 182 (1968), 259–261; translation, Soviet Math. Dokl. 9 (1968), 1130–1132.
147. O. Körner. Die Masse der Geschlechter quadratischer Formen vom Range $\leqslant 3$ in quadratischen Zahlkörpern, Math. Ann. 193 (1971), 279–314.
148. O. Körner. Bestimmung einklassiger Geschlechter ternärer quadratischer Formen in reellquadratischen Zahlkörpern, Math. Ann. 201 (1973), 91–95.
149. O. Körner. Integral representations over local fields and the number of genera of quadratic forms, Acta Arith. 24 (1973/74), 301–311.
150. T. Kreid. On representations of natural numbers as sums of 2 squares, Demonstratio Math. 9 (1976), 367–378.
151. M. Kula and L. Szczepanik. Quadratic forms over formally real fields with 8 square classes, Manuscripta Math. 29 (1979), 295–303.
152. J. Lagrange. Décomposition d'un entier en somme de carrés et fonctions multiplicatives, Séminaire Delange-Pisot-Poitou 14 (1972/73), Exp. 1.
*153. T. Y. Lam. *The Arithmetic Theory of Quadratic Forms*. Reading, MA: Benjamin, 1973.
154. L. C. Larson. A theorem about primes proved on a chessboard, Math. Mag. 50 (1977), 69–74.
155. A. Lax and P. Lax. On sums of squares, Linear Algebra Appl. 20 (1978), 71–75.
156. P. A. Leonard and K. S. Williams. Forms representable by integral binary quadratic forms. Acta Arith. 24 (1973/74), 1–9.
157. H. von Lienen. The quadratic form $x^2 - 2py^2$, J. Number Theory 10 (1978), 10–15.
158. G. A. Lomadze. Representations of numbers by certain quadratic forms in 6 variables I, II (in Russian; Georgian summary), Thbilis. Sahelmc. Univ. Šrom. Mekh.-Math. Mecn. Ser. 127 (1966), 7–43, 129 (1968), 275–297.
159. G. A. Lomadze. The number of representations of numbers by the forms $x_1^2 + 3x_2^2 + 36x_3^2$ and $x_1^2 + 12x_2^2 + 36x_3^2$ (in Russian; Georgian summary), Sakharth. SSR Mecn. Akad. Moambe 51 (1968), 25–30.
160. G. A. Lomadze. The representation of numbers by certain binary quadratic forms, Izv. Vysš. Učebn. Zaved. Mat. 102 (11) (1970), 71–75.
161. G. A. Lomadze. Formulae for the number of representations of numbers by all primitive positive ternary diagonal quadratic forms that belong to one-class genera (in Russian; Georgian summary), Sakharth. SSR Mecn. Akad. Math. Inst. Šrom. 40 (1971), 140–179.
162. G. A. Lomadze. On the representations of integers by positive ternary diagonal quadratic forms I, and II (in Russian), Acta Arith. 19 (1971), 267–305, 387–407.
163. G. A. Lomadze. The number of representations of numbers by quadratic forms of 4 variables (in Russian; Georgian summary), Sakharth. SSR Mecn. Akad. Math. Inst. Šrom. 40 (1971), 106–139.
164. G. A. Lomadze: The behavior of the derivative of the theta functions under linear substitutions (Russian. Georgian summary). Thbilis. Univ. Šrom A 4 (146) (1972), 15–27.
165. F. Lorenz. Quadratische Formen und die Artin-Schreier Theorie der formal reellen Körper, Colloque sur les formes quadratiques (Montpellier, 1975), *Bull Soc. Math. France Suppl. Mém.*, No. 48 (1976), 61–73.
166. H. W. Lu. A system of matrix Diophantine equations, Sci. Sinica 22 (1979), 1347–1361.

167. I. S. Luthar. A generalization of a theorem of Landau, Acta Arith. 12 (1966/67), 223–228.
168. A. V. Malyshev. On the representation of integers by positive definite quadratic forms (in Russian), Trudy Mat. Inst. Steklov 65 (1962).
169. K. I. Mandelberg. A note on quadratic forms over arbitrary semi-local rings, Canad. J. Math. 27 (1975), 513–527.
170. K. I. Mandelberg. On the classification of quadratic forms over semi-local rings, J. Algebra 33 (1975), 463–471.
171. Yu. I. Manin. On Hilbert's 11th problem (in Russian), *Hilbert's Problems* (in Russian), Moscow: Izdatel'stvo "Nauka" 1969, 154–158.
172. M. Marshall. Round quadratic forms, Math. Z. 140 (1974), 255–262.
173. J. Martinet. Sommes de carrés, Séminaire Théorie des Nombres 1970–71, Univ. Bordeaux I, Talence, Exp. 26.
174. G. Maxwell. A note on Artin's Diophantine conjecture, Canad. Math. Bull. 13 (1970), 119–120.
175. J. Meyer. Präsentation der Einheitsgruppe der quadratischen Form $F(X) = -X_0^2 + X_1^2 + \cdots + X_{18}^2$, Arch. Math. (Basel) 29 (1977), 261–266.
176. H. Möller. Imaginär-quadratische Zahlkörper mit einklassigen Geschlechtern, Acta Arith. 30 (1976), 179–186.
177. L. J. Mordell. On the equation $ax^2 + by^2 + cz^2 = 0$, Monatsh. Math. 53 (1951), 323–327.
178. L. J. Mordell. The representation of integers by three positive squares, Michigan Math. J. 7 (1960), 289–290.
179. L. J. Mordell. The representation of a number by some quaternary quadratic forms, Acta Arith. 12 (1966/67), 47–54.
180. L. J. Mordell. The Diophantine equation $dy^2 = ax^4 + bx^2 + c$, Acta Arith. 15 (1968/69), 269–272.
181. L. J. Mordell. On the magnitude of the integer solutions of the equation $ax^2 + by^2 + cz^2 = 0$, J. Number Theory 1 (1969), 1–3.
182. L. J. Mordell. The minimum of a single ternary quadratic form, J. London Math. Soc. (2) 2 (1970), 393–394.
183. C. Moser. Représentation de -1 comme somme de carrés d'entiers dans un corps quadratique imaginaire, Enseign. Math. (2) 17 (1971), 279–287.
184. C. Moser. Représentation de -1 comme somme de carrés dans un corps cyclotomique quelconque, J. Number Theory 5 (1973), 139–141.
185. Y. Motohashi. On the number of integers which are sums of 2 squares, Acta Arith. 23 (1973) 401–412.
186. T. Nagell. Sur une catégorie d'équation diophantienne insoluble dans un corps réel (English summary), Norske Vid. Selsk. Skr. (Trondheim), 1971, No. 4.
187. T. Nagell. Sur la représentabilité de zéro par certaines formes quadratiques (English summary), Norske Vid. Selsk. Skr. (Trondheim), 1972, No. 6.
188. T. Nagell. Sur la résolubilité de l'équation $x^2 + y^2 + z^2 = 0$ dans un corps quadratique, Acta Arith. 21 (1972), 35–43.
189. T. Nagell. Sur la représentation de zéro par une somme de carrés dans un corps algébrique. Acta Arith. 24 (1973), 379–383.
190. Y. Nakamura. On Pfister dimensions of some non-formally real fields, Bull. Ibaraki Univ. Ser. A Math. No. 10, May 1978, 77–79.
191. N. Nobusawa. On the distribution of integer solutions of $f(x, y) = z^2$ for a definite binary quadratic form f, Canad. Math. Bull. 10 (1967), 755–756.
192. D. Nordon. Zéros communs non-singuliers de deux formes quadratiques, Acta Arith. 30 (1976), 107–119.
193. B. Novák. Mean value theorems in the theory of lattice points with weights, I, II, Comment. Math. Univ. Carolinae 8 (1967), 711–733, 11 (1970), 53–81.
194. B. Novák. Ueber Gitterpunkte mit Gewichten in mehrdimensionalen Ellīpsoiden, *Number Theory* (Colloquium J. Bolyai, Math. Soc. Debrecen, 1968). Amsterdam: North-Holland, 1970, 165–179.
195. R. W. K. Odoni. On norms of integers in a full module of an algebraic number field and the distribution of values of binary integral quadratic forms, Mathematika 22 (1975), 108–111.
196. O. T. O'Meara. The construction of indecomposable positive definite quadratic forms, J. Reine Angew. Math. 276 (1975), 99–123.
197. T. Ono and H. Yamaguchi. On Hasse principle for division of quadratic forms, J. Math. Soc. Japan 31 (1979), 141–159.

198. G. Pall. Discriminantal divisors of binary quadratic forms, J. Number Theory 1 (1969), 525–533.
199. G. Pall. The number of representations of a binary quadratic form as a sum of 2, 4 or 8 squares, *Proceedings of the Washington State University Conference*. On *Number Theory* (Washington State Univ., Pullman, WA, 1971). Pullman, WA: Department of Mathematics, Washington State University, 1971, 174–181.
200. G. Pall. Some aspects of Gaussian composition, Acta Arith. 24 (1973), 401–409.
201. C. J. Parry. Primes represented by binary quadratic forms, J. Number Theory 5 (1973), 266–270.
202. M. Peters. Ternäre und quaternäre quadratische Formen und Quaternionenalgebren, Acta Arith. 15 (1968/9), 329–365.
203. M. Peters. Die Stufe von Ordnungen ganzer Zahlen in algebraischen Zahlkörpern, Math. Ann. 195 (1972), 309–314.
204. M. Peters. Quadratische Formen über Zahlringen, Acta Arith. 24 (1973), 157–164.
205. M. Peters. Summen von Quadraten in Zahlringen, J. Reine Angew. Math. 268/269 (1974), 318–323.
206. M. Peters. Einklassige Geschlechter von Einheitsformen in totalreellen algebraischen Zahlkörpern, Math. Ann. 226 (1977), 117–120.
207. M. Peters. Darstellung durch definite ternäre quadratische Formen, Acta Arith. 34 (1977/78), 57–80.
208. M. Peters. Definite binary quadratic forms with class-number one, Acta Arith. 36 (1980), 271–272.
209. H. Pfeuffer. Bemerkungen zur Berechung dyadischer Darstellungsdichten einer quadratischen Form über algebraische Zahlkörper, J. Reine Angew. Math. 236 (1969), 219–220.
210. H. Pfeuffer. On a conjecture about class numbers of totally positive quadratic forms in totally real algebraic number fields, J. Number Theory 11 (1979), 188–196.
211. A. Pfister. Quadratic forms over fields, *1969 Number Theory Institute, Proceedings of Symposia in Pure Mathematics*, v.20 (State Univ. of New York, Stony Brook, NY, 1969), Providence, RI: American Mathematical Society, 1971, 150–160.
212. A. Pfister. Neuere Entwicklungen in der Theorie der quadratischen Formen, Jahresber. Deutsch. Math.-Verein 74 (1972/73), 131–142.
213. A. Pfister. Quelques problèmes des formes quadratiques sur les corps, Bull. Soc. Math. France Suppl. Mém. No. 48 (1976), 89–91.
214. A. Pizer. Class number of positive definite quaternary forms, J. Reine Angew. Math. 286/287 (1976), 101–123.
215. P. A. B. Pleasants. The representation of primes by quadratic and cubic polynomials, Acta Arith. 12 (1966/67), 131–153.
216. N. Plotkin. The solvability of the equation $ax^2 + by^2 = c$ in quadratic fields, Proc. Amer. Math. Soc. 34 (1972), 337–339.
217. E. V. Podsypanin. Sums of squares of square-free numbers (in Russian), Zap. Naučn. Sem. Leningrad. Otdel. Mat. Inst. Steklov (LOMI) 33 (1973), 116–131.
218. E. V. Podsypanin. On the representation of the integer by positive quadratic forms with square-free variables, Acta Arith. 27 (1975), 459–488.
219. P. Ponomarev. Class number of definite quaternary forms with non-square discriminant, Bull Amer. Math. Soc. 79 (1973), 594–598.
220. P. Ponomarev. Arithmetic of quaternary quadratic forms, Acta Arith. 29 (1976), 1–48.
221. V. V. Potockii. A. sharpening of the estimate of a certain trigonometric sum (in Russian), Izv. Vysš. Učbn. Zaved. Mat. 82 (3) (1969), 42–51.
222. A. Prestel. A local–global principle for quadratic forms, Math. Z. 142 (1975), 91–95.
223. L. N. Pronin. Integral binary Hermitian forms over the field of quaternions (in Russian), Vestnik Har'kov Gos. Univ., 1967, No. 26, 27–41.
224. A. R. Rajwade. Note sur le théorème des 3 carrés, Enseign. Math. (2) 22 (1976), 171–173.
225. L. Rédei. Ueber die Pellsche Gleichung $t^2 - du^2 = -1$, J. Reine Angew. Math. 173 (1935), 193–221.
226. H. P. Rehm. On a theorem of Gauss, concerning the number of integral solutions of the equation $x^2 + y^2 + z^2 = m$, Paper No. 4 in [260].
227. P. Revoy. Une note sur les formes quadratiques en characteristique 2 (Colloque sur les formes quadratiques, 2, Montpellier, 1977), Bull. Soc. Math. France, No. 59 (1979), 13–31.
228. P. Ribenboim. 17-ème problème de Hilbert, Bol. Soc. Brasil. Mat. 5 (1974), 63–77.
229. P. Ribenboim. 17-ème problème de Hilbert, *Report of Algebra Group*, Queen's Univ.,

Kingston, Ontario 1973/74, Queen's Papers Pure and Appl. Math. No. 41, 1974, 148–164.
230. P. Ribenboim. Hilbert's 17th problem (in Spanish), Bol. Mat. 8 (1974), 56–73.
231. P. Ribenboim. Quelques développements récents du 17-ème problème de Hilbert, *Journées arithmétiques de Bordeaux* (Conf. Univ. Bordeaux, Bordeaux 1974), Astérisque No. 24–25. Société Mathématique de France, 237–241.
232. G. J. Rieger. Zum Satze von Landau über die Summe aus 2 Quadraten, J. Reine Angew. Math. 244 (1970), 198–200.
233. L. G. Risman. A new proof of the 3 squares theorem, J. Number Theory 6 (1974), 282–283.
234. G. Rosenberger. Zu Fragen der Analysis im Zusammenhang mit der Gleichung $x_1^2 + \cdots + x_n^2 - ax_1 \cdots x_n = b$ (English summary), Monatsh. Math. 85 (1978), 211–283.
235. S. S. Ryškov. On the reduction theory of positive quadratic forms, Dokl. Akad. Nauk SSSR 198 (1971), No. 5; translation, Soviet Math. Dokl. 12 (3) (1971), 946–950.
236. S. S. Ryškov. On complete groups of integral automorphisms of positive quadratic forms, Dokl. Akad. Nauk SSSR 206 (1972), No. 3; translation, Soviet Math. Dokl. 13 (1972), 1251–1254.
237. S. S. Ryškov. The reduction of positive quadratic forms of n variables in the sense of Hermite, Minkowski, and Venkov (in Russian), Dokl. Akad. Nauk SSSR 207 (1972), 1054–1056; translation, Soviet Math. Dokl. 13 (1972), 1676–1679.
238. S. S. Ryškov. On the theory of the reduction of positive quadratic forms in the sense of Hermite and Minkowski (in Russian), Zap. Naučn. Sem. Leningrad. Otdel. Mat. Inst. Steklov (LOMI) 33 (1973), 37–64.
239. G. Sansone. Il sistema diofanteo $N + 1 = x^2, 3N + 1 = y^2, 8N + 1 = z^2$, Ann. Mat. Pura. Appl. (4) 111 (1976), 125–151.
240. W. Scharlau. A historical introduction to the theory of integral quadratic forms, *Proceedings of the Conference on Quadratic Forms*. Queen's Papers in Pure and Appl. Math. No. 46. Kingston, Ontario: Queen's University, 1977, 284–339.
241. D. B. Schapiro. Hasse principle for the Hurwitz problem, J. Reine Angew. Math. 301 (1978), 179–190.
242. H. Schmidt. Eine diophantische Aufgabe des Leonardo von Pisa, Bayer. Akad. Wiss. Math.-Natur. Kl.-Sitzungsber. 1975, 189–195 (1976).
243. W. Schwarz. Ueber B-Zwillinge II, Arch. Math. (Basel) 23 (1972), 408–409.
244. S. M. Shah. On triangular numbers which are sums of 2 triangular numbers, Vidya 9 (1966), 161–163.
245. C. L. Siegel. Zur Theorie der quadratischen Formen, Nachr. Akad. Wiss. Göttingen Math.-Phys. Kl., 1972, No. 3, 21–46.
246. T. Skolem. On the Diophantine equation $ax^2 + by^2 - cz^2 = 0$. Rend. Mat. Appl. 5 (11) (1952), 88–102.
247. I. Sh. Slavutsky. On representations of numbers by sums of squares and quadratic fields, Arch. Math. (Brno) 13 (1977), 23–40.
248. C. Small. Sums of 3 squares and level of quadratic fields, *Proceedings of the Conference on Quadratic Forms, 1976*. Queen's Papers in Pure and Appl. Math. No. 46. Kingston, Ontario: Queen's University, 1977, 625–631.
249. R. A. Smith. On $\sum_n r_2(n) r_2(n + a)$, Proc. Inst. Sci. India Part A 34 (1968), 132–137.
250. R. A. Smith. The average order of a class of arithmetic functions over arithmetic progressions with application to quadratic forms, J. Reine Angew. Math. 317 (1980), 74–87.
251. W. Sollfrey. Note on sums of squares of consecutive odd integers, Math. Mag. 4 (1968), 255–258.
252. H.-J. Stender. Zur Parametrisierung reell-quadratischer Zahlkörper, J. Reine Angew. Math. 311/312 (1979), 291–301.
253. E. Stevenson. On the representation of integers by p-adic diagonal forms, J. Number Theory 12 (1980), 367–371.
254. M. I. Stronin. Integral points on circular cones (in Russian) Izv. Vysš. Učebn. Zaved. Mat. 87 (8) (1969), 112–116.
255. L. A. Sulakvelidze. The representation of numbers by certain positive ternary forms (in Russian; Georgian and English summaries). Sakharth. SSR Mecn. Akad. Moambe 88 (1977), 21–24.
256. P. P. Tammela. On the reduction theory of positive quadratic forms (in Russian), Dokl. Akad. Nauk SSSR 209 (1973), 1299–1302; translation, Soviet Math. Dokl. 13 (1973), 651–655.

257. P. P. Tammela. On the reduction theory of positive quadratic forms (in Russian), Zap. Naučn. Sem. Leningrad. Otdel. Mat. Inst. Steklov 50 (1975). 6–96, 195.
258. O. Taussky. Two facts concerning rational 2 × 2 matrices leading to integral ternary forms representing zero, Paper No. 1 in [260].
259. O. Taussky. Automorphs of quadratic forms as positive operators, Inequalities III (Proceedings of the Third Symposium, Univ. of California, Los Angeles, 1969, Dedicated to the memory of Th. Motzkin). New York: Academic Press, 1972.
*260. O. Taussky (editor). *Selected Topics on Ternary Forms and Norms* (Seminar in Number Theory, Calif. Institue of Technology, Pasadena, Calif., 1974/75). Pasadena: California Institute of Technology, 1976.
261. G. Tennenbaum. Une fonction arithmetique liée aux sommes de 2 carrés, Séminaire Theorie des Nombres 1975/76, Univ. Bordeaux I, Talence, 1976, Exposé 2.
262. G. Terjanian. Un contre-example à une conjecture d'Artin, C. R. Acad. Sci. Paris Sér. A.-B. 262 (1966), A612.
263. U. P. Tietze. Zur Theorie quadratischer Formen über Hensel-Körpern, Arch. Math. (Basel) 25 (1974), 144–150.
264. U. P. Tietze. Zur Theorie quadratischer Formen der Charakteristik 2, J. Reine Angew. Math. 268/269 (1974), 388–390.
265. J.-D. Thérond. *Détermination Effective de Certaines Décompositions en Somme de 2 Carrés.* Thèse, Univ. of Montpellier, Montpellier, 1970.
266. S. Uchiyama. A five-square theorem, Publ. Res. Inst. Math. Sci. 13 (1977/78), 301–305.
267. P. L. Varnavides. The non-homogeneous minima of a class of binary quadratic forms, J. Number Theory 2 (1970), 333–341.
268. P. L. Varnavides. Non-homogeneous binary quadratic forms, Ann. Acad. Sci. Fennica, Ser. A I, No. 462 (1970), 28 pages.
269. T. V. Vephradze. The representation of numbers by positive Gaussian binary quadratic forms (in Russian; Georgian, and English summaries), Sakharth. SSR Mecn. Akad. Math. Inst. Šrom. 40 (1971), 21–58.
270. E. B. Vinberg. The unimodular integral quadratic forms (in Russian), Funkcional. Anal. i Priložen. 6 (1972), 24–31.
271. A. B. Voroneckii. On the question of one-class genera of positive binary quadratic forms (in Russian) Zap. Naučn. Sem. Leningrad. Otdel. Mat. Inst. Steklov (LOMI) 67 (1977) 144–155, 226.
272. A. B. Voroneckii and A. V. Malyšev. Simultaneous representations of a pair of numbers by sums of integers and their squares (in Russian), Trudy Mat. Inst. Steklov 142 (1976), 122–134, 269.
273. A. R. Wadsworth. Similarity of quadratic forms and isomorphism of their function fields, Trans. Amer. Math. Soc. 208 (1975), 352–358.
274. B. L. van der Waerden. On the first and second quadratic form, Paper No. 2 in [260].
275. B. L. van der Waerden. Das Minimum von $D/(f_{11}f_{22}\cdots f_{55})$ für reduzierte positive quinäre quadratische Formen, Aequationes Math. 2 (1969), 233–247.
276. C. Waid. Generalized automorphs of the sum of 4 squares, J. Number Theory 3 (1971), 468–473.
277. A. Z. Walfisz and A. A. Walfisz. Ueber Gitterpunkte in mehrdimensionalen Kugeln: Part I, Jahresber. Deutsch. Math.-Verein 61 (1958), Abt. 1, 11–31. Part II, Acta Arith. 6 (1960), 115–136. Part III, Acta Arith. 6 (1960), 193–215. Part IV, *Number Theory and Analysis* (Papers in Honor of E. Landau). New York: Plenum Press, 1969, 307–333.
278. A. A. Walfisz. Ueber die simultane Darstellung zweier ganzer Zahlen durch quadratische und lineare Formen, Acta Arith. 35 (1979), 289–301.
279. R. Ware. A note on quadratic forms on pythagorean fields, Pacific J. Math. 58 (1975), 651–654.
280. R. Warlimont. Ueber die k-ten Mittelwerte der Klassenzahl primitiver binärer quadratischer Formen negativer Diskriminante, Monatshefte der Mathematik 75 (1971), 173–179.
281. G. L. Watson. Transformations of a quadratic form which do not increase the class-number: Part I, Proc. London Math. Soc. (3) 12 (1962), 577–587. Part II, Acta Arith. 27 (1975), 171–189.
282. G. L. Watson. The least common denominator of the coefficients of a perfect quadratic form. Acta Arith. 18 (1971), 29–36.
283. G. L. Watson. The number of minimum points of a positive quadratic form, Dissertationes Math. Rozprawy Mat. 84 (1971), 42 pages.

284. G. L. Watson. One-class genera of positive ternary quadratic forms: Part I, Mathematika 19 (1972), 96–104.. Part II, Mathematika 22 (1975), 1–11.
285. G. L. Watson. One-class genera of positive quaternary quadratic forms, Acta Arith. 24 (1973/74), 461–475.
286. G. L. Watson. One-class genera of positive quadratic forms in at least 5 variables, Acta Arith. 26 (1974/75), 309–327.
287. G. L. Watson. Existence of indecomposable positive quadratic forms in a given genus of rank at least 14, Acta Arithm. 35 (1979), 55–100.
288. G. L. Watson. Determination of a binary quadratic form by its values at integer points, Mathematika 26 (1979), 72–75.
*289. A. Weil. Elliptic Functions according to Eisenstein and Kronecker, *Ergebnisse der Mathematik und Ihrer Grenzgebiete*. Berlin, Heidelberg, New York: Springer-Verlag, 1976.
290. H. Wild. Die Anzahl der Darstellungen einer natürlichen Zahl durch die Form $x^2 + y^2 + z^2 + 2t^2$, Abh. Sem. Hamburg. Univ. 40 (1974), 132–135.
291. K. S. Williams. Forms representable by an integral positive definite binary quadratic form, Math. Scand. 29 (1971), 73–86.
292. K. S. Williams. Representation of a binary quadratic form as a sum of two squares, Proc. Amer. Math. Soc. 32 (1972), 368–370.
293. J. Wojcik. On sums of 3 squares, Colloq. Math. 24 (1971), 117–119.
294. J. Wojcik. On rational automorphs of quadratic forms, Colloq. Math. 30 (1974), 69–88.
295. D. Wolke. Moments of the number of classes of primitive quadratic forms with negative discriminants, J. Number Theory 1 (1969), 502–511.
296. D. Wolke. Moments of class numbers (II), Arch. Math. (Basel) 22 (1971), 65–69.
297. D. Wolke. Moments of class numbers (III), J. Number Theory 4 (1972), 523–531.
298. R. T. Worley. Non-negative values of quadratic forms, J. Australian Math. Soc. 12 (1971), 224–228.
299. H. Zassenhaus. Gauss' theory of ternary quadratic forms, an example of the theory of homogeneous forms in many variables, with applications, Paper No. 3 in [260].
300. W. Zelidowicz. On the equation $x^2 + y^2 - z^2 = k$ (in Polish), Fasc. Math. No. 4 (1969), 79–90.
301. T. M. Zuparov. A multidimensional additive problem with a growing number of terms (in Russian; Uzbek summary) Izv. Akad. Nauk UzSSR Ser. Fiz.-Mat. Nauk 13 (1969), 21–27.

Addenda

1. E. Grosswald. Partitions into squares, L'Enseignement Mathématique 30 (1984), 223–245.
2. T. Y. Lam *The algebric theory of quadratic forms*. Menlo Park, CA: Addison-Wesley Publ. Co., 1973.
3. M. Kneser. Klassenzahl quadratischer Formen, Jahresbericht der Deutschen Mathematiker Vereinigung 61 (1958), 76–88.
4. W. Magnus. Ueber die Anzahl der in einem Geschlecht enthaltenen Klassen von positiv-definiten quadratischen Formen. Mathematische Annalen 114 (1937), 465–475 and 115 (1938), 643–644.

Author Index

Subject Index